The Nano–Micro Interface

Edited by
Hans-Jörg Fecht and Matthias Werner

The Nano–Micro Interface

Bridging the Micro and Nano Worlds

Edited by
Hans-Jörg Fecht and Matthias Werner

WILEY-VCH

WILEY-VCH Verlag GmbH & Co. KGaA

Edited by

Prof. Dr. Hans-Jörg Fecht
University of Ulm
Faculty of Engineering
Materials Division
Albert-Einstein-Allee 47
89081 Ulm
Germany

and

Forschungszentrum Karlsruhe
Institute of Nanotechnology (INT)
P.O. Box 3640
76021 Karlsruhe
Germany

Dr. Matthias Werner
NMTC
Soorstrasse 86
14050 Berlin
Germany

■ This book was carefully produced. Nevertheless,
editors, authors and publisher do not warrant
the information contained therein to be free of
errors. Readers are advised to keep in mind that
statements, data, illustrations, procedural details
or other items may inadvertently be inaccurate.

Library of Congress Card No.: applied for
A catalogue record for this book is available from
the British Library.

**Bibliographic information published
by Die Deutsche Bibliothek**
Die Deutsche Bibliothek lists this publication
in the Deutsche Nationalbibliografie; detailed
bibliographic data is available in the Internet at
<http://dnb.ddb.de>.

© 2004 WILEY-VCH Verlag GmbH & Co. KGaA,
Weinheim

Printed in the Federal Republic of Germany
Printed on acid-free paper

Cover design   SCHULZ Grafik-Design,
Fußgönheim
Composition   K+V Fotosatz GmbH, Beerfelden
Printing   betz-druck GmbH, Darmstadt
Bookbinding   Schäffer GmbH, Grünstadt

ISBN   3-527-30978-0

Preface

The key topic of this book "The Nano-Micro-Interface" (NAMIX) intends to bridge the gap between microsystem technology and nanotechnology. Micro- and nanotechnologies are becoming key technologies having a significant impact on the development of new products and production technologies for nearly all industrial branches.

This newly emerging field includes the use of nanotechnology effects to achieve a better device performance or to create completely new devices (bottom-up). The trend to a continuous miniaturization and the corresponding increase in the density of integration is a challenge to the processes and materials in use (top down). Therefore, this exciting area at the interface between the micro- and nanoworld is gaining more and more interest from the fundamental point of view as well as for industrial applications. One of the key scientific and commercial problems to be solved is the development of novel functional structures of superior performance by controlling the atomic or molecular structure on a scale between 1 and 100 nm.

Nanotechnology Comprises

- All products with a controlled geometry size of at least one functional component below 100 nanometres in one or more dimensions that makes physical, chemical or biological effects available which cannot be achieved above the critical dimension(s) ($\leq$100 nm) without a loss of performance.
- Equipment for analytical or manipulatory purposes that allows controlled fabrication, movement or measurement resolution with a precision below 100 nanometres.

According to this definition, a nanotechnology product contains at least one functional component which should fulfill one of the above-mentioned boundary conditions. Obviously, only in a few cases does such a product consist of nanoscale building blocks alone without any macroscopic element. Since the value of the nanotechnology contribution to such a product is difficult to estimate, it is only possible to consider the market price value of the end product. This has implications

The Nano–Micro Interface: Bridging the Micro and Nano Worlds
Edited by Hans-Jörg Fecht and Matthias Werner
Copyright © 2004 WILEY-VCH Verlag GmbH & Co. KGaA, Weinheim
ISBN: 3-527-30978-0

for the determination of the overall market size. Therefore, a "nanotechnology product" is defined as the smallest unit with a functional nanotechnology component that can be commercially sold in the marketplace. For example, the functional component of a magnetic disk drive is the read/write head that is based on the GMR (Giant Magneto Resistance) effect. However, the smallest unit that is commercially available is the magnetic disk drive and not the read/write head itself. Therefore, market figures are most often based on the market price of the smallest commercially available units with functional nanotechnology components.

Most fundamental physical properties change if the geometry size in at least one dimension is reduced to a critical value below 100 nanometres, depending on the material itself. For example, this allows tuning of the physical properties of a macroscopic material if the material consists of nanoscale building blocks with controlled size and composition. Every property has a critical length scale, and if a nanoscale building block is made smaller than the critical length scale, the fundamental physics of that property changes. By altering the sizes of those building blocks, controlling their internal and surface chemistry, their atomic structure, and their assembly, it is possible to engineer properties and functionalities in completely new ways.

Nanoparticles and nanomaterials possess radically different phenomena and behaviours, as compared to their larger scale counterparts. Such mechanisms include quantum effects, statistical time variations of properties and their scaling with structure size, dominant surface and interface interactions and absence of defects in the nanocrystals. These nanoparticles and nanomaterials have unique mechanical, electronic, magnetic, optical, and chemical properties, opening the door to enormous new possibilities of engineered nanostructures and integrated nanodevice designs, with application opportunities in information and communications, biotechnology and medicine, photonics and electronics. Examples include developments in very high-density data storage, molecular electronics, quantum dots and spintronics. Atomic or molecular units, with their well-known subatomic structure, offer the ultimate building blocks for a bottom-up, atom-by-atom synthesis and, in some cases, self-assembly manufacturing. Advanced nanostructured materials such as high purity single wall carbon nanotubes are being considered for microelectronics, sensors, thermal management for micro- and optoelectronics, and flat panel displays.

This is the first book picking up these emerging technology trends and compiling contributions from 25 authors and international research groups. It addresses the interface between micro- and nanotechnology with a strong focus on synergy effects provided by the combination of both. The book's contributions cover the entire range of basic technology aspects with a strong focus on potential applications. Moreover, business aspects such as potential markets, roadmaps, transnational networking, and investment opportunities are some of the key topics as well.

Many users are already unknowingly using effects based on nanotechnology. A case in point is sunscreen with a high protection factor, the effect of which is based on nanocristalline titanium oxide. Nanocristalline titanium oxide provides a high protection factor without having a negative impact on the transparency and biocompatibility of the sun cream. Only through a low particle size may this effect be obtained. Another example is the Giant Magneto Resistive Effect (GMR), which is found in virtually any hard disk drive of a computer as a read/write head. The currently high storage densities may only be obtained through use of this nanotechnology effect.

The subject of Nano-Micro Interface ranges from nanomaterials through electronic to biological systems. Using nanotechnology effects in combination with microtechnology is about to open up a considerable market potential. It is remarkable in this context that the European Union, and Germany in particular, is playing an outstanding role, next to the United States and Japan, in the field of micro- and nanotechnology. Germany ranks among the top three in the world both in MST and nanotechnology. This is borne out by technology indicators as well as by the number of publications and the number of patents per country.

Applications in Microsystems

Microsystems, including microelectromechanical systems (MEMS), bioMEMS, nanoelectromechanical systems (NEMS), optical, electronic, and electrochemical microsystems, hold the promise of a new class of multifunctional devices and systems for many applications ranging from advanced computing, chemical and biological analysis/detection, drug delivery/discovery, tissue engineering, chemical and materials synthesis, to energy conversion and storage. New advanced micro-

Effects of nanomaterials and applications due to the reduced dimension.

Effects	*Applications*
Higher surface to volume ratio, enhanced reactivity	Catalysis, solar cells, batteries, gas sensors
Lower percolation threshold	Conductivity of materials
Increased hardness with decreasing grain size	Hard coatings, thin protection layers
Narrower bandgap with decreasing grain size	Opto-electronics
Higher resistivity with decreasing grain size	Electronics, passive components, sensors
Increased wear resistance	Hard coatings, tools
Lower melting and sintering temperature	Processing of materials, low sintering materials
Improved transport kinetics	Batteries, hydrogen storage
Improved reliability	Nanoparticle encapsulated electronic components

systems with integrated nanometer-scale structures and functions present a multidisciplinary challenge.

The performance of such microsystems also depends on the understanding of the properties on both the nano- and microscales. Recently, the Review Committee of the National Nanotechnology Initiative in the U.S. recommended: "Revolutionary change will come from integrating molecular and nanoscale components into high order structures… To achieve improvements over today's systems, chemical and biologically assembled machines must combine the best features of the top-down and bottom-up approaches."

An overview of the effects and applications of the reduced dimensionality of nanomaterials is listed in the table. Furthermore, the addition of nanoparticles to an otherwise homogenous material can lead to a change in the macroscopic material behaviour. Most material properties may be changed and engineered dramatically through the controlled size-selective synthesis and assembly of nanoscale building blocks.

September 2004

Hans-Jörg Fecht
Matthias Werner

Contents

The Nano–Micro Interface: Bridging the Micro and Nano Worlds
Edited by Hans-Jörg Fecht and Matthias Werner
Copyright © 2004 WILEY-VCH Verlag GmbH & Co. KGaA, Weinheim
ISBN: 3-527-30978-0

**Asia–Pacific Nanotechnology: Research, Development, and
Commercialization** *35*
Lerwen Liu

New Approach to Improve the Piezoelectric Quality of ZnO Resonator Devices by Chemomechanical Polishing *181*
Jyrki Molarius, Martin Kulawski, Tuomas Pensala, and Markku Ylilammi

Self-Assembled Semiconductor Nanowires *195*
Theodore I. Kamins

3D Nanofabrication of Rutile TiO_2 Single Crystals with Swift Heavy-Ions *207*
Koichi Awazu, Ken-ichi Nomura, Makoto Fujimaki, and Yoshimichi Ohki

III Applications

Bio-Inspired Anti-reflective Surfaces by Imprinting Processes *263*
Thomas Sawitowski, Norbert Beyer, and Frank Schulz

Preparation and Properties of MgO–Ni(Fe) Nanocrystalline Composites *281*
Oldřich Schneeweiss, Naděžda Pizúrová, Yvonna Jirásková, and Tomáš Žák

Nanocrystalline Oxides Improve the Performances of Polymeric Electrolytes *289*
Silvia Licoccia and Enrico Traversa

**Optimized Electromechanical Properties and Applications of Cellular Polypropylene,
a New Voided Space-Charge Electret Material** *303*
Michael Wegener and Werner Wirges

List of Contributors

Santos F. Alvarado
IBM Zurich Research Laboratory
Säumerstraße 4
CH-8803 Rüschlikon
Switzerland

Koichi Awazu
CAN-FOR
National Institute
of Advanced Industrial
Science and Technology
Tsukuba Central 4
305-8562
Japan

M. R. Baghbanan
University of Toronto
Department of Materials Science and
Engineering
Toronto, M5S 3E4
Canada

Marie-Isabelle Baraton
University of Limoges
Faculty of Sciences
SPCTS – UMR CNRS 6638
123 Avenue Albert Thomas
87060 Limoges
France

Murielle Batude-Thibierge
University of Stirling Innovation Park
Institute of Nanotechnology
6 The Alpha Centre
Stirling FK9 4NF
United Kingdom

Norbert Beyer
AlCove Surfaces GmbH
Am Wiesenbusch 2
45966 Gladbeck
Germany

Markus Böhm
University of Siegen
Institute for Microsystem Technology
Hölderlinstrasse 3
57068 Siegen
Germany

Federica Bondioli
Department of Materials
and Environmental
Engineering
University of Modena
and Reggio Emilia
Via Vignolese 905/a
41100 Modena
Italy

The Nano–Micro Interface: Bridging the Micro and Nano Worlds
Edited by Hans-Jörg Fecht and Matthias Werner
Copyright © 2004 WILEY-VCH Verlag GmbH & Co. KGaA, Weinheim
ISBN: 3-527-30978-0

STEFFEN CHEMNITZ
Ruhr-Universität Bochum
Fraunhofer Institutszentrum
Schloss Birlinghoven
Biomolecular Information Processing
53754 Sankt Augustin
Germany

CEDRIC CHEUNG
University of Toronto
Department of Materials Science and
Engineering
184 College Street, Wallberg Building
Toronto, Ontario M5S 3E4
Canada

MARCO CUCINELLI
Interstate University of Applied
Science
Institute for Microsystems NTB
Werdenbergstrasse 4
9471 Buchs
Switzerland

ALEX DOMMANN
Interstate University of Applied
Science
Institute for Microsystems NTB
Werdenbergstrasse 4
9471 Buchs
Switzerland

UWE ERB
University of Toronto
Department of Materials Science and
Engineering
184 College Street, Wallberg Building
Toronto, Ontario M5S 3E4
Canada

DIETMAR EHRHARDT
University of Siegen
Institute for Microsystem Technology
Hölderlinstrasse 3
57068 Siegen
Germany

STEPHAN ERTL
GFD-Gesellschaft für Diamantprodukte
mbH
Lise-Meitner-Strasse 13
89081 Ulm
Germany

HANS-JÖRG FECHT
University of Ulm
Faculty of Engineering
Materials Division
Albert-Einstein-Allee 47
89081 Ulm
Germany

ALEXANDER FISCHER
University of Siegen
Institute for Microsystem Technology
Hölderlinstrasse 3
57068 Siegen
Germany

ANDRÉ FLÖTER
GFD Gesellschaft
für Diamantprodukte mbH
Wilhelm-Runge-Strasse 13
89081 Ulm
Germany

RALF-PETER FRANKE
University of Ulm
Department of Biomaterials
ZIBMT
Albert-Einstein-Allee 47
89081 Ulm
Germany

THOMAS FRIES
FRT, Fries Research & Technology
GmbH
Friedrich-Ebert-Strasse
51429 Bergisch Gladbach
Germany

MAKOTO FUJIMAKI
Waseda University
Shinjyuku
Tokyo 169-8555
Japan

STANISLAW GIERLOTKA
High Pressure Research Center
of the Polish Academy of Science
Sokolowska 29/37
01-142 Warsaw
Poland

PETER GLUCHE
GFD Gesellschaft für
Diamantprodukte mbH
Wilhelm-Runge-Strasse 11
89081 Ulm
Germany

LARISA GRIGORJEVA
University of Latvia
Institute of Solid State Physics
8 Kengaraga str.
1063 Riga
Latvia

EWA GRZANKA
High Pressure Research Center
of the Polish Academy of Science
Sokolowska 29/37
01-142 Warsaw
Poland

SHARIFAH BEE ABD HAMID
University of Malaysia
Postgraduate Studies & Research
50603 Kuala Lumpur
Malaysia

DARIUSZ HRENIAK
Institute for Low Temperature
and Structural Research
of the Polish Academy of Science
Okolna 2
50-950 Wroclaw
Poland

YVONNA JIRÁSKOVÁ
Academy of Sciences
of the Czech Republic
Institute of Physics of Materials
Zizkova 22
61662 Brno
Czech Republic

THEODORE I. KAMINS
Hewlett Packard Laboratories
Quantum Science Research
1501 Page Mill Road
Palo Alto, CA 94304
USA

JÜRGEN KOGLIN
FRT, Fries Research & Technology
GmbH
Friedrich-Ebert-Strasse
51429 Bergisch Gladbach
Germany

VOLODYMYR KOZIY
University of Siegen
Institute for Microsystem Technology
Hölderlinstrasse 3
57068 Siegen
Germany

MARTIN KULAWSKI
VTT Centre for Microelectronics
P.O. Box 1208
02044 VTT-Espoo
Finland

CRISTINA LEONELLI
University of Modena
and Reggio Emilia
Department of Materials
and Environmental
Engineering
Via Vignolese 905/a
41100 Modena
Italy

Silvia Licoccia
Università di Roma Tor Vergata
Dipartimento di Scienze e Tecnologie
Chimiche
Via della Ricerca Scientifica 1
00133 Rome
Italy

Lerwen Liu
Nanotechnology Consultant
Nanotechnology Research Institute
National Institute of Advanced
Industrial Sciene and Technology
1-1-1 Umezono
Tsukuba City Ibaraki 305-8568
Japan

Witold Lojkowski
High Pressure Research Center
of the Polish Academy of Science
Sokolowska 29/37
01-142 Warsaw
Poland

Hiromichi Maeno
Mitsui & Co., Ltd.
8F Sumitomo Fudosan Hamacho
Building
3-42-3 Nikonbashi Hamacho
Chuo-ku
Tokyo 103-0007
Japan

Lhadi Mehari
CERAMEC R&D
64 Avenue de la Libération
87000 Limoges
France

Matthias Meyer
FRT, Fries Research & Technology
GmbH
Friedrich-Ebert-Strasse
51429 Bergisch Gladbach
Germany

Donats Millers
University of Latvia
Institute of Solid State Physics
8 Kengaraga str.
1063 Riga
Latvia

Jyrki Molarius
VTT Centre for Microelectronics
P.O. Box 1208
02044 VTT-Espoo
Finland

Peter Müller
IBM Zurich Research Laboratory
Saeumerstrasse 4
8803 Rueschlikon
Switzerland

Marc-Aurele Nicolet
California Institute of Technology
Pasadena, CA 91125
USA

Ken-ichi Nomura
Waseda University
Shinjyuku
Tokyo 169-8555
Japan

Yoshimichi Ohki
Waseda University
Shinjyuku
Tokyo 169-8555
Japan

Agnieszka Opalinska
High Pressure Research Center
of the Polish Academy of Science
Sokolowska 29/37
01-142 Warsaw
Poland

Bogdan Palosz
High Pressure Research Center
of the Polish Academy of Science
Sokolowska 29/37
01-142 Warsaw
Poland

Gino Palumbo
Integran Technologies Inc.
1 Meridian Road
Toronto, Ontario M9W 4Z6
Canada

Tuomas Pensala
VTT Centre for Microelectronics
P.O. Box 1208
02044 VTT-Espoo
Finland

Nadezda Pizurová
Academy of Sciences
of the Czech Republic
Institute of Physics of Materials
Zizkova 22
61662 Brno
Czech Republic

Adam Presz
High Pressure Research Center
of the Polish Academy of Science
Sokolowska 29/37
01-142 Warsaw
Poland

Edward Reszke
Ertec-Poland
Rogowska 146/5
54-440 Wroclaw
Poland

Mihail C. Roco
U.S. National Science and Technology
National Science Foundation
4201 Wilson Blvd.
Arlington, VA 22230
USA

Laura Rossi
IBM Zurich Research Laboratory
Säumerstraße 4
CH-8803 Rüschlikon
Switzerland

Otilia Saxl
Stirling University
Institute of Nanotechnology
Innovation Park
Stirling FK9 4NF
Scotland/UK

Thomas Sawitowski
AlCove Surfaces GmbH
Am Wiesenbusch 2
45966 Gladbeck
Germany

Heiko Schäfer
University of Siegen
Institute for Microsystem Technology
Hölderlinstrasse 3
57068 Siegen
Germany

Robert Schlögl
Fritz-Haber-Institut der
Max-Planck-Gesellschaft
Faradayweg 4–6
14195 Berlin
Germany

Torsten Schmidt
Genthe-X-Coatings GmbH
Im Schleeke 27–31
38642 Goslar
Germany

Oldrich Schneeweiss
Academy of Sciences
of the Czech Republic
Institute of Physics of Materials
Zizkova 22
61662 Brno
Czech Republic

Frank Schulz
AlCove Surfaces GmbH
Am Wiesenbusch 2
45966 Gladbeck
Germany

Konstantin Seibel
University of Siegen
Institute for Microsystem Technology
Hölderlinstrasse 3
57068 Siegen
Germany

Tomasz Strachowski
High Pressure Research Center
of the Polish Academy of Science
Sokolowska 29/37
01-142 Warsaw
Poland

Wieslaw Strek
Institute for Low Temperature
and Structure Research
of the Polish Academy of Science
Okolna 2
50-950 Wroclaw
Poland

Enrico Traversa
Università di Roma Tor Vergata
Dipartimento di Scienze e Tecnologie
Chimiche
Via della Ricerca Scientifica 1
00133 Rome
Italy

K. Venkat Rao
KTH The Royal Institute of Technology
Department of Materials Sciences
10044 Stockholm
Sweden

Michael Wegener
University of Potsdam
Department of Physics
Am Neuen Palais 10
14469 Potsdam
Germany

Matthias Werner
NMTC
Soorstraße 86
14050 Berlin
Germany

Werner Wirges
University of Potsdam
Department of Physics
Am Neuen Palais 10
14469 Potsdam
Germany

Markku Ylilammi
VTT Centre for Microelectronics
P.O. Box 1208
02044 VTT-Espoo
Finland

Tomás Zák
Institute of Physics of Materials
Academy of Sciences
of the Czech Republic
Zizkova 22
61662 Brno
Czech Republic

1
Nanotechnology Research Funding and Commercialization Prospects

U.S. National Nanotechnology Initiative: Planning for the Next Five Years

Mihail C. Roco

1
Introduction

Nanoscience and nanotechnology are opening up a new era of integrated fundamental research at the nanoscale, a more coherent science and engineering education, economic nanoscale manufacturing of products, and an enabling foundation for improving human capabilities and societal outcomes in the long term. The U.S. National Nanotechnology Initiative (NNI) is a visionary program that coordinates 17 departments and independent agencies [1–5] with a total budget of U.S.$ 961 million in the fiscal year 2004. An overview of the main research and development (R&D) themes, outcomes in the first two years of the initiative, and plans for the future are presented. At least 35 countries have initiated national activities in this field, partially stimulated by the NNI vision and plans.

Priority in funding in 2004 is oriented to:
- research to enable the nanoscale as the most efficient manufacturing domain;
- innovative nanotechnology solutions to biological, chemical, radiological, and explosives detection and protection;
- development of instrumentation and standards;
- nanobiosystems;
- the education and training of a new generation of workers for the future industries;
- societal implications; and
- partnerships to enhance industrial participation in the nanotechnology revolution.

Priority nanoscale science and technology goals in the next five years are in currently exploratory areas of research (including nanomedicine, energy conversion, food and agriculture, realistic simulations at the nanoscale, molecular nanosystems), in areas transiting to technological innovation (nanostructured materials, nanoelectronics, catalysts, and pharmaceuticals, development of tools for measurement and simulation), and in areas to advance broad societal goals (such as better understanding of nature and life, increasing productivity in manufacturing, interdisciplinary education, improving human performance, and sustainable develop-

The Nano–Micro Interface: Bridging the Micro and Nano Worlds
Edited by Hans-Jörg Fecht and Matthias Werner
Copyright © 2004 WILEY-VCH Verlag GmbH & Co. KGaA, Weinheim
ISBN: 3-527-30978-0

ment). Societal and educational implications, including environmental research, will increase in importance in NNI as nanotechnology products and services reach the market.

Several generations of nanotechnology products are expected to evolve from relatively simple nanostructures for products such as coatings and hard metals, to active components such as nanoscale transistors, and then nanosystems with new architectures. This chapter shows how miniaturization, self-assembling from molecules up, and multiscale architectures lead to the integration of nano- and microcomponents into system applications.

2
Government R&D Investments

The worldwide nanotechnology R&D investments reported by government organizations has increased more than six-fold from U.S.$ 430 million to about U.S.$ 3 billion between 1997 and 2003 (Tab. 1 and Fig. 1). At least 35 countries have initiated national activities in this field, partially stimulated by the U.S. NNI.

Scientists have opened a broad net that does not leave any major research area untouched in the physical, biological, materials, and engineering sciences. Industry has gained confidence that nanotechnology will bring competitive advantages to both traditional and emerging fields, and significant growth is noted in small businesses, large companies, and venture capital firms. The annual global impact of products where nanotechnology will play a key role was estimated in 2000 to exceed U.S.$ 1 trillion by 2015, which would require about 2 million nanotechnol-

Tab. 1 Estimated government nanotechnology R&D expenditures during 1997–2003 (in U.S.$ millions/year).

Region	1997	1998	1999	2000	2001	2002	2003
West Europe	126	151	179	200	~225	~400	~600
Japan	120	135	157	245	~465	~700	~810
USA [a]	116	190	255	270	422 (465) [b]	600 (697) [b]	774 (862) [b]
Others	70	83	96	110	~380	~550	~800
Total	432	559	687	825	1492	2347	2984
(% of 1997)	100	129	159	191	346	502	690

Explanatory notes: West Europe includes countries in EU and Switzerland; the rate of exchange U.S.$1 = 1.1 Euro until 2002; U.S.$1 = 1 Euro in 2003; Japan rate of exchange U.S.$1 = 120 yen in 2002; others include Australia, Canada, China, Eastern Europe, FSU, Israel, Korea, Singapore, Taiwan and other countries with nanotechnology R&D.

a) A financial year begins in the United States on 1 October of the previous calendaristic year, six months before most other countries.

b) Denotes the actual budget recorded at the end of the respective fiscal year. Estimations use the nanotechnology definition as defined in NNI [1]; this definition does not include MEMS, and includes the publicly reported government spending.

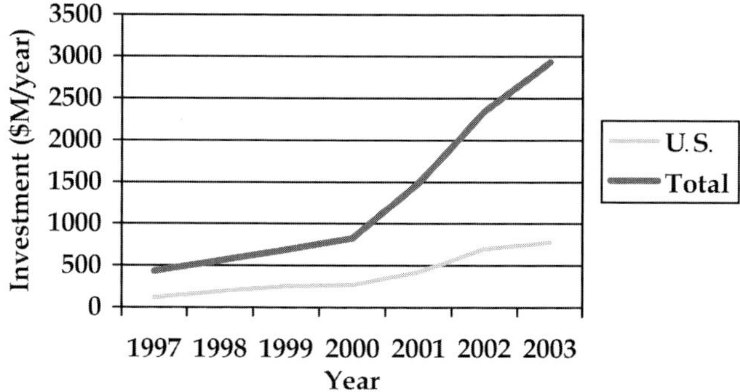

Fig. 1 Government investments in nanotechnology during 1997–2003.
Upper curve = total worldwide including USA; lower curve = USA.

ogy workers [3]. This estimate was based on the analysis of existing R&D activities in industry in the U.S., Japan, and Western Europe. One notes that U.S.$ 1 trillion represents about 10% of the U.S. GDP in 2003. If one would extrapolate the previous experience, where for each information technology worker another 2.5 jobs are created in related areas, nanotechnology has the potential to create 7 million jobs overall by 2015 in the global market. Also, if one considers the impact of information technology of increasing U.S. productivity more than 1% per year in 1990s (that is roughly half of the overall productivity growth of about 2.1% in the 1990s), a similar or possibly larger impact is expected from nanotechnology. This is because the impact is broader than a new generation of electronic hardware once nanotechnology is reaching a critical mass in knowledge and commercial markets. One may note that the initial estimates for information technology significantly under-estimated its long-term positive implications (because of successive and non-scalable qualitative changes) and over-estimated the negative effects (beginning with the risk of macroscale robots that would take over the world). By envisioning the potential synergism of many fields contributing to nanotechnology and various phases of its introduction, a similar scenario would be possible at an even more pronounced scale.

The U.S. has initiated a multidisciplinary strategy for development of science and engineering fundamentals through the NNI based on a long-term vision [1]. The estimated Federal Government budget for nanotechnology research U.S.$ 961 million in fiscal year 2004, and the request is U.S.$ 982 million for the fiscal year beginning in October 2004 [5]. Japan [6–7], the European Community (Fig. 2) [8] and more recently China [9] have initiated broad programs, and their current plans look up to five years ahead. Other countries, including Korea [10], Taiwan [11], Australia [12], Canada, Eastern Europe, Israel, India, and Singapore have encouraged their own areas of strength, several of them focusing on fields of the potential markets. Their rate of increase in government spending in the last year is higher than the sum of all other three areas (U.S., Japan, Western Europe). Differences among countries are observed in the research domains they are aiming for,

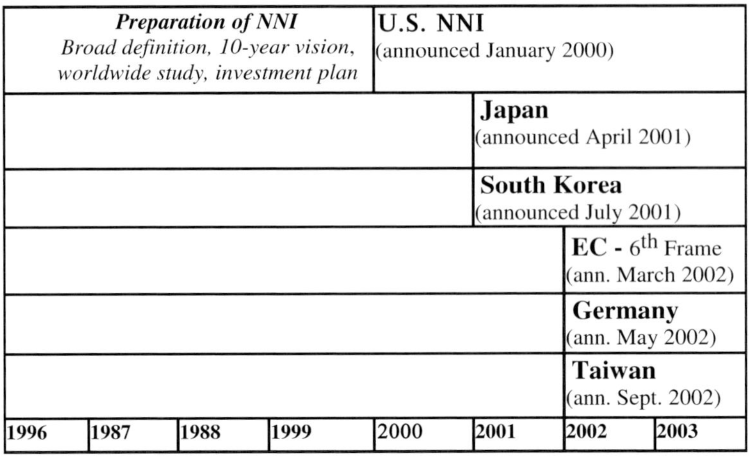

Preparation of NNI *Broad definition, 10-year vision, worldwide study, investment plan*	**U.S. NNI** (announced January 2000)		
		Japan (announced April 2001)	
		South Korea (announced July 2001)	
			EC - 6th Frame (ann. March 2002)
			Germany (ann. May 2002)
			Taiwan (ann. Sept. 2002)
1996	1987	1988	1999	2000	2001	2002	2003

Fig. 2 Comprehensive nanotechnology research programs with funding exceeding U.S.$ 100 million/year by national governments or EC, announced after 2000.

Tab. 2 Contribution of key agencies to NNI.

Federal department or agency	FY 2000 Actual ($M)	FY 2001 Actual ($M)	FY 2002 Actual ($M)	FY 2003 Actual ($M)	FY 2004 Current Plan ($M)	FY 2005 Request ($M)
National Science Foundation (NSF)	97	150	204	221	254	305
Department of Defense (DOD)	70	125	224	322	315	276
Department of Energy (DOE)	58	88	89	134	203	211
National Institutes of Health (NIH)	32	40	59	78	80	89
National Institute of Standards and technology (NIST)	8	33	77	64	63	53
National Aeronautics and Space Administration (NASA)	5	22	35	36	37	35
Environmental Protection Agency (EPA)	–	6	6	5	5	5
Homeland Security (TSA)	–	–	2	1	1	1
Department of Agriculture (USDA)	–	1.5	0	0	1	5
Department of Justice (DOJ)	–	1.4	1	1	2	2
Total	270 (100%)	465 (172%)	697 (258%)	862 (319%)	961 (356%)	982 (364%)

the level of program integration into various industrial sectors, and in the time scale of their R&D targets.

The actual U.S. NNI budget in fiscal year (FY) 2003 was U.S.$ 862 million and current plan in FY 2004 is U.S.$ 961 million (Tab. 2). The budget decreases in FY 2004 request noted at NASA (National Aeronautics and Space Administration) and DOD (Department of Defense) in the FY 2005 may be explained by the reassignment of applied nanotechnology projects to the respective areas of relevance instead of NNI. The state and local organizations committed additional funds for infrastructure, education, and commercialization of more than half of the NNI investment in 2002.

The NNI centers and networks of excellence encourage long-term system-oriented projects, research networking, and shared academic users' facilities. These nanotechnology research centers play an important role in the development and utilization of specific tools, and in promoting partnerships (Tabs. 3, 4).

The research outcomes are not proportional to the investments because of research productivity, various components of the infrastructure, and culture. For example, the timeline of the patents recorded with U.S.PTO (U.S. Patent and Trade Office) is shown in Fig. 3. That office receives domestic and foreign applications as being the main target for investors because the U.S. provides the largest single market.

Tab. 3 NNI centers and networks of excellence.

Center name	Institution
NSF	
Nanoscale Systems in Information Technologies, NSEC (Nanoscale Science and Engineering Center)	Cornell University
Nanoscience in Biological and Environmental Engineering	Rice University
NSEC	
Integrated Nanopatterning and Detection, NSEC	Northwestern University
Electronic Transport in Molecular Nanostructures, NSEC	Columbia University
Nanoscale Systems and their Device Applications, NSEC	Harvard University
Directed Assembly of Nanostructures, NSEC	Rensselaer Polytechnic Institute
Nanobiotechnology, Science and Technology Center	Cornell University
DOD	
Institute for Soldier Nanotechnologies	MIT
Center for Nanoscience Innovation for Defense	UC Santa Barbara
Nanoscience Institute	Naval Research Laboratory
NASA	
Institute for Cell Mimetic Space Exploration	UCLA
Institute for Intelligent Bio-Nanomaterials & Structures for Aerospace Vehicles	Texas A&M
Bio-Inspection, Design and Processing of Multi-functional Nanocomposites	Princeton
Institute for Nanoelectronics and Computing	Purdue

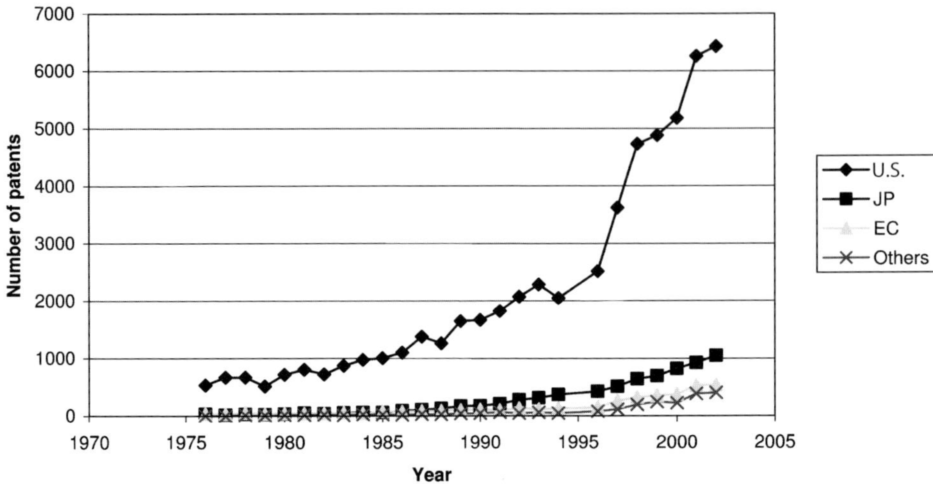

Fig. 3 Number of nanotechnology patents per four regions (1976–2002). The leading ten countries in 2002 were: U.S., 6425 patents; Japan, 1050; France, 245; UK, 100; Korea, 87; Taiwan, 86; Netherlands, 66; Australia, 61; Switzerland, 55; Italy, 44. The survey was taken using the USPTO database in April 2003 [13].

Tab. 4 NNI R&D user facilities, 2003.

Center name	Institution
NSF	
National Nanofabrication Users Network (NNUN; this network has been re-competed in 2004 with an expanded role and increased budget under the National Nanotechnology Infrastructure Network)	Cornell University Howard University Stanford University Pennsylvania State University UCSB
Network for Computational Nanotechnology	Purdue University University of Illinois Stanford University University of Florida University of Texas, El Paso Northwestern University Morgan State University
DOE	
Center for Functional Nanomaterials	Brookhaven National Laboratory
Center for Integrated Nanotechnologies	SNL and LANL
Center for Nanophase Materials Sciences	Oak Ridge National Laboratory
Center for Nanoscale Materials	Argonne National Laboratory
Molecular Foundry	Lawrence Berkeley National Laboratory

Nanotechnology is growing in an environment where international interactions accelerate in science, education, and industrial R&D, while industrial competitiveness difficulties are surfacing at national and industry consortia levels. Government investments in nanotechnology have jump-started the development of nanoscale science and engineering. Government activities should equally prepare society for introduction of new technologies and products, as well as future unexpected consequences of nanotechnology such as health and environmental concerns.

Acknowledgments

This chapter is based on a keynote presentation the 2003 NAMIX Conference in Berlin on 26 May 2003 and an updated information in [5]. Opinions expressed here are those of the author and do not necessarily reflect the position of NSET (the U.S. National Science and Technology Council's subcommittee on Nanoscale Science, Engineering and Technology) or NSF (National Science Foundation).

References

1 Roco M.C., Williams R.S., and Alivisatos P. (eds.), 1999, *Nanotechnology Research Directions*, U.S. National Science and Technology Council, Washington, DC (also Kluwer Academic Publishers, Boston, 2000, 316 pages). http://nano.gov

2 Roco M.C., *National Nanotechnology Investment in the fiscal year 2004*. Budget Request by the President, 2003, AAAS Report 28: R&D in FY 2004, Washington, DC, pp. 255–264. http://nano.gov

3 Roco M.C. and Bainbridge W. (eds.), *Societal implications of nanoscience and nanotechnology*. National Science Foundation Report (also Kluwer Academic Publishers, Boston, 2001, 370 pages).

4 Roco M.C. and Bainbridge W.S. (eds.), *Converging technologies for improving human performance*, National Science Foundation and Department of Commerce Report, June 2002 (also Kluwer Academic Publishers, Boston, 2003). http://www.nsf.gov/nano

5 U.S. National Nanotechnology Initiative official web site: http://www.nano.gov

6 Yamaguchi Y. and Komiyama K., Structuring knowledge project in nanotechnology: Materials program launched in Japan. *J. Nanoparticle Res.* **2001**, *3*, 105–110.

7 Government of Japan: *The Science and Technology Basic Plan (2001–2006)*, Tokyo, 30 March 2001.

8 EC, Sixth R&D Framework, Draft 6, November 2002, Brussels, Thematic Area 3 "Nanotechnology and Advanced Materials", Luxembourg.

9 Bai C., Progress of Nanoscience and Nanotechnology in China, *J. Nanoparticle Res.* **2001**, *3*, 251–256.

10 Lee J.W., Overview of nanotechnology in Korea – 10 year blueprint, *J. Nanoparticle Res.* **2002**, *4*, 473–476.

11 Lee C.K., Wu M.K., and Yang J.C., A catalyst to change everything: MEMS/NEMS – a paradigm in Taiwan's nanotechnology. *J. Nanoparticle Res.* **2002**, *4*, 377–386.

12 Braach-Maksvytis V., Nanotechnology in Australia – towards a national initiative. *J. Nanoparticle Res.* **2002**, *4*, 1–7.

13 Huang Z., Chen H., Yip A., Ng G., Guo F., Chen Z.-K., and Roco M.C., Longitudinal patent analysis for nanoscale science and engineering: country, institution and technology field. *J. Nanoparticle Res.* **2004** (in press).

Technological Marketing for Early Nanotechnologies

Murielle Batude-Thibierge

1
Introduction

Today's descriptions of nanotechnology markets and trends contain a lot of estimates or "guesstimates". Information is not reliable, customers and suppliers have few references, and consequently only subjective market pictures are available. But if you are as rigorous with market data as you are with scientific and technological data, you will be able to demonstrate your vision.

The purpose of this chapter is to give the scientific audience a synopsis of marketing and management tools adapted to nanotechnological innovations.

When describing future nanotechnology market segments, nearly every author (from top political officials, through scientists, to venture capitalists) has a different way of mapping the industrial scene. Many use basic industrial groupings (which they call sectors, instead of market segments) but this is inadequate. Expert market analyzers face a difficult situation: there is a need to assess the commercial impact of nanotechnologies rapidly and in a pragmatic manner. The need is to evaluate the overall impact of nanotechnology, but the only information available is some micro-market studies, because no reliable data has been collected and markets are yet to be invented.

In these conditions the temptation is to analyze the available technology, but describe it as an analysis of the market. This is not valid, so when reading such studies you must be careful not to fall into the trap of confusing the offer with the demand. To illustrate the difference think about the following example.

It takes a long time to assess, in a particular business case, what technical problem an innovation might solve, who might be interested, how much potential purchasers might pay, and if they are willing to pay. It is much easier to look at the Internet and find the sales forecasts of a few big manufacturers who hope that such an innovation could work. But if you do that, you are not doing a marketing study (determining demand): you are analyzing what is on offer.

You also need to understand the difference between a validation marketing study and an exploratory marketing study. In the former, you know your future customers and you examine several scenarios, whereas in the latter, the problem is fuzzy: you do not know the market your innovation will reach and you try to

The Nano–Micro Interface: Bridging the Micro and Nano Worlds
Edited by Hans-Jörg Fecht and Matthias Werner
Copyright © 2004 WILEY-VCH Verlag GmbH & Co. KGaA, Weinheim
ISBN: 3-527-30978-0

work this out. An exploratory market study is more creative; it also acts as a first anchor with potential customers when you interview them. Professor Millier of the Lyon Graduate School of Management developed a method for undertaking an exploratory market study [1].

If a chemical or pharmaceutical multinational asks how nanotechnology could renew their product lines, the research is difficult. This is not the supplier looking for markets but the purchaser looking for nanotechnology offers that could impact their product portfolio. First, note that it is likely that nanotechnology could also renew their processes, but the purchaser gives marketing researchers no data about processes. Second, when assessing the impact of nanotechnology on the product range, the study can only be superficial. The word "nanotechnology" might be a great managerial trick: a strong incentive for science to restructure around nanoscience, and for ministries to work hand in hand. If so, it is also a marketing trap. Trying to assess how nanotechnology could affect the product range of a large company would require hundreds of single marketing studies. The problem is too wide. Such a poor brief is likely never to be answered reliably, unless people realize what it requires and allocate sufficient time and money. (An assessment of the cost of a marketing study is included later in this chapter). To do this, it is necessary to grasp what it means to analyze markets that do not yet exist.

In the first part of this chapter we present a method for studying markets that do not yet exist. This method can be adapted also to make it easier to study existing markets for nanotechnology innovations.

We will then show how a good exploratory marketing study enables us to formulate a good marketing strategy and prepare its implementation because it has:

- converged effort and thinking by giving functional unity and coherence to your project,
- given you a means of controlling the introduction of your innovation in the market, and
- created a guide for smart resource allocation.

1.1
Managerial Synthesis with Recommendations

The real aim of marketing is to possess markets, not just to sell products, and elegant marketing consists of defining your idea of the whole cake (MacKenna [9]).

The "whole cake" is something you make: it does not exist before your product goes on sale. The "big and ready market" is a myth. Therefore you need to think about how you are going to create a demand for your product early in the innovation process and how you will drive your institution towards commercializing your innovation successfully.

In order to create the demand market, you need to map it out and choose which market segments you are going to target, being aware of the risks for your organization. Only then can you say how you are going to address them.

To give you the essence of what is contained in this chapter, we will start by asking a few questions:

- Do you identify industrial sectors as market segments? If you do not know, you probably do.
- Do you identify the offer market with the demand? That is, do you say "the global market for glucosensors is so much, so the market for my nanoglucosensor is the same"? This is not valid, because there might be some customers who do not need the performance you will expect them to pay for, some who need these sensors to work under some constraints your sensors cannot sustain, some who cannot afford it, and so on.
- Do you look at the Internet (or something similar), compile data, and declare "here is my market!". This does not give you primary data. You do not know how reliable secondhand data is or how it was collected. You should not build a marketing strategy based on secondhand data because you will lack all the necessary information that could be gathered by visiting the customer.

Our main recommendations are the following:

- Differentiate between industrial sectors and market segments.
- Differentiate between the offer market and the demand market.
- Differentiate between the applications of your innovation and its functions.

Our general advice on how to study markets that do not yet exist comprises these key points:

- Observe downstream instead of what you already know: your technology is upstream.
- Use primary data acquired in the field by visiting the customer instead of using secondhand data.
- Do not be afraid to start by collecting qualitative information and then converting it to quantitative data. Our marketing methodology tries to quantify as much as possible and gives you tools to do so.
- Look for contacts (conferences, publications, patents holders, databases) and experts who know the technique and the market (to save time).
- Exchange technicians between nanotechnology supplier and customer until integration is completed.

This marketing methodology has the virtue of giving common marketing tools and vocabulary to a team and of involving a collective effort that will strengthen any management decision based on its results.

1.2
Working Definitions

1.2.1 Nanotechnology Product

Nanotechnology is the creation and utilization of materials, devices, and systems through the control of matter at the nanometer length scale [2]. It involves not only working on a smaller scale, but also using new properties that appear at this scale. Sellable products do not consist exclusively of nanoparts, but the marketer will evaluate the market for a new smaller device exploiting nanocomponents.

1.2.2 Innovation

Innovation can be incremental: marginally improving the product, but it is still based on something the customer knows and can refer to. Innovation can also disrupt the customer's work and purchasing habits. The customer might have to learn how to implement and ensure maintenance.

1.2.3 Technological Function

Technological functions are often expressed in terms of performance, of improvement for a given process (my technology "improves, increases, allows").

1.2.4 Application

The application of a technology is a technical problem at the customer. The compilation of customers' technical problems can be used to discover generic problems that the innovation can try to address.

Note that technological functions and their applications are different. By the end of the marketing study we want to pair them up, when they are called "market segments", but this is a result: it is not given from the start, otherwise there would not be a need to analyze the market.

1.2.5 Market Segmentation

Segmentation of the demand market means dividing it into homogeneous groups according to their behavior towards our nanotechnology innovation, and thus to the key success factors.

1.2.6 Translation Process

This is the dialectical process occurring during the diffusion of an innovation, which consists of shaping the right message to efficiently communicate with the market so that the innovation is adopted. This is the theory of Callon and Latour. Setting the vocabulary is a key to a successful translation process.

1.2.7 Collective Learning Process

A collective learning process occurs when a group tries to capitalize on its experience as a group. It means that each party is conscious of playing a role with regards to the general intellectual enhancement of the group and knows its share of responsibility in assessing if the messages are efficiently conveyed between members. Such a willing attitude is needed in nanotechnology in order collectively to build the message that will sell nanotechnological products and services. The "collective learning process" should include the engineers, researchers, marketers, and customers.

1.3
Setting the Scene

We imagine ourselves to be a small nanotechnology company characterized by having a good technical base but rather poor commercial know-how. More than 80% of nanocompanies in Europe are small companies, defined as having revenues below US\$ 10 million. In comparison the USA has fewer small companies but more medium-sized companies [3].

Suppose you are in such a company with a nanotechnology project. The project is not necessarily at an advanced stage but even at an early stage you want to be able to answer these questions.

- Do I really master the whole technology of my offer?
- Should I continue with this project?
- Should I sell it?
- Should I wait until I improve the technology?
- Should I commercialize it right now?
- In this case, how can I exploit it to the maximum?
- What range of products can be derived from it?
- Who are the first customers who will open the market for me?
- How do I present my innovation to them?
- Which market segments will I then be able to address?
- Who are my competitors?

For most nanotechnology products (apart from nanotools, for the characterization and manipulation of nanostructures) we cannot set targets such as trying to increase the turnover from X to Y within one year and maintain an annual $Z\%$ growth. The reason for this is that most nanotechnology innovation projects are in a transitory state. The transitory state is a turbulent and intermediate state in an innovation's life cycle: between its research life and its life as a product. At this stage, especially for nanotechnologies, there is a proliferation of perceived applications due to the power of generic discoveries. One cannot afford to tackle all the applications at once. The marketer must guide the nanotechnology project towards a few lucrative applications.

To select them and to identify the market segments the marketer needs a methodology. The methodology presented here does not focus on competitors (which would be the center of a marketing approach in a highly competitive environment) but on the demand market and the customer. Technological innovations are not (or should not be) comparable with what competitors could do, making it irrelevant to insist too much on this aspect of a marketing study.

Most nanotechnology products imply a business-to-business relationship or a business-to-research laboratory relationship. We are talking about industrial marketing, that is we deal with group purchasing with power over suppliers. Soon some nanotechnology developments might imply a business-to-consumer relationship that will necessitate communicating through the media. This study focuses on marketing towards industry and research labs.

1.4
"Raison d'être" of Marketing, Especially for Nanotechnologies at Early Stages

Working at the nanometer scale not only allows more precise control of materials but also reveals new properties such as surface effects, molecular forces, thermal vibration, quantum effects, and nanoscale damage mechanisms. These new properties make nanotechnology potentially an extremely disruptive technology. Nanotechnologies have the potential to revolutionize existing markets and create whole new ones. This is the right time to try to exploit nanotechnology to the maximum because standards have not yet been agreed that could block the way for dramatic developments.

A broad range of supply methods and working methods (for businesses and individual consumers) will evolve, but this will not happen smoothly; it will happen when some companies identify the right points at which to force their way into the market. To catch the full potential of this revolution one has to dramatically

- stretch ones imagination,
- refuse determinism,
- recall the essence of customer needs,
- envisage, anticipate, and accelerate the function–application pairings
  that have the best chance of success.

This is marketing's mission for nanotechnology.

1.5
"Raison d'être" of Management Thinking and Strategic Planning
for Nanotechnologies

Between the nanotechnology research center trying to create spin-off companies, and the multinational trying to exploit nanotechnology in their current products, there are differences in means, interest (probably sometimes divergence of interests), and culture. For example, the nanotechnology research center is likely to be less at ease with integration of marketing and strategic thinking than the multinational. However, those centers have the same critical mass of research tools needed to achieve quality nanotechnological breakthroughs as multinationals.

From a management point of view, being able to switch from small entities to big corporations, and from public or semi-public research centers to multinationals, is much needed. Let us illustrate the diversity of behaviors this implies: in Europe, some nanotechnology research centers are private, but most are public institutions that have only recently stepped into the business world. They have been working hand in hand with industry for a long time, but have not previously expected any revenue from this work, so they did not value their research in the same way that a private research lab would have done. The manager of such an establishment faces a broad range of issues arising from this history and culture.

The number of nanotechnology start-ups (defined as companies with nanotechnology as their core business) is growing worldwide. There has been an exponential growth of the number of nanotechnology companies founded in the last ten years. Roughly 66% of all active nanotechnology companies in the world have emerged since 1990 [3]. In small entities like this, having people who know what is at stake on different fronts as diverse as legal, organizational behavior, competition, entrepreneurial finance, political lobbying, and so on is crucial. There are few people able to deal with the culture of all those different parties because one has to have been sensitized to a broad range of business issues.

In a multinational, implementing nanotechnology (unless it is already part of daily work) means being able to create the need for change within the organization and to sustain the change. Strategic thinking is also a good advocate of nanotechnology. Nanotechnology will reveal technological impact once a solution for mass fabrication is found (which is not the case for most cases). There is an opportunity for a concentration of the key success factors: companies will make economies of scale, have a renewed value chain, they will have a new approach to their industrial infrastructure, and they will have access to new markets that would not exist without nanotechnology.

Nanotechnology has government support and lobbying *is* possible. For example, "Ambri", a small Australian start-up in the field of medical diagnosis, obtained great financial support from the Australian government.

The managers' mission is to exploit nanotechnology for the achievement of the business strategy, to deal with all kinds of people involved in nanotechnology developments (from public to private organizations) and, in small spin-offs, to anticipate on all fronts.

1.6
Problematic Nanotechnologies

Owing to the characteristics of nanotechnologies and the general challenges of launching a high-tech revolution, there are many difficulties along the way. Nanotechnologies are not just high-tech innovations. They are complex. Business people and scientists have to acquire knowledge additional to their initial background, especially when working on nanotechnology innovations in the biotechnology field. A common and efficient vocabulary has to be found. Do you know the difference between a nanate and a nanite, for example? (Solution at the end of the chapter.)

Benchmarking in nanotechnology is a very complex task because it is necessary to compare heterogeneous industries, sciences, and techniques and to deal with different entities (university labs, company labs, and so on). A prerequisite for

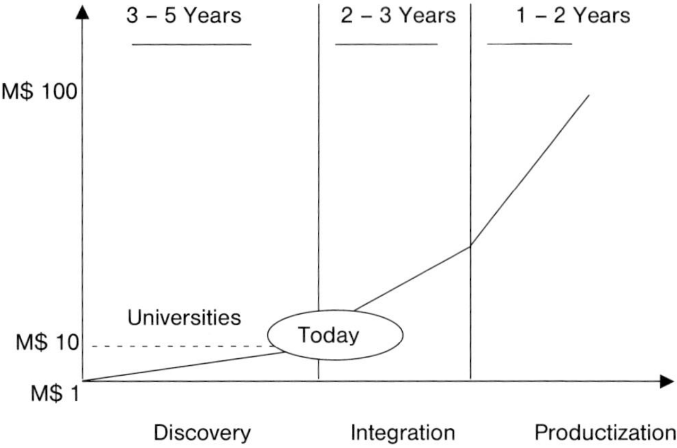

Fig. 1 Phases of a project in nanotechnology [2].

nanoscience is the characterization of nanostructures, which involves clean rooms, microscopes, and simulation software, and therefore only players with a critical mass of tools can keep in the race. Nanotechnologies take time to develop: Fig. 1 shows the phases of a project in nanotechnology.

1.6.1 Discovery

In this phase annual expenditure is over US$ 1 million per year. Partners at this stage are: universities, national labs, government groups such as research funds, project agencies, or chamber of commerce in the US.

1.6.2 Integration

Integration costs are about US$ 10 million per year. Try to find engineering solutions because you have already solved the science. Look for various manufacturing partners to take the financial risk or look for government sponsorship.

1.6.3 Exploitation

Turning an idea into a product costs about US$ 100 million a year and only large companies can play this game. Nanotechnologies are not just high-tech innovations because they imply a multiplicity of actors. The techno-economic networks are still in the process of being built at a global scale. They involve:

- the researchers (labs, research centers, *sensitive to customer focus or not*),
- the engineers (companies, *sensitive to customer focus or not*),
- the marketers (companies, *scientists or not*),
- the policy makers (state, *scientists or not*),
- the customers (military, businesses, individuals, *scientists or not*), and
- the financiers (venture capitalists, banks, *scientists or not*).

And last but not least, there is a risk of technological backlash, like for GM food. With regards to Rogers thesis [4], there are many bad points for nanotechnologies.

- There is no symbolic advantage of nanotechnology (declaring you do nanotechnology only works with a certain elite, which not everyone belongs to).
- Nanotechnology is not compatible with the value system of the target (not sure at all that society will not react like with GM).
- Nanotechnology is complex (what can be done to make it easy to understand, what can be done to make it easy to use?).
- Nanotechnology is not divisible. (Are samples relevant? What can be done to demonstrate effectiveness to the customer? Is it light? That is, is it possible to integrate just a part of it in the customer supply chain?)
- Nanotechnology is intangible some of the time.

2
Marketing for a Nanotechnological Innovation

2.1
Marketing Study Budget

If you intend to be as rigorous with market data as with scientific data you must plan how much this will cost. An exploratory market study can cover questions dealing more with market entry for innovation introduction or technological competition and so on. Once you have assessed the object of analysis, "taken a picture of the problem" before the study (in order to compare it with the picture after the market study) you can start planning the marketing actions required. Tab. 1 shows market research and the marketing budget.

Tab. 1 Market research and the marketing budget [1].

Marketing study budget	8–15 k Euros	15–45 k Euros	45–90 k Euros
Expert interview	one on the phone	2–3	3–8
Face to face interview	3 interviews	5–20	20–50
Phone interview	15–20	10–30	20–100
Time: months per man	1–2	3–5	5–10

2.2
Collecting Information for Marketing Tools

Marketing is the window of a business or a research center aiming at commercialization; it proposes to investigate both outside (commercial and technological landscape) and inside (center resources and ability to employ them) and resulting from this confrontation it shapes an optimized offer.

To look outside you will build up an interview guide, not a questionnaire, which means that you will allow the interviewee to be freer during the phone or face-to-face interview. In business-to-business marketing or business-to-research laboratory marketing, you and your customer do create the first function–application fits and make a breach into a new market. Therefore you have to let imagination have a bit of space during the interview process.

Hint: progressively guide the interview towards the object of analysis, that is your innovation, then expose the innovation characteristics and then open conversation again to stimulate the customer's creativity and make him/her suggest applications

Hint: exposing the innovation characteristics, a few observations.
- Methods regularly used in education (drawings and calculations) are often forgotten in a business environment, but business is the place for thinking and communicating.
- Often designs and pre-prepared models (powerpoint presentations and graphs that pop up all at once) are unconscious ways of avoiding explanations: we all avoid cooperating – it is tiring!
- It is better to do live drawings in front of the audience, accompanied with body motions and words (use overhead projector with transparent slides and color pens).
- Skipping calculations by just giving the results or formulae is another way of avoiding explaining. Calculations acquire real relevance if their goal is clear, every step is explained, and the result is given in context.

It is also relevant to mention the frequent practice of the conference call. Native English speakers tend to believe that everyone understands English well, but this can be a costly assumption. The idea of setting the idioms in English and then communicating them broadly could help reduce these costs. It is important that the communities reach agreement on the vocabulary to be used, with everyone having an input. Examples of idioms that help understanding are "top-down and bottom-up approaches" or "nanates and nanites".

As the interview guide aims at collecting information to feed the marketing tools (technical analysis and economical analysis), one needs to include three elements: the customer's technical problem, behavior towards our innovation, and commercial expectations.

2.3
Technical Analysis

Here we give a feel for the structure of the technical analysis, shown in Fig. 2.

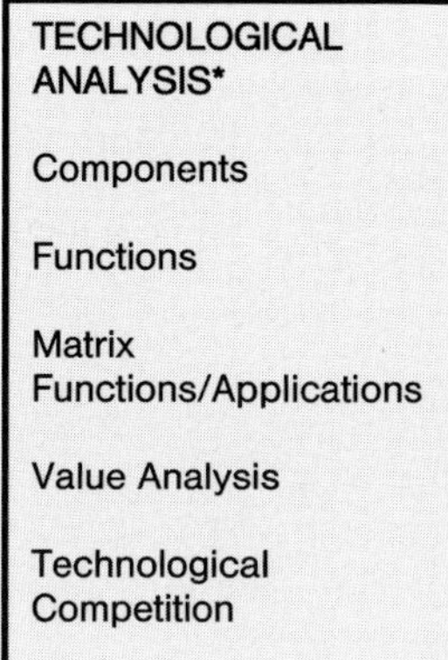

Fig. 2 The structure of technical analysis [1].

2.3.1 Components

Understanding of the innovation should be broken down into simple elements in order to make sure you master each component. After undertaking this step of the analysis you will be able to answer the basic question: "My technology is made of …". It is likely that your technology is not, in fact, only composed of physical components but also involves the equipment required to make it, theoretical knowledge, and know-how. So one has to recognize and value the full range of skills. Bearing this in mind, start by establishing the material and abstract components from conception to technology transfer, maintenance, and recycling (see Tab. 2).

Tab. 2 The component elements of a nanotechnological innovation

Technical element	Key?	Basic?	Level of mastery	Position amongst competition
Nano	✓		N/A (table designed for a given company)	N/A (Table designed for a given company)
Micro		✓		
System for integration of nano- and microtechnology	✓			

2.3.2 Functions

One needs to explain clearly what the technology can do. This means that you want to be able to answer simply the question: "My technology is made for . . . ."

- Thinking process: determine (service functions/constraints functions/architecture functions/esteem); classify (dominant or induced functions); link functions to performance (quantitative formulation).
- Marketing tool: arborescence. Example of metal oxide nanoparticles [3].
- F 1: allow better, finer polishing.
- F 2: acute detection sensitivity as sensors.
- F 3: increase reaction rate as catalysts.
- F 4: increase UV protection and are transparent.

In general, functions are often expressed in terms of performance improvement for a given process ("improve, increase, allow" and so on).

2.3.3 Matrix of Functions and Applications

- Goal: determine the offer that could be declined starting from the innovation.
- Thinking process: by interviewing customers we can pass from particular uses to generic applications, which we define as the types of problems we can solve for customers. Thanks to the previous step we already have the functions of our technology so we are naturally tempted to draw the function/application matrix to find out if some pairing are possible. The function/application pairs are possible market segments: this function of our technology could solve this application.
- Marketing tool: function/application matrix.

Example of nanoparticles in general: this is heretic, in a methodology sense because it deals with too broad a concept, but it gives an idea in a way that is easy to understand [3].

- Nanoparticles can determine the rate of reaction, the place, and timing: applications as drug delivery agents
- Better surface stability in air and aqueous environments: applications as coatings
- Better UV absorption: applications as cosmetics
- Better control of viscosity and thermal expansion: applications as lubricants
- Lighter, more resistant, thinner: applications as advanced materials.

Then to complete the matrix you would add for each market segment a note about what performance level should be achieved and what level of control you have.

2.3.4 Value Analysis

- Goal: to optimize the set of components necessary to obtain the requested functions, offer rationalized steps, and use for mastering costs.

- Thinking process: first agree the functions to deliver with the community, then define the necessary components from supply to delivery.
- Marketing tool: double entry table with components and functions.

2.3.5 Technological Competition

- Goal: try to solve customer's technical problem or needs better than others can.
- Thinking process: analyze direct and indirect competition on the function we propose.
- Marketing tool: again use the Functions/Applications matrix by inserting in the boxes the competitors' names as well as the performances they reach.

2.4
Commercial Analysis

This part aims at getting the best commercial effect with the less commercial effort (Fig. 3). We will not review all that Professor Millier proposes; in particular, we will not discuss the analysis of the competition. Commercial analysis is probably what is often lacking in nanotechnology: being developed in an environment where the focus is on technology, the "easy" part is missing. Unfortunately the purchase decision depends mostly on the non-technical part of the offer. So, as well as technological leadership, both commercial know-how and the commercial part of the offer are important.

Thanks to the previous technical part, we now know where we want to go technology wise: we have selected those market segments that we want to work on and we have defined them in terms of how the function fits the application. For each of the market segments we now need to analyze the economic and commercial environment relevant to the introduction of our innovation. This means being able to characterize this commercial environment: structurally and functionally. Structurally: who is playing there? With whom? Functionally: how will this evolve? Who is a decision maker? Who is a trend maker? Hence, whom should we first convince with our innovation?

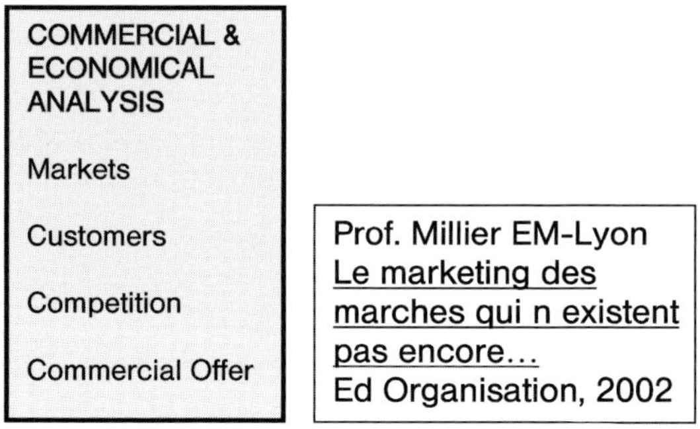

COMMERCIAL &
ECONOMICAL
ANALYSIS

Markets

Customers

Competition

Commercial Offer

Prof. Millier EM-Lyon
Le marketing des
marches qui n existent
pas encore...
Ed Organisation, 2002

Fig. 3 Commercial analysis [1].

2.4.1 Influence Matrix and Drive/Dependence Matrix

If we can convince the group of customers that has most influence on the rest of the industry, we save time and money. These matrixes aim at analyzing which group of customers should be addressed first.

2.4.2 Customer Behavior Towards Innovation

Motivation differs between a customer with industry logic and a customer with technical logic. The first type might be less likely to invest time in codeveloping, and reluctant to adopt until your innovation has had some success. Whereas if you target a problem that a research lab faces, they are likely to invest time and money for it to work. In the case of an industrial customer, let us review on which sensitive cords we can play [5]: Tab. 3 shows motivation for behavior towards innovation.

2.4.3 Analysis of Key Commercial Success Factors

Each customer group needs to be addressed with a special commercial strategy. There is a whole spectrum of commercial know-how:

- general characteristics of the company (core activity, positioning, image),
- sales and commercialization (customer focus, quality, availability),

Tab. 3 Motivation for behavior towards innovation [5].

Commercial motivations	Social motivations
Better serve customers	Reduce head count
Make products customers are asking for	Save jobs
Reach new markets, new customers	Answer a staff request
Change image	Reduce nuisance
Boost top line (revenues)	
Competition motivations	**Legal motivations**
Get ahead of competitors	Norms, laws, fiscal or political hint,
Catch up with competitors	
Financial motivations	**Environmental motivations**
Reduce investment, inventory	Ecology, natural constraints
Produce at less cost	(snow, dryness, rains,)
Increase profitability	
Obtain subsidies	
Obtain decrease in assurance	
Industrial motivations	**Strategic motivations**
Lead a quality policy	Independence towards a monopoly
Increase productivity	Suppliers' nationality
Use a tool to the maximum	Diversification
Find a unique solution to a technical problem	

- capacity to make the customer evolve (customer understanding, innovations in the pipeline),
- service (staff training), and
- financial elements (price).

2.5
Defining a Price

You should not base your pricing policy on competitive products, because your product is significantly different (and your competitors might not have set the price at the best level). Without a reference pricing is very difficult. The best way is to find out how much profit your nanotechnological innovation would bring to the customers. To evaluate the profit made by the customer you need to have them use it for a while. So initially you will sell at symbolic price or lease it; then you can undertake a cost comparison study [6].

2.5.1   The Cost Comparison Study: Step 1

This is a quick and easy way to identify whether the customer can reduce their costs thanks to your innovation. You build a base cost walk (exposure of each base cost situation before and after) using current facility data and nanotechnology innovation implementation costs. The study targets the following cost areas:

- former supply cost per activity,
- associated costs such as warehousing, services, transportation, and
- lost revenue resulting from former defect rate.

This enquiry can be done through a standardized questionnaire, which sales representatives should help fill in. The resulting table pinpoints the areas where the nanotechnology system would reduce the monthly costs.

2.5.2   The Cost Comparison Study: Step 2

Once it has been established that the innovation could save money, a further analysis is required. This should include: capital and related recurring costs of the nanotechnology equipment and labor, less labor and consumable costs that might be eliminated by the implementation over a 10 year period. The study targets the following cost areas:

- current operation budget,
- current procedure volume and supply costs,
- current central department operations,
- defect correction spending,
- transportation,
- system acquisition cost,
- implementation cost,

- travel, and
- anticipated reduction in operations management.

2.6
Quantified Diagnostic and Simulations

Even when we have started thinking about how to set a price, we are still evaluating the market. Then we should compile the results of our technical and commercial analyses. For each market segment found we need to assess the technical risk and the commercial risk. Thus, for each market segment, we quantify parameters that come into play to assess both those risks separately. We end up with two matrixes: one containing all the segments represented as circles (size representing market size) positioned depending on their level of technical risk and the other containing all the segments represented as circles positioned depending on their level of commercial risk. Then it becomes easy to synthesize technological analysis and commercial analysis into a unique matrix with all the segments positioned according to both their technical and commercial risk. The final matrix is shown in Fig. 4.

For most nanotechnology innovations, we are likely to observe that most segments are located in the top right hand corner of the diagram, representing high technological and commercial risks.

2.6.1   Simulations on the Diagnostic Matrix

We can try to rank the probability of each segment for bad technical results. By assuming that we become able to control one or more of the technical stumbling blocks, we can see how each market segment is affected and moves down the dia-

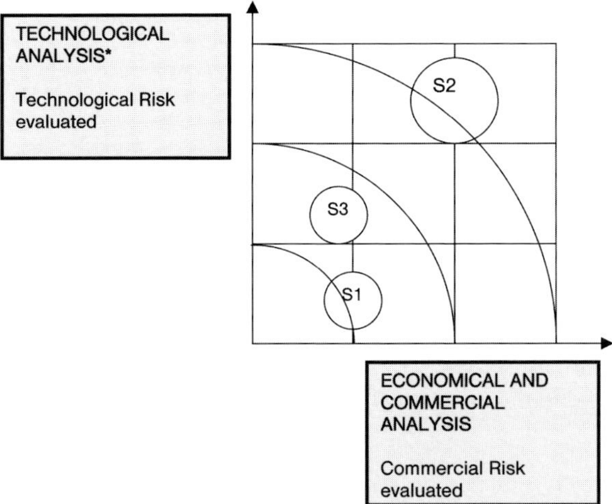

Fig. 4 Final marketing matrix.

gram. If we find a powerful parameter that drags nearly all segments down, we are likely to decide to work on this technical aspect first.

2.6.2 Exploitation of the Diagnostic Matrix

We can evolve from segment to segment, trying to make them reachable from a technical and a commercial point of view until we have exploited the full range of the applications we had forecast, always checking that our company has the means to undertake the actions.

3
Management Thinking and Strategic Planning for Small Nanotechnology Businesses

Any entrepreneur has to be a strategic planner. Strategic planning means being able to formulate a strategy resulting from an in-depth and structured analysis. But strategic management is also about taking into account right from the start that people will have to implement it. This understanding should influence the content of the strategy, the level of collaboration during its formulation, and the means used to communicate it.

A nanotechnology entrepreneur must embody the attraction of their vision for their innovation. Hence, it is necessary to master the translation process across a variety of backgrounds, cultures, and companies, and to make adopters feel the need for implementing and securing change.

3.1
Strategic Planning: from Segment Action Plan to Business Action Plan

Thanks to the marketing matrix simulations, we have chosen an efficient order in which to address the market segments. Additionally, we know which customers to contact in the first place from the drive/dependence matrix. Basically the action for this business activity is planned. One has to check it is coherent with the whole business strategy and resources, but the main work is done.

3.2
Co-Developing with a Big Player

If you intend to co-develop with a big player, it will be helpful for you to understand the methods and management tools that they use to manage change. Here I explain the method known as "Six Sigma" used by General Electric (GE) and the management tools used to lead change inside this organization.

3.2.1 The Six Sigma Method to Stimulate a Disruptive Change

The Six Sigma method was first developed at Motorola [7] and General Electric (GE) has taken to it with enthusiasm: "Six Sigma: The Way we Work". You need to know about the Six Sigma methodology because it entails project management beyond company boundaries. You and your client need to share quality standards and vocabulary if you are to convince them to adopt your innovation. We can not explain Six Sigma in detail here, but basically if you use the CTQ approach, you will considerably gain their attention (CTQ = Critical To Quality issues: matters that have been identified as significant for satisfying the needs and expectations of the consumers).

GE leaders do not hesitate to talk about the Six Sigma Quality Attitude, which consists of the following mottos.

- The Customer is the Final Judge of Product or Service Quality.
- Understanding Customer Needs is a Prerequisite to Improving Operations.
- Meeting Customer Needs at the Lowest Possible Cost is the Only Way to Compete Successfully.

A Six Sigma project can involve customers, and the methodology works for any type of project whether financial, marketing, engineering, or operations. Two methodologies are used: DMAIC (Define Measure Analyze Improve Control) for existing product or service improvement and DMADV (Define Measure Analyze Design Verify) for new product and service introduction. Nanotechnological innovations are likely to fall under both of them equally.

3.2.2 Management Tools for Change

If you can get your nanotechnology team to participate in a business experience such as a GE workout, they will better understand what is at stake for your customer. Here we show some management tools used to accelerate a change across the company either when facing an acute and urgent problem or in a more general way [8].

Workouts are powerful problem solving processes. Even if constantly evolving, the key concepts are speed, simplicity, and self-confidence. Participation in any workout is determined by the nature of the problem to be solved. Participants may be any level, any function, and from inside/outside the organization. Clients and suppliers often join workout sessions to the benefits of their own business as well as GE's. All ideas are valued; everyone who participates has a stake in the outcome. Results, action-oriented plans, and recommendations are communicated directly to business leaders for their decisions and support. Follow-up is part of the process.

Speed teams are teams that use the workout method to generate solutions to problems. These "speed" or "action" team are so called because they are able to make decisions and implement solutions within the workout sessions. The groups do not need to refer their decision to any other person within the organi-

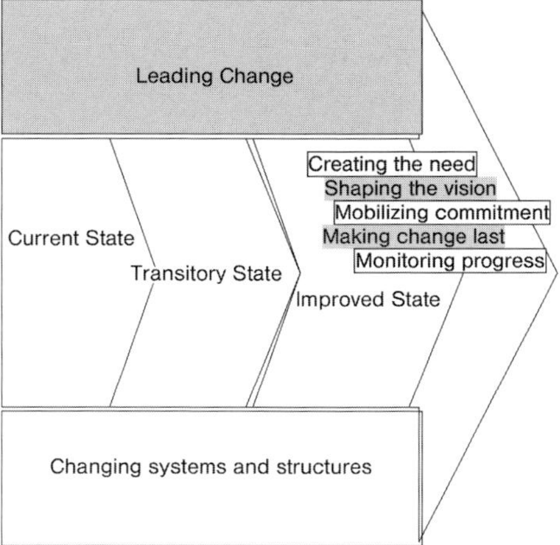

Fig. 5 The change acceleration process.

zation: the decision makers are in the speed team. "Speed" or "action" means that action is taken immediately. In addition the topics being worked on usually relate to speeding up company processes and procedures.

Change acceleration processes is the management tool that is especially relevant to an organization trying to integrate nanotechnology into their business thinking habits. Here is how GE explains this process:

> *In many ways, leading any organization to higher levels of success is about effectively managing change. Beyond quality (technical) change, a critical element of any successful change is the degree to which the individuals and/or teams support the change and to what extent. CAP consists in tools to build acceptance within the organization in order to reach the most ambitious goals.* [8]

The process should be used with suppliers and customers alike. Insist on expectations of levels of performance, volume of business, and quality of work.

Quick market intelligence what GE calls best practice in quickly getting information from the field, and this is vital when you lead a disruptive change. QMI is a way for a large business to respond to issues, competitive intelligence, new ideas, and other information as a small business would. QMI meetings are scheduled on a regular basis with representatives from around the world. Cross-functional team members share their insights on the issues being presented, which should relate to at least one of the following areas:

– competition,
– pricing,
– quality assurance,
– customer satisfaction, and
– lessons learnt.

3.3
Mastering the Translation Process

In a small nanotechnology business where you do not have all the means you would want to commercialize and diffuse your innovation, there is a need to be more clever than others with regards to the translation process. It is not likely to be expensive: it is much more of a mindset and the implementation of a value system. In communications it is very easy for any specialist (either scientist or marketer) to lose the understanding of the less expert counterpart. Thus the culture of being valued because you efficiently conveyed a message is key. We so often meet people who want to rub in the message they know better than you.

3.3.1 Example

Professor Williams of Hewlett-Packard's Palo Alto Quantum Research labs gave an account of his team's experience at the *Asia Pacific Nanotechnology Forum* in February 2002. We report their conclusions here.

Let us set the scene: 27 people are working in basic research. Professor Williams needs to master the invention process: How to invent? Of course there is some general advice such as "think architecture first" and so on, but in this chapter we are more interested in the team building aspect. Their method was most importantly networking and, at the same time, analyzing their own collective learning process. It took them two years of adaptation to understand what each was saying, but then, only then "they started inventing like crazy" [2].

We propose a scenario. The team is composed of specialists either in science or business (because very few people have a double culture of engineering and business, but this will be the case in a few years).

- First, we already experience a lack of researchers.
- Second, the needs in nanotechnology will pretty soon overrun the number of trained students. In the USA grants for molecular biology are much greater than those for nanotechnology: the NNI receives half a billion dollars per year, but the life sciences budget is US$ 10 billion per year: 20 times more! So naturally undergraduate students are flowing into life sciences at the cost of maths, engineering and physics.
- Third, it takes 10 years for each specialty to be mastered (10 years for engineering and 10 years for business as testified a californian venture capitalist who assured he had to work 10 years in engineering and 10 years in purely commercial functions to be qualified enough to work into venture capitalism for nanotechnologies – example of the venture capitalist of Alameda).

Thus we have to share knowledge more than before. A collective learning process must occur within nanoteams to capitalize on experience. This means that every party must be conscious of playing a role with regards to the general intellectual enhancement of the group and know its share of responsibility in assessing if the messages are efficiently conveyed between members.

3.3.2 **Nanotechnology Team Building**

Effective team building uses basic game rules, such as those used by GE in training exercises.

- Open minded: "We are all beginners". We have been able to see at the atomic level for less than 30 years, and on a historical time scale we are just at the beginning of nanoworld exploration.
- Curiosity: "The fields of technology and applications are larger than the current knowledge".
- Creative thinking skills: "think outside the box". Do not allow your ideas to be constrained by any knowledge you currently have on the topic.
- Interactivity: "This is my chance, it is my opportunity to learn".
- Challenge each other.
- Have fun!

Becoming really good at doing that is cheap and can give a strong competitive advantage to nanotechnology teams.

3.4
Formulating a Strategy

A suitable strategy should be short, simple, realistic, and be communicated to every concerned party. The strategy should encompass the corporate strategy, strategy for each business activity, and strategy for each market segment.

For a business to be successful in the long term, an innovation is not sufficient, however good it is. You also need a clear goal when you put together your corporate identity. Do you want to make money (income), to make your business big and reach the whole world (growth), or do you want to improve the lot of your stakeholders (social added value)? These goals are all good, but you need to be conscious of your goal because it should be reflected in your strategy.

The formulation of your strategy should also reflect the value system you want to put in place in your organization. If, as we have suggested, you would like to see your team be a living example of mastery of the translation process, you will probably have to make it clear and to invent some tools to encourage it.

3.5
Implementing the Strategy

The best strategies are those that are implemented, not those that stay in the cupboard. If you have analyzed the exploitation of your innovation in the ways that we have explained here, there is a very good chance that you will be successful. Your people have been involved in deciding which market segments to go for and how to go for them. Hence they should support the decision when choosing, for example, to tackle a market segment that is riskier but potentially bigger. People will know where the business is going and will act accordingly.

4
Conclusions

From the beginning of this chapter we have said that we would adopt a different strategy from what is commonly done with regards to segmenting the possible nanotechnology markets. We believe the best way to control the hype is to undertake exploratory micro-market studies and then compile the results to get a bigger picture.

We have seen that a marketing study can be undertaken and deliver a quantified diagnostic, even for markets as new as the ones nanotechnology can create. This methodology reduces the complexity of the innovation situation as it puts down rationally all possible data. It also permits good communication as it makes great use of graphs and diagrams. The necessary involvement of technology and marketing people into this market segmentation helps to base strategy on technical and economic risk assessment and enhances the chances of buy-in from the team.

We have underlined that no innovation can replace entrepreneurial drive and a company mission, and that strategic thinking had to rise above the innovation that justified creation of a company in the first place.

We have exposed some good managerial practices and methodologies that can be shared across company barriers especially when co-developing with a multinational. We have especially insisted on the need for nanotechnology innovation teams to master the translation process, and to be conscious of their own collective learning.

5
Appendix

Definition of Nanates and Nanites

Nanates are passive applications of nanotechnology. They are already in the market place. Some of them have been in the market place for a very long time (for example photographic film, which exploits chemical reactions of nanoscale particles in a nanocomposite). They are easy to make: there are not many entry barriers. For example, in nanocomposites you mix two components with different properties and get a new material with the properties of both components. So you can obtain toughness and hardness by "mixing" polymers and ceramics, which is useful for plastics, baskets, and tires.

Nanites are active applications of nanotechnology. They are active systems, in which information is actively transferred into a particular nanoscale system or between a nanoscale system and its surroundings (Fig. 6).

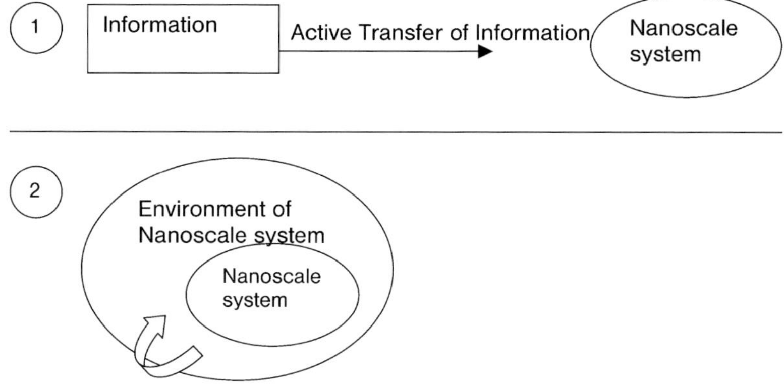

Fig. 6 Definitions of nanates and nanites.

Acknowledgments

The author acknowledges the creation and authorship of the methodology to her professor, Professor P. Millier, professor of Marketing and Innovation at EM-Lyon Graduate School of Management.

References

1 MILLIER P., *L'étude des Marches qui n'éxistent pas Encore*. Paris: Editions. d'Organisation, **2002**.

2 WILLIAMS S., in speech at the *Asia Pacific Nanotechnology Forum*, Tsukuba, Japan, Feb **2002**.

3 FECHT H., WERNER M., Nanotechnology. *Companies and Markets – Finding Hidden Pearls*, Ulm, Germany, **2003**.

4 ROGERS E.M., *Diffusion of Innovations*, 4th ed. New York: The Free Press, **1995**.

5 MILLER P., *Marketing the Unknown: Developing Market Strategies for Technical Innovations*. Wiley, **1999**.

6 General Electric. *Comparison of cost studies*, Engineering staff documentation.

7 Six Sigma training – *Quality Taining*, DMAIC Binder.

8 General Electric, Abbreviations Glossary, *CAP – Change Acceleration Process Section*.

9 McKENNA R., Harvard Business Review, *Marketing is Everything*, **1991**.

Asia–Pacific Nanotechnology:
Research, Development, and Commercialization

Lerwen Liu

1
Nanotechnology Funding in the Asia–Pacific Region

The Asia–Pacific (AP) region is advancing to becoming the most ambitious and dynamic nanotechnology region in the world. There have been significant changes in science and technology policy making in AP countries since the announcement of the US National Nanotechnology Initiative (NNI) in January 2000. Governments in the AP region nominated nanotechnology as one of the priority areas in science and technology planning. The budget for nanotechnology research and development (R&D) has been increased substantially and more strategically allocated. The total public spending in the AP region is about US$ 1.4 billion for 2003 (70% of which is from Japan, assuming 100 yen equals 1 US$) and private investments are increasing. The appreciation of the importance of nanotechnology R&D has been growing in various industries and businesses in the region.

The commercialization of micro- and nanotechnology has become a key focus in government and corporate R&D strategy particularly in advanced countries such as Japan. Manufacturing facilities for microelectromechanical systems (MEMS) are being established across Asian countries including Japan, Korea, and Taiwan. Nanotechnology in particular is the buzzword in science and technology policy making all over Asia. In most Asian countries the definition of nanotechnology includes MEMS.

Japan, one of the most technologically advanced countries, has been investing in nanotechnology since the mid-1980s with various national programs (typically with a period of 5–10 years). Its government funding for nanotechnology per capita has been the highest in the AP region and the world. Its funding for 2002 has been increased about 20–30% from 2001. The funding for 2003 will be similar to that in 2002, but there is a stronger focus on the convergence of nanotechnology and biomedical areas. Countries such as China, South Korea, and Taiwan have increased drastically their nanotechnology spending since 2001.

China has planned to spend 2–2.5 billion RMB (US$ 250–300 million) within the current five-year plan (2001–2005). More aggressive initiatives are about to be launched as China wishes to match the funding of other more advanced countries

The Nano–Micro Interface: Bridging the Micro and Nano Worlds
Edited by Hans-Jörg Fecht and Matthias Werner
Copyright © 2004 WILEY-VCH Verlag GmbH & Co. KGaA, Weinheim
ISBN: 3-527-30978-0

such as South Korea. A National Nanotechnology R&D Center is currently being built near Beijing University, TsingHua University, and the Chinese Academy of Sciences (CAS), which is expected to be finished by 2004. A Nanotechnology Industry Base was built in Tianjin (about 100 km east of Beijing), and is fully operational since 2003.

South Korea has committed 2.391 trillion won (US$ 2 billion) over a 10-year period (2001–2010). The increase in the government spending in nanotechnology for 2002 compared with 2000 is about 400%. One of the goals of its National Nanotechnology Initiative is to make Korea first in the world in certain competitive areas and to develop niche markets for industrial growth. Korea has a clear focus on a number of core technologies such as Tera-level Integration of Electronic Devices. The *Year 2002 Plan for Implementing Nanotechnology Development* was launched together with two new Frontier Research Programs: *Development of Nanostructured Materials Technologies*, and *Development of Nanoscale Mechatronics & Manufacturing Technologies*. Each of the programs is funded with 100 million for the next 10 years. In addition to the Frontier Research Programs for Nanotechnology, the Korean government has launched "Core", "Basic", and "Fundamental" nanotechnology research programs whose total research budget is about US$ 20 million every year for the next 6–9 years. To establish the infrastructure, a nanofabrication center was set up last year at the Korea Advanced Institute of Science and Technology (KAIST) in Daejoen Science City, where most government research labs are located. US$ 165 million has been allocated for this center over 9 years (2002–2010). The Government has recently formulated the *2003 Action Plan for Nanotechnology Development*, which includes the *Presidential Decree and Enforcing Regulations* for implementing the *Nanotechnology Development Promotion Act*. The aim of this Act is to prepare a solid research basis for nanotechnology and to encourage industrialization of matured nanotechnology. The Korean government also allocated US$ 380 million (19% of the total nanotechnology spending) to the *National Nanoindustrialization Program* including an Industrial R&D Fund and a Venture Capital Fund.

Taiwan's *National Initiative on Nanoscience & Technology* is a 6-year plan with a total spending of US$ 620 million from 2003 to 2008. Its strategy and programs are very similar to the US National Nanotechnology Initiative. It aims to achieve academic excellence and create innovative industrial applications through the establishment of common core facilities and education programs. The *Academic Excellence Program* covers the topics of: basic research on the physical, chemical, and biological properties of nanostructures; synthesis, assembly, and processing of nanomaterials; R&D of probes and manipulation techniques; design and fabrication of interconnections, interfaces, and systems of functional nanodevices; development of MEMS/NEMS technology; and nanobiotechnology.

Taiwan places important emphasis on nanotechnological education. Its education program includes: establishing interdisciplinary nanoscience and technology curricula at college and high-school level; enhancing basic science education from high school to college; promoting international collaboration and personnel exchange; recruiting talents from abroad including personnel from mainland China; and promoting academic–industry collaborating research and personnel exchange.

Other countries in the AP region such as Australia, Hong Kong, India, New Zealand, Singapore, Malaysia, and Thailand have launched their own nanotechnology programs and initiatives. Tab. 1 shows funding information. Note that the funding in each country may vary in its definition of nanotechnology.

Tab. 1 Funding comparison for Asia–Pacific countries in 2003.

Country	Population	Funding (5y)	Priority	Policy coordination
Australia	19.2 million	100 M	Bio, IT	Common Core Facilities
China	1.2 billion	300 M+	Mtl, ME	National and Regional Centers
Hong Kong	6.7 million	30 M	Mtl, IT, Egy	Centers of Excellence
India	1.0 billion	20 M+	Mtl, MEMS	Centers of Excellence
Korea (South)	48.3 million	1 B	Mtl, Ets, Bio	National Centers and Core Facilities
Malaysia	21.8 million	23 M+	Mtl	National Centers and Core Facilities
New Zealand	4 million	50 M	Mtl, Ets	Centers of Excellence
Singapore	4.2 million	60 M	Mtl, Ets, Bio	Centers of Excellence
Taiwan	21.5 million	500 M	Mtl, Ets, MEMS	Common Core Facilities
Thailand	62 million	25 M	Mtl, MEMS	National Centers

Bio = Biomedical, Egy = Energy, Ets = Electronics, IT = Information technology, ME = Molecular electronics, Mtl = Nanomaterials

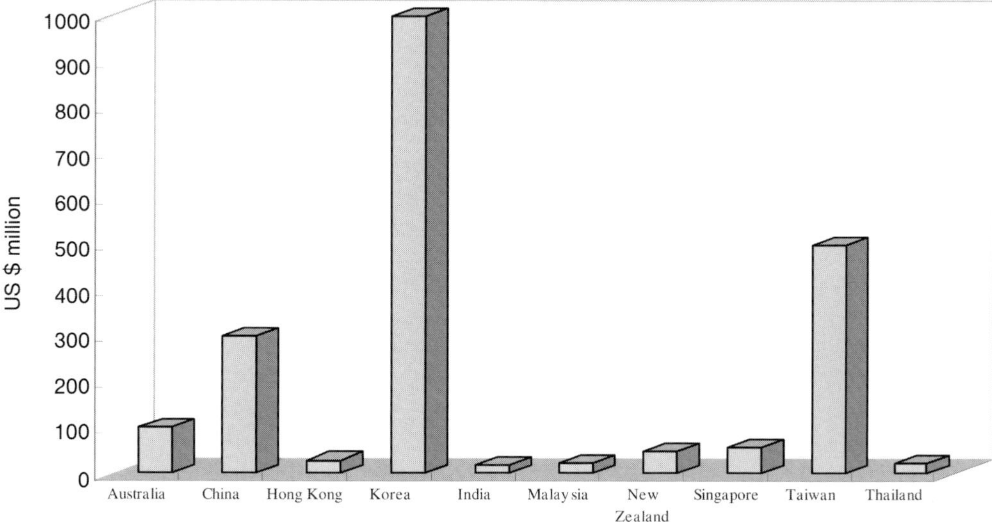

Fig. 1 Funding comparison for Asia–Pacific countries for 2003–2007.

Tab. 2 Funding comparison for EU, Germany, Japan, and USA for 2001–2003. Note that the USA funding figure does not include microelectronics or MEMS.

Country	2001	2002	2003
EU	300	450	700
USA	422	695	774
Japan	781	1044	1044
Germany	150	200	230

All amounts in US $ millions

Fig. 1 compares nanotechnology funding for 2003–2007 among AP countries including China, Korea (south), Hong Kong, India, Malaysia, New Zealand, Singapore, Taiwan, and Thailand.

Tab. 2 shows the global comparison in nanotechnology funding for EU, USA, Japan and Germany during 2001-2003. The unit is US$ and it is assumed 100 yen = 1 US$ and 1 Euro = 1 US$. The funding rose substantially globally since 2002.

Fig. 2 is a plot of EU, Japan and USA funding figures for 2001–2003.

The total funding for the AP region including Japan has risen particularly substantially since 2003, as shown in Tab. 3 and Fig. 3, and will continue to increase when developing Asian countries increase their science and technology spending.

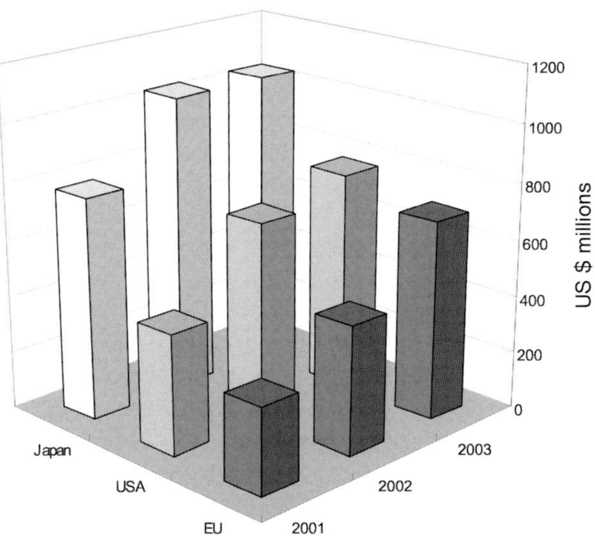

Fig. 2 Funding comparison for EU, Japan, and USA for 2001–2003. Note that the USA funding figure does not include microelectronics or MEMS.

Tab. 3 Funding comparison for AP, EU, and USA for 2002–2004. Note that the figures for 2004 for both AP and EU are not confirmed.

Country	2002	2003	2004
EU	450	700	700
USA	695	774	845
Asia	900	1400	1400

All amounts in US$ millions

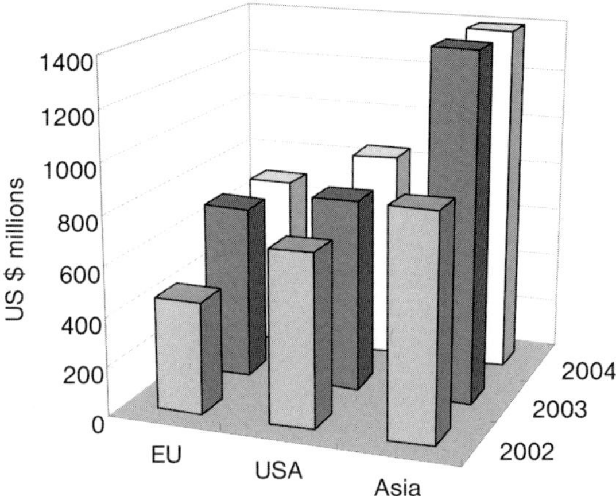

Fig. 3 Funding comparison for AP, EU, and USA for 2002–2004.
Note that the figures for 2004 for both AP and EU are not confirmed.

2
Commercialization Efforts

The Japanese government has made a strong commitment in nanotechnology to creating new industries as an important policy for revitalizing the Japanese economy. The government adopted at the end of last year the New Industry Development Strategy (NIDS) for Nanotechnology and Materials in the Council on Economic and Fiscal Policy (CEFP). The Ministry of Economy, Trade and Industry (METI), which is the key ministry in Japan supporting Japanese industries, launched in 2003 the following R&D programs for economic revitalization:

- Nanotechnology & Materials – 20 projects, 11.6 billion yen (US$ 116 million).
- IT + Nanotechnology & Materials – 23 projects, 22.4 billion yen
  (US$ 224 million).
- Life Science + Nanotechnology & Materials – 6 projects, 3.3 billion yen
  (US$ 33 million).

Tab. 4 Japan market projections of 5 key industries in 2010.

Nano industries	Trillion yen	US$ billions
Innovation materials	0.6–1.4	6–14
Nano environment and energy	0.9–1.7	9–17
Nano Bionic	0.6–0.8	6–8
Network and nano devices	17–20	170–200
Nano metrology and manufacturing	0.8–2.2	8–22

Japan has been very strong in the fine ceramics industry (occupying over half of the world market share) and high-resolution electron microscopy (occupying 60–70% of the world market share). As for the new nanotechnology industry, Japan is expecting the market size of the five key areas that Japan expects to lead the world by 2010 shown in Tab. 4.

MEMS technology R&D is included in the nanotechnology program in most Asian countries. In Japan, METI has just launched the New Manufacturing Technology Program MEMS Project (US$ 20 million for 2003–2005) targeting device realization of RF-MEMS, optical MEMS, and ultra-small MEMS sensors. The government has also established a policy to overcome the problems in Japanese MEMS development such as lack of MEMS engineers (several hundreds compared with 5000 in EU), lack of education programs, over-focused on specialized industry, lack of venture companies, poor standardization, and poor networks.

Both Japan and Taiwan have competitive MEMS foundries. There are more than 10 foundries in Japan including Olympus, Omron, Matsushita Electric, and Sumitomo Metal. Taiwan has established the Taiwan MEMS Industry Alliance with about nine foundries and 10 other MEMS start-ups. The Alliance aims to provide a platform for the exchange of technical and up-to-date market information, to set industry standardization, to integrate the existing technologies, and provide road mapping. It also provides intellectual property management and legalization services as well as international trade services and advice. Its members include Asia Pacific Microsystems, Inc., Walsin Lihwa Corp., Micro Base Technology Corp., and Neostones Microfabrication Co., Ltd.

Countries such as India have emerging MEMS industries, and Thailand, China and Singapore have competitive MEMS facilities and research activities.

3
Private Investment in Nanotechnology

In the business sector, the two top trading houses in Japan, Mitsui & Co. and Mitsubishi Corporation, have established new nanotechnology business divisions two years ago and are actively engaging in nanotechnology R&D, facilitating commercialization and investment in nanotechnology. Mitsui & Co. in particular has started the process of building a global nanotechnology enterprise. See the chapter by Maeno, Section I, for details of Mitsui nanotechnology strategy and the state of the development. The Japanese top industries such as NEC, Hitachi, Fujitsu, NTT, Toshiba, Sony, Sumitomo Electric, Fuji Xerox, and others are continuing nanotechnology R&D efforts and taking more aggressive measures to accelerate the commercialization of their R&D. The Korean companies Samsung and LG and other Korean industries are aggressively investing their effort in nanotechnology R&D and commercialization. The Taiwanese semiconductor foundries such as TSMC and UMC are aggressively pursuing semiconductor nanoelectronics. Venture Capital firms such as Innovation Engine (Japan), Apax Globis Partners & Co. (Japan), and Juniper Capital Ventures Pte Ltd (Singapore) have invested in nanotechnology start-ups in Asia. Firms such as Cranes Software International Ltd (India), Good Fellow Group (HK), and so on have invested in nanotechnology venture companies in Asia. The Australian Macquarie Bank and Pacific Dunlop are the main investors in the first nanotechnology company, AMBRI, listed on the Australian Stock Exchange.

Investment hotspots in Asia include: MEMS, optoelectronics, memory, carbon materials, diagnostic tools, drug delivery systems, measurement and characterization tools, display technologies, and coating.

4
Advantages of Collaborating with Asians

The advantages of collaborating with Asian countries are:

- abundant human resources,
- excellent facilities (Japan, Korea),
- advanced technology (Japan, Korea, Taiwan), and
- dynamic technology and strong market growth.

Over the AP region, governments, industries, and business sectors have indicated strong ambition and effort in pushing their countries into the nanofuture. The AP region is becoming an exciting place for global partnerships and business opportunities in nanospace.

Unlike in the European Union, there is not yet an "Asian Commission" to coordinate Asian nanotechnology R&D and networks. However, there has been an increasing awareness of the necessity of working together. At the bottom-up level, there have been various collaborations in research and now even in business. The Japanese government got together Asian government representatives for the Nanotech 2003 + Future meeting, 26–28 February 2003, sponsored by the Japanese government to discuss the state of development in each country and from this an Asian nanotechnology network is on the way.

The following appendix provides additional information about nanotechnology policy and programs in Australia, China, Hong Kong, India, Malaysia, New Zealand, Singapore, and Thailand.

5
Appendix

Australia

The Australia Research Council (ARC, the key funding agency for Australian Science and Technology focusing on fundamental science, www.arc.gov.au) received an additional A$ 736.4 million over 5 years for doubling ARC competitive grants funding. Under the National Competitive Grants Program, four priority areas for ARC 2003 funding were announced in January 2002, which are:

- nanomaterials and biomaterials,
- genome-phenome research,
- photon science and technology, and
- complex/intelligent systems.

A$ 170 million has been allocated from 2003 to support projects and centers for up to 5 years. A$ 90 has been allocated for 5 years for the ARC Center of Excellence (COE) program starting 2003 for eight centers located around Australia.

The nanotechnology related COE include:

- quantum computer technology,
- quantum-atom optics,
- advanced silicon photovoltaics and photonics, and
- ultrahigh-bandwidth devices for optical systems.

In addition, there is over A$ 45 million investments committed by State Governments, venture capitals, and other business investors.

China

There are over 50 universities, 20 institutions within CAS, and over 100 companies active in nanotechnology R&D in China, according to the Chinese Ministry of Science and Technology (MOST) survey. The short-term strategy of China nanotechnology is to integrate nanotechnology with the traditional industries and develop products with competitive quality and performance that would benefit the consumers and change the quality of their daily lives. To provide a platform for the commercialization of nanotechnology, China is establishing the Industry Base and Engineering Center near Beijing and Shanghai areas. In particular, the nanotechnology Industry Base was built in TianJin, a harbor city about 100 km from Beijing. The applied research of the main R&D organizations in the Beijing area moved to TianJin last year when the nanotechnology Industry Base became operational. The long-term strategy of China nanotechnology is to strengthen basic science and enhance the global competitiveness of nanotechnology in China. The government has allocated 270 million RMB (US$ 33 million) for building the National Research Center for Nano Science and Technology. The Center will integrate China's top R&D institutions such as the Chinese Academy of Sciences, Beijing University, TsingHua University, Fudan University, Jiaotong University, Nanjing University, and the East China University of Science and Technology, to perform better coordinated and world glass scientific research in nanotechnology.

The Chinese Academy of Sciences (CAS) has the largest research institution network in China and the world. The CAS Nanotechnology Engineering Center Co., Ltd (CASNEC) was founded in November 2002 as a platform for accelerating the commercialization of nanotechnology in CAS. The main investor of the company is the Good Fellow Group (a HK listed company), which takes 55% of the company share. The total amount of investment is about 50 million RMB (US$ 6 million). The CAS has 20% shares, scientists at CAS have 15% in total and YongFeng High Tech Park has 10%.

The founding of CASNEC is to reorganize the nanomaterials and nanotechnology industrialization within CAS using top-down management. The CASNEC is also viewed as a role model for promoting a national industrial base for nanomaterials and nanotechnology in YongFeng High-Tech Park in Beijing. It will occupy 7000 square meters of land in YongFeng High-Tech Park centrally located in Zhong-Guan-Cun area, which is the Beijing Science and Technology R&D and industry center.

The business concept of CAS Nanotechnology Engineering Center Co. Ltd (CASNEC) is to provide a technology-transfer platform for the CAS R&D. Its main revenue sources come from technology licensing, mid-scale production, and consulting services. For industrial scale production, CASNEC cooperate with large manufacturers, which are well established in the related industry and market.

CASNEC recently signed a technology licensing agreement with ERDOS Group, the biggest cashmere manufacturer in China that occupies about 30% national market.

The ERDOS Group sells 2 million pieces of cashmere shirt every year. Their annual sale is about 3 billion RMB (US$ 362 million).

Compared with other venture business, CASNEC has the advantages of:

- direct access to the R&D results of CAS,
- access to a pool of excellent human resources of the experienced scientists and engineers including those retired from CAS,
- receiving stable financial support from government, and
- CASNEC consists of 26 PhDs, 112 Master degree personnel, 3 MBAs, and 7 technicians.

Hong Kong

In Hong Kong, nanotechnology R&D funding mainly becomes two sources: the Research Grant Council (RGC) and the Innovation and Technology Fund (ITF). The RGC funds mainly the basic research at universities and the ITF funds "midstream" and "downstream" research at universities and industries to promote technology advancement and enhance competitiveness of existing industry and creating new industry in Hong Kong. The RGC and ITF communicate with each other and coordinate their funding programs to avoid overlaps. The ITF started its strategic nanotechnology programs in 2001 after the Legislative Council approved the Development of Nanotechnology Initiative on 31 October 2001. The total funding during 1998–2002 was about US$ 20.6 million plus 2.3% industry funds. Most recently funding of two key nanotechnology centers has been approved: one is the Hong Kong University of Science and Technology (HKUST), which received US$ 7.3 million and the other is the Polytechnic University of Hong Kong (PolyU), which received US$ 1.6 million. The total funding for the two key nanotech centers PolyU and HKUST cost total US$ 8.9 million for 2003–2004.

India

India, a country with 1 billion population, is making its way in to the nanotechnology era. The Indian government has already started the Nano Science and Technology Initiative. Various funding agencies such as the Department of Science and Technology (DST) and the University Grants Commission (UGC) have launched large nanoscience research programs. The main nanoscience research has been conducted at institutions such as the Indian Institute of Science (Bangalore), Indian Institute of Technology (Madras, Chennai, Kharagpur, Mumbai, New Delhi), Central Electronics Engineering Research Institute (Pilani), University of Pune, Solid State Physics Laboratory (Delhi), Tata Institute of Fundamental Research (Mumbai). Recently a number of institutions have also started more coordinated nanoscience and technology research, they include Raman Research Institute (Bangalore), National Chemical Laboratory (Pune), Central Glass and Ceramic Research Institute (Jadavpur), University of Delhi, and the University of Hyderabad.

Three years ago the government launched a US$ 15 million/5 yr National Program on Smart Materials coordinated by five government agencies involving 10 research centers across India with a key focus on MEMS technology. The nanomaterials topic has been included in this program and more funding is expected in expanding the program. Last year the Department of Science and Technology launched the National Nanotech Program with total funding of US$ 10 million committed over the next 3 years. The Indian Institute of Science (IISc) was awarded US$ 1.0 million for its Nanoscience Research Center. IISc is known as the Knowledge Hub of India.

Indian nanoscience and technology covers a wide spectrum of topics, which include MEMS, nanostructure synthesis and characterizations, DNA chips, nanoelectronics (transistors, quantum computing, optoelectronics), nanomaterials (CNT, nanoparticles, nanopowder, nanocomposites) and so on.

Similar to China, the Indian science and technology and business network is worldwide spread. However, one of the problems that impedes international cooperation with Indian scientists and the business community is the difficulty of getting a visa to most countries.

Unlike China, the English speaking Indian population makes Indian science much more accessible to the western world, drawing investment and global cooperation opportunities. For example, the IndiaNano is a platform recently established by the US and Indian US Community in the Silicon Valley together with Indian R&D community trying to coordinate the Indian academic, corporate, government, and private laboratories, entrepreneurs, early-stage companies, investors, IP, joint ventures, service providers, start-up ventures, and strategic alliances.

There has been some recent interesting development in the nanotechnology industry in India. Private companies began investing in the R&D laboratories at university and government institutions. In the past, companies have largely shied away from investing in research. Universities and national research centers have worked in isolation. This lack of synergy and cooperation between the two sectors has prevented the growth of inventive technology. Private companies have, for the most part, worked with university labs in consulting mode where a short-term interaction has been sought for solving a well-defined problem, mostly of troubleshooting nature. This sort of interaction has never blossomed into a relationship with a long-term vision for research driven product or technology development. However, the CranesSci MEMS Lab, India's first privately funded MEMS Research Laboratory joint venture between Indian Institute of Science and Cranes Software International Ltd (CSIL) in Bangalore, was established precisely to create a new culture in micro- and nanotechnology business in India through creating synergy between public research institution and private industries. The CSIL is a listed company at Bombay stock exchange with market capital of about US$ 20 million and it is a leading player in the high-end scientific and engineering software products and solutions. The laboratory was created with a unique philosophy. It believes in science with conscience with the vision of production based on business, social obligations, and education. Not only that, but the laboratory transforms MEMS technology to the market place but also it stresses on IP rights, strategic need of the country and mankind as well as providing other researchers with infrastructure knowledge management in MEMS technology.

Malaysia

In Malaysia nanotechnology is categorized under the Strategic Research (SR) of Intensification of Priority Research Areas (IPRA) program in the Eight Malaysia Plan (2001–2005) funded by the Ministry of Science, Technology and Environment (MOSTE). The SR projects have the following requirements: are within 60 months, future competitive socioeconomic environment or new breakthrough, multi-institutional, multi-disciplinary, industry linkages, and commercialization potential.

A budget of RM 1 billion (US$ 263 million) was allocated for IPRA in the Eight Malaysia Plan. Funding of the Strategic Research is 35% of the IPRA program and it includes four evenly distributed areas, which are:

- design & software technology,
- specialty fine chemical technology,
- optical technology, and
- nanotechnology and precision engineering.
  The funding for nanotechnology and precision engineering over the next 5 years is around US$ 23 million for a country with about 20 million population. Note that in Taiwan with a population of about 21.5 million, the national committed spending for nanotechnology over the next 6 years is US$ 620 million.

The areas for nanoscience research include nanophotonics, nanobiosystems, nanoelectronics, nanostructured materials, and nanometrology. Its short-term strategy is to

- identify researchers in various areas of nanoscience with specific expertise,
- upgrade and equip nanoscience laboratories with state-of-the-art facilities, and
- prepare a comprehensive human resource development program for training nanoscientists.

The long-term strategy is to

- nurture a nanoscience research culture among researchers,
- develop a world-class national nanoscience laboratory in Malaysia, and
- produce renowned nanoscientists.

New Zealand

The main activities of nanoscience and technology in New Zealand have recently been coordinated in the MacDiarmid Institute for Advanced Materials and Nanotechnology, New Zealand's premier research organization concerned with high quality research and research education in materials science and nanotechnology. It is founded upon principles of inter-disciplinary cooperation and led by two Universities, Victoria University of Wellington and the University of Canterbury, with partner organizations Industry Research Ltd (IRL) and Institute of Geological & Nuclear Sciences (IGNS) Ltd, as well as research groups at Massey and Otago Universities. The Institute builds upon pre-existing and dynamic research collaborations, exceptional scientific and engineering capability, outstanding leadership, unparalleled international networking, strong industry and business linkages and longstanding expertise in training world-class graduate students. Of the Principal Investigators of the Institute, nine are Fellows of the Royal Society of New Zealand, and six have been awarded prestigious RSNZ science medals. The MacDiarmid Institute is led by its Director, Professor Paul Callaghan FRS, and its Deputy Director, Dr Richard Blaikie.

The institute focuses on materials and technologies that are attracting world-wide attention, including: nanoengineered materials and devices, optoelectronic activity in semiconducting materials and devices, superconductors, conducting polymers, carbon nanotubes, sensing and imaging systems, functional materials and coatings, energy storage materials, light harvesting and photochemical materials, soft materials, biomaterials and complex fluids.

Singapore

The key funding agency in Singapore for nanoscience and technology is the Agency for Science, Technology and Research (A*STAR). The A*STAR Nanotechnology Initiative started in September 2001. Their approach is to seed focused nanotechnology research as part of Singapore's continuing effort to build on accumulated capabilities and the promotion of innovations in areas that fuel Singapore industries.

A*STAR develops nanotechnology research programs through existing capabilities development programs at the:

- Institute of Materials Research and Engineering – photonics, advanced materials.
- Institute of Microelectronics and Data Storage Institute – semiconductors, electronics, storage.
- Institute of Bioengineering and Nanotechnology – bionanotechnology.

Their efforts are focused on the advancement of the technological envelopes in industries key to Singapore, such as electronics, chemical and biomedical industries. The Singapore Economic Development Board is another funding agency for supporting industry applications R&D particularly funding nanotech start-ups and international joint ventures.

Taiwan

The Taiwan MEMS national programs started in 1996 funded by both the National Science Council (NSC) and the Ministry of Economy Affair (MOEA). From 1998, NSC established three main MEMS national centers, which are the Northern, Central and Southern Region MEMS Research Centers to build up MEMS R&D common facilities and core technologies in Taiwan. From 2003, MEMS government program has been imbedded in the National Science and Technology Program on Nanotechnology. Taiwan MEMS focuses on information technology, industrial processes/devices, communication, consumer electronics, semiconductors, and biomedical technology.

Taiwan MEMS has shifted from R&D to commercial production now. Taiwan MEMS foundry business started in 2000 and currently there are nine foundries in Taiwan. The total investment for MEMS foundries in Taiwan is about

US$ 211 million up to now. The accumulated investment in MEMS business in Taiwan is about half a billion US$. The Asia Pacific Microsystems Inc. (APM) is a leading MEMS foundry, founded in August 2001 with over 200 employees currently, in providing microsystems solutions.

Starting with over US$ 50 millions of capital investment, the company is pursuing the MEMS fabrication technology and servicing inkjet, smart transducer, wireless, optical, and bio-MEMS applications. They provide one-stop-shop services including design support, process development, volume production, packaging, assembly, and testing for the customers. APM took over a 5-inch wafer foundry for making CMOS chips from Winbond Electronics Corp., and transformed it into an advanced 6-inch CMOS compatible MEMS facility in Hsinchu Science-based Industrial Park. APM is targeting at 8000 wafers/month by 2003. The major shareholders of APM include Chi Mei Industrial Co., Mobiletron Electronics Co., Universal Microelectronics Co., Wintek Corp., an affiliate of Acer Group, and 30% venture capital investment. Details about APM can be found at www.apmsinc.com.

Thailand

The top funding organization for Science and Technology is the Ministry of Science and Technology (MOST). The National Science Technology Development Agency (NSTDA), which is under the MOST, is supporting three main national centers, which are the National Center for Genetic Engineering and Biotechnology (BIOTEC), National Metal and Materials Technology Center (MTEC), and the National Economic and Computer Technology Center (NECTEC).

A National Nanotechnology Center (NANOTEC) has been proposed recently and it aims to:

- identify and focus on niche areas in nanotechnology, thus enhancing Thailand's competitiveness,
- assemble and produce a critical mass of researchers and educators on nanotechnology,
- act as a national coordinating body between academia, industry and government.

The proposed budget is about US$ 25 million for the period 2004–2008 with 300 personnel. The R&D areas of focus are advanced polymers, nanocarbon, nanoglass, nanometal, nanoparticles, nanocoating, nanosynthesis with applications to the industries of automotive, foods, energy, environment, medicine, and health. Currently in Thailand there are 14 laboratories in six universities and five laboratories in two government agencies with about 100 researchers. The current research areas in nanoscience have been mainly in nanoparticles: quantum dot devices, carbon nanotubes, nanocoating, and MEMS.

Cooperation with Small- and Medium-Sized Enterprises Boosts Commercialization

Torsten Schmidt

1
The Company

Genthe-X-Coatings GmbH (GXC) applies nanotechnological antifog coatings onto the inside surface of automotive headlamps and foglamps in order to prevent visible condensation. The company was founded in the year 2000 by nanotechnology inventors and medium size business entrepreneurs. The nanotechnological functionality was developed by the inventors but the adaptation and customizing to meet the full set of specification for automotive mass production was accomplished in cooperation with the Dr. Genthe GmbH & Co (abbreviated: Genthe Glas). Dr. Torsten Schmidt joined the company in September 2002 as Chief Executive Officer after serving 18 years in leadership positions in global specialty and fine chemicals corporations. He holds a PhD in chemical engineering from the Technical University of Berlin. GXC is one of the leading suppliers of transparent antifog coating technology on headlamps and foglamps for the automotive industry.

2
Scope

Small- and medium-sized enterprises (SME) play an important role comprising economic as well as social and psychological aspects in German society. This economic field is constituted by economically and legally independent businesses. The primary trait is a close relationship between a person, such as the entrepreneur, and an economic unit strongly influencing its market behavior and performance in ways such as [1]:

- identity of ownership and personal responsibility for the enterprise's activities,
- identity of ownership and personal liability for the entrepreneur's and the enterprise's financial situation,
- personal responsibility for the enterprise success or failure, and
- personal relationship between employer and employees.

The Nano–Micro Interface: Bridging the Micro and Nano Worlds
Edited by Hans-Jörg Fecht and Matthias Werner
Copyright © 2004 WILEY-VCH Verlag GmbH & Co. KGaA, Weinheim
ISBN: 3-527-30978-0

Tab. 1 IFM classification of company size.

Classification	Personnel	Annual revenue (EUR million)
Small	< 10	< 0.5
Medium-sized	10 to 499	0.5–50
Large	> 500	> 50

In quantitative terms the following classification (Tab. 1) is used by the IfM Institute for Mittelstandsforschung, Bonn [1].

The European Union chose to apply the following classification (Tab. 2) from 1 January 2005 onwards [2].

Based on the aforementioned definition, IfM investigated SME in Germany in 1999. They reported these findings:

- about 70% of employees have their occupation in SME,
- SME create approx. 60% of the gross value added,
- SME create approx. 45% of all revenues subject to VAT,
- SME carry out approx. 45% of gross investments, and
- SME carry 13% of German R&D expenditures [3].

Performing R&D activities requires a critical mass and occupies an excessive number of personnel the smaller the size of the enterprise. Following this train of thought, SME have the tendency to curb these activities in recession periods. SME increasingly respond in order to sustain the level of product development necessary for survival in the market place by entering into R&D cooperation or outsourcing R&D.

Comparing corporations and SME the first finding is that new products and processes are developed by both but the number as well as success of R&D activities increase with increasing size of the enterprise. But even more striking is the scope of the R&D activity, while corporations mostly target so-called defensive projects, which maintain market position, SME focus their projects on creating and expanding new market niches. SME do not exploit their full innovation potential due to imminent constraints such as access to financial as well as human re-

Tab. 2 EU classification of company size.

Classification	Personnel	Annual revenues (EUR million)	Total assets (EUR million)
Micro	< 10	< 2	< 2
Small	< 50	< 10	< 10
Medium-sized	< 250	< 50	< 43

sources. In a particular way, German SME suffer from a lack of equity, which is an impediment to the acquisition of fresh funds for R&D projects. If it comes to organizational issues, the strength of SME might also contribute to its weakness: the organizational focus on the entrepreneur who owns and drives the business. The entrepreneur personalizes the company and develops the product into an innovation and technology leadership position in niches. Hence, he has the trust of the customer who in return shares vision, demands, and unmet needs with him. This is a strength, as it enables concentrating state-of-the art customer as well as technology knowledge in the decision making process. Therefore it allows SME to act with velocity; in other words, SME are less prone to "paralysis by analysis". In pulling through the project, SME are probably not built around robust and self-supporting innovation management, such as R&D stage gate processes [4]. Neither are they typically connected to R&D networks, which allow access to external know-how. Protecting intellectual property also is a weakness of SME as they typically protect their innovation by commercial confidentiality rather than applying for patents.

The automobile industry consists of the original equipment manufacturers (OEM) (which are large global corporations), first-tier suppliers, the so-called system integrators (which are corporations), and second-tier suppliers, the so-called component suppliers (which to a large extent are SME). This also applies to the third tier, the so-called raw material suppliers. In the German business environment of today, the automotive industry is the key innovation driver with expenditures of Euro 16.2 billion, which represent 39.1% of the total German R&D expenditures in the year 2000 [5]. In 2002, Volkswagen, BMW, and DaimlerChrysler have ploughed back 5%, 4.8%, and 4.1% respectively of their revenues into R&D. First-tier suppliers carry an even higher share of R&D expenditures: on average 6.2% [6]. In a thrust towards pushing R&D activities up the value chain [8], SME in the automotive have a demand for product and process innovations from external sources.

In the light of these facts and figures, the following three propositions were developed, which are supported by the actual case of Genthe-X-Coatings GmbH, Goslar, Germany.

3
Proposition: In Current Technology Markets, Commercial Success of New Product Ideas Evolves from Cooperation

The majority of new product ideas are developed following a product push attitude, without a deep understanding of customer needs. Successful SME have developed the competency to determine the real customer needs within strong customer supplier networks. By cooperating with such enterprises, the inventor gains access to this critical success factor.

Case Study

In modern design, headlights and foglights at the front of the car, as well as rear lights, constitute a major element of the distinguished image of an automobile. Major automotive companies were faced with the problem that after start-of-the-art production, condensation on the inside surface of the lenses, so called fogging or misting, occurs. This creates esthetic as well as safety issues giving rise to an increased number of customer complaints. Genthe Glas [9] has been producing high precision glass for the lighting, technical, and optical industries for more than 75 years and, hence, is a player in the automotive lighting market. Due to its market position Genthe Glas was informed by prestigious OEM of the real customer need to prevent visible condensation (Fig. 1).

Surface engineering and coating technology have become important means of upgrading systems, by bypassing relevant problems or expanding material limits. This is demonstrated by the use of hard coatings for polycarbonate headlamp finishing. Automotive lighting has many functions for vehicle design and safety aspects and is approaching new technologies such as advanced front lighting (AFS) systems or the use of LED applications. Thus new solutions on different substrate materials are required, reflecting a bundle of technical demands and following the well-known basics such as low costs, high reproducibility, reliability, and long-term stability. Antifog effects, self-cleaning or easy to clean functions, impact resistance, and diffusion barriers on glass and plastics are accessible through nanotechnology and will be some of these challenges for the future. The response to new materials can be skeptical, because of experience, but GXC succeeded in introducing such a new technology in the automotive market within two years and has been in series production of antifog coatings since 2001. This technology gives an antifog effect on glass, by using a sol-gel material with nanoparticular structures applied by approved and low-cost spray processing. The benefits of avoiding fogging in today's mainly transparent headlamp lenses are enhancing the spectrum for headlamp designs, and are enabling stable light efficiencies,

Fig. 1 Condensation in car headlamp.

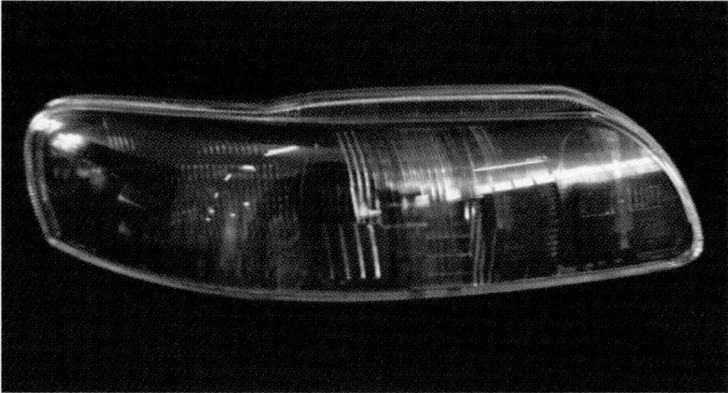

Fig. 2 New halogen headlamp.

hence, conformity with safety aspects even under difficult ambient conditions. The antifog treatment comprises a coating for glass that has a spontaneous water-spreading effect based on a hydrophilic inorganic–organic matrix consisting of a nanoparticle-network supported by diffusible hydrophilic additives. The coatings on glass provide contact angles for water of below $10°$, an excellent long-term stability, high transparency, good adhesion, and a functionality accomplished by very thin coatings (approx. 1 µm). The coatings on glass are state-of-the-art. Currently GXC adapts these systems to polymer substrates. The transition of the antifog materials to other markets and the introduction of further products with customer-tailored functions and applications are foreseen following a medium-term schedule.

In 1997, Genthe Glas got in touch with researchers who later founded the Nano-X GmbH [10]. These researchers had developed a lab-scale sol-gel coating varnish, which decreased the contact angle for water and were looking for a channel to commercialize their invention. Both companies cooperated in developing a solution that complied with the specification of a premium automobile OEM in early 2000 (Figs. 2 and 3).

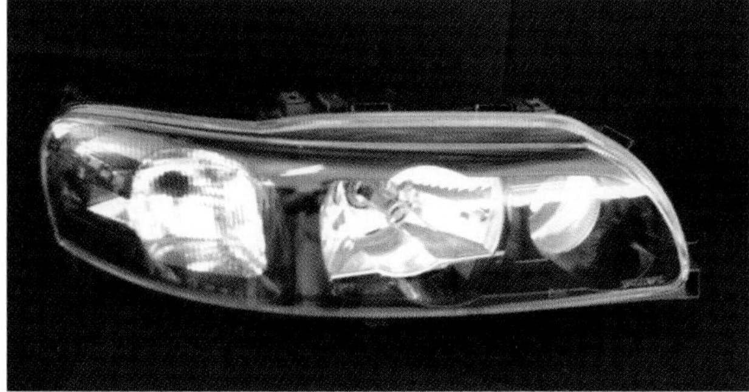

Fig. 3 New headlamp.

4

Proposition: Innovations Get on a Fast Track to Market if Implemented in SME

Because of their pragmatic decision making processes as well as their customer intimacy, SME enable the inventor in the early phases of feasibility testing and market entrance to match his idea with reality.

Case Study

By virtue of established quality procedures, Genthe Glas drove the validation process of the antifog coating receiving the exclusive order for the coating of the glass lenses for large Volvo sedans and station wagons. In the later part of the year 2000, Genthe-X-Coatings GmbH (GXC) [7] was founded by the owners of Genthe Glas and the majority share holders of Nano-X. The financing package was put together by Genthe Glas including private equity and bank credits allowing GXC to invest in a fully automated series production line (Fig. 4). It was preceded by an optimization of the coating material and the substrate material in a batch mode in response to the demands of the automotive customers. The automatic coating line is a prototype fulfilling the required technical and commercial objectives. Additional investments were made into GXC logistics and quality systems, and the losses of the first year of operation were covered by the Genthe Glas side of the founders.

In summary the process is characterized by:
- high flexibility regarding the coating material, substrate material and geometry, and process parameters,
- marketable coating costs,
- high reproducibility, and low scrap rates.

The nanotechnological coating has been in operation since 2001 without a single failure in the field.

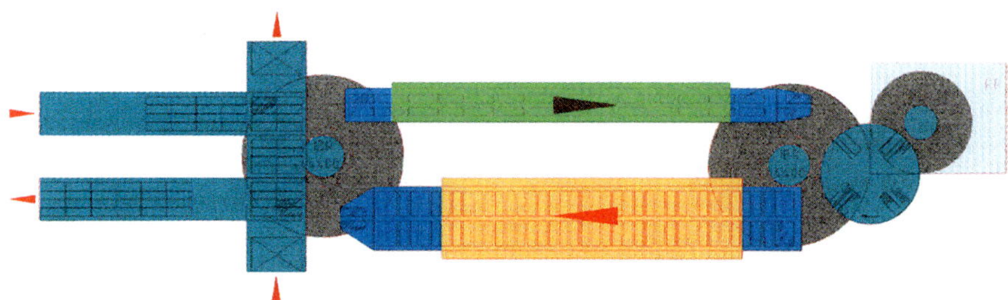

Fig. 4 Diagram of the automatic coating line comprising: material handling, quality check, pretreatment, spray coating, and curing.

5

Proposition: Competencies of SME and Inventors are Complementary Rather than Competing

In cycles of economic slow-down, many SME are prone to curb their R&D expenditures. Partnering with inventors can help to protect their product pipeline from otherwise drying out. In return, inventors gain access to industrial experience such as financial and quality management and a proven network of customer relationships.

Case Study

GXC focused itself on transforming nanotechnology into series production by identifying customer needs and developing tailor-made coating process solutions. The playing field was defined by automobile, safety, and optics markets (Tab. 3).

GXC is positioned at the interface between materials research and industrial mass production. The current business is focused on enabling design of headlamps without risk of end-use customer complaints and, hence, helps to drive down the cost of customer returns. The nanotechnological route promises higher robustness than other in-kind or not-in-kind technologies. The key competence of GXC is the adaptation and modification development of nanotechnological materials ("tweaking") to meet the requirements of automotive mass production with respect to functionality, quality assurance, and environmental friendliness. In future, the emphasis will be on selling process technology and materials. Coating services will be offered to cover small series or pre-series volumes.

Tab. 3 Functions and applications of nanocoatings on glass.

Functionality	Automobile	Optics	Safety
Antifog	Glass headlight	Lens systems	Helmets
	Polycarbonate headlight lenses	Displays	Protective masks
	Glass foglamp lenses	Ophthalmic glasses	Skiing apparel glasses
	Polycarbonate foglamp lenses		Diving goggles
	Polycarbonate rearlamp lenses		
	Instrument covers		
	Driver Support Systems		
Impact resistance	Glass foglamp lenses		
	Glass truck headlights		
Easy-to-clean	Polycarbonate rear lamp lenses	Display covers	
	Polycarbonate headlight lenses	Cellular phone covers	
	Rear mirror	Watch covers	
	Driver Support Systems	Ophthalmic glass	

In order to keep GXC focused on its core competencies such as coating process development as well as keeping its organization lean, key material development projects were outsourced allowing GXC to keep the new product pipeline filled without excessive build-up of R&D staff.

The key project within GXC's business plan to 2007 is coating of high-volume plastic lenses. Amongst others, this new product commercialization will enable GXC to grow at a compounded average rate of 25% CAGR into a stand-alone future, boosted by venture capital.

Beyond the business plan projections there are other growth opportunities inside and outside the automotive industry. The innovation roadmap of the German automobile industry [8] comprises features that might be enabled or supported by nanotechnological coatings such as hydrophobic surfaces, sensors, cameras, driver support systems (e.g., night vision) (Tab. 3). Outside the automobile applications possibilities exist in the following functionalities:

- antifog
  antiscratch
- chemical resistance
- environmental resistance (sun, pollution)
- mechanical resistance
- dirt-repelling, self-cleaning
- electrical conductivity
- heating
- electromagnetic interference (EMI)
- antistatic.

The sol-gel process is a novel route for synthesis of coating materials. In this, the "sol" is a colloidal solution, marked by a homogeneous distribution of particles with a diameter of around 1 nm in a dispersing agent. The "gel" component is a two-phase solid featuring a grid-phase next to a dissolved phase in a colloidal manner.

In the sol-gel process, the materials first go through a gel-phase and subsequently through a sol-phase. As in every polymerization reaction, completely new combinations and grid-structures are formed. One of the big advantages of the sol-gel process is the liquid sol-phase, which facilitates the application of the material to complex surface structures. In the subsequent curing process, the gel-phase molds the homogenous film into a long lasting, strong coating.

Owing to their thin layers nanotechnological coatings are designed to keep the optical properties and transparency of the substrates while adding one or more performance features to the substrate. If thickness is below the wavelength of the light, the coating layers are highly transparent and do not interfere with the optical properties of the substrate. This is one of the reasons why nanotechnological coatings offer great possibilities to the optical and lighting industries as well as construction industries.

6
Summary

The current business cycle characterized by the cash crunch is a fertile ground for partnership between inventors and small and medium size businesses.

Acknowledgments

I thank Mr. Sven Probst for his contributions to the technology sections of this chapter.

References

1 H.-E. HAUSER, IfM Bonn 2000, http://www.ifm-bonn.org

2 http://europa.eu.int/comm/enterprise/enterprise_policy/sme_definition/index_en.htm

3 vdi nachrichten, Berlin, 07.03.2003.

4 R.G. COOPER, Top oder Flop in der Produktentwicklung. Wiley-VCH, Weinheim, 2002.

5 J. KLUGE, 3. VDA-Mittelstandstag, 21.03.2003, Mainz, VDA Frankfurt.

6 Frankfurter Allgemeine Zeitung, 12 June 2003.

7 http://www.gxc-coatings.de

8 Jahresbericht 2002, VDA, Frankfurt.

Rapid Commercialization of Nanotechnology in Japan: from Laboratory to Business

Hiromichi Maeno

1
Background: Japan at the Crossroads

If you want to get a sense of the economic situation in Japan, ask a taxi driver. A typical answer has been "not good" or "terrible" for several years. Now it has changed into "dying" or "paralyzed". Personal bankruptcies hit a grim record of more than 200 000, and around 30 000 people took their own lives last year. Our government has often explained that the economy will recover steadily after a sharp readjustment, but the words are not believed by the people. A taxi driver said with dismay and anger that he was willing to work very hard, but he did not know what to do or how to do it.

People are conditioned to think that not taking risks is safer for individuals. This can lead to an irresponsible mood like an epidemic, rather than sensible restraint, although people hardly become aware of their own sloppiness. If so, society might fall into a bad situation and then there is no question that it leads to hazards for individuals consequently.

Today's downward competitiveness might be an inevitable result from this approach, which I can hardly ignore as a part of the serious reasons, because underneath is a bad social situation.

Many corporate scandals have erupted recently. Such business paralysis might be a remnant of the old success model. Japan may be at a crossroads. I wonder if the economic mess might be a sign of change heralding an economic revival. I suggest that a new model for decision making is very important for this renewal. We have imagined a different way of conducting business to inspire sincere people to build great businesses.

2
Motivation and Strategy: Shake up Unique People

We have come out openly with a new business model on our own initiative to motivate new business, new industry, and new markets from science and technology with the acknowledged risk that there are many difficulties before success is achieved.

The Nano–Micro Interface: Bridging the Micro and Nano Worlds
Edited by Hans-Jörg Fecht and Matthias Werner
Copyright © 2004 WILEY-VCH Verlag GmbH & Co. KGaA, Weinheim
ISBN: 3-527-30978-0

The first key factor in the new business model is to promote the right people on their merit, but it is not just a meritocracy. We need to make a new culture for good communications with a minimum of restraint, with sincerity, integrity, and no sycophancy in our new organization. Most of our staff have joined from well-established large manufacturers, and they want to break with the obsolete style. We want our staff to bring their capacity into full play in their working life. Our organization has spirit and excitement, which most other large corporations seem to have lost. What is the optimal way to achieve critical mass for business success from nanotechnology? Nanotechnology is such a wide category, from single molecules to multiple products, that it can not be managed by conventional vertical organization.

The second key factor in the new business model for achieving success from nanotechnology is to use an interdisciplinary approach to overcome limited knowledge and sometimes complacency. The range of nanotechnology is so broad that it seems like a completely new business should be built up from basic science, but it would be an uphill fight for anyone to achieve success from basic research. There are two major difficulties to overcome: choice of project and collaboration.

Thus the third key factor for success is the selection of worthwhile jobs and working with reliable partners. These are both necessary pre-requisites for success. It is almost as critical to know what is not important as to know what is vital for any research group, while deep and flexible collaboration with others is necessary for timely results. Otherwise nanotechnology will not be able to attract investment and become widely applied, because an awesome task tackled at a basic science lab is apt to fail to achieve commercialization due to isolation.

3
Research and Development of a New Idea

We have established such a new facility where we hope to stimulate and challenge the spirit of the people who work there. The venture R&D company is divided into four nanotechnology segments:

– nanostructures,
– nanomaterials,
– devices such as photonics and MEMS,
– biomedical and organic self-assembly systems.

The group is called "XNRI" , where X symbolizes across the broad field and NRI is Nanotechnology Research Inc. XNRI is a new venture R&D group, established almost 2 years ago by Mitsui's seeding fund of around US$ 100 million over 3–5 years plus other priming funds such as government subsidies. We have learned from recent economic havoc that an old fashioned style of management might fail. So we are struggling to build a new company culture through candid discussion. We retain the best aspects of the traditional Japanese system, such as a respect for craftsmanship, but reject any lingering haughtiness, which estranges people from becoming excited about their work.

4
Pump Priming and Leadership

The government has pump-primed developments in nanotechnology by subsidizing academic research and providing venture capital for small and medium-sized enterprises (SME). I am not sure if this is sufficient to restore the economy, but we are sure that leadership is most important for success. We have established our new enterprise at the 'Nanotech Park' in Tsukuba, where collaborative work is done between academic researchers and our XNRI group researchers. Generally speaking, in the past collaborative work was superficial and slow. In the new business model the XNRI group manager will oversee the progress of each project rigorously, and not accept any trivial or sluggish work. This system may offer mutual benefits by saving time, workload, and costs by keeping an overview. Supervision by XNRI will, we hope, wean inexperienced scientists from their immature behavior. This strategy is designed to stimulate productive activity that will lead to commercial success.

5
Nanotechnology Activities

5.1
Bio Nanotec Research Institute, Inc. (BNRI): Zeolite Membranes

Inorganic membranes for separation of liquids and gases at the molecular level are expected to contribute significantly to physical and chemical processes that are important to the progress of green chemical technology (Fig. 1). At BNRI, we are

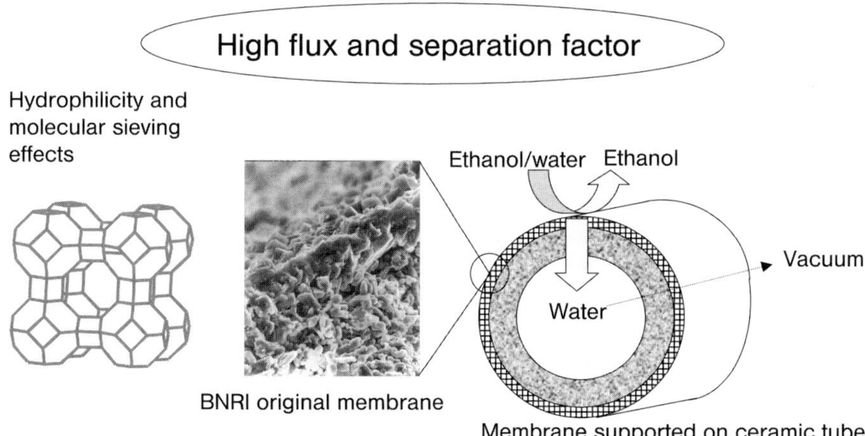

Fig. 1 Inorganic membrane technology. Synthesis and application of zeolite membranes to separation of biomass ethanol and organic chemicals.

actively involved in basic research and development of commercial products in the areas of environmental protection and energy-efficient systems, as well as the creation of more simple, efficient, and less costly processes that can be utilized in industrial chemical processes. As an example, we present a case for separation of biomass ethanol by a zeolite membrane.

"Gashol" is a fuel made by adding dehydrated ethanol to gasoline for automobiles: it is expected to have a significant global impact. Implementation of this technology has begun in the USA and Europe. The use of Gashol is expected to double or triple by 2010. In December 2002 the Japanese government announced that Japan would join the world in this endeavor. Ethanol produced by fermentation of biomass requires purification. This is currently done by multiple distillation including processes to handle azeotropy, which are complicated and costly. We have developed a highly efficient zeolite membrane that can provide high purity ethanol with high flux and high separation factor. The membrane system offers simple, safe, and less energy-consuming production of ethanol, which can be used for automobiles, power plants, and elsewhere. Further developments in zeolite membranes for other applications are also pursued at BNRI.

5.2
Carbon Nanotech Research Institute, Inc. (CNRI): Clean Single-Walled Carbon Nanotubes

CNRI successfully has developed high purity and clean single-walled carbon nanotube, which are produced by catalytic chemical vapor deposition using alcohol (ACCVD process). The single-walled carbon nanotubes contain almost no amorphous carbon, as shown in Fig. 2. The TEM photograph shows a clean surface

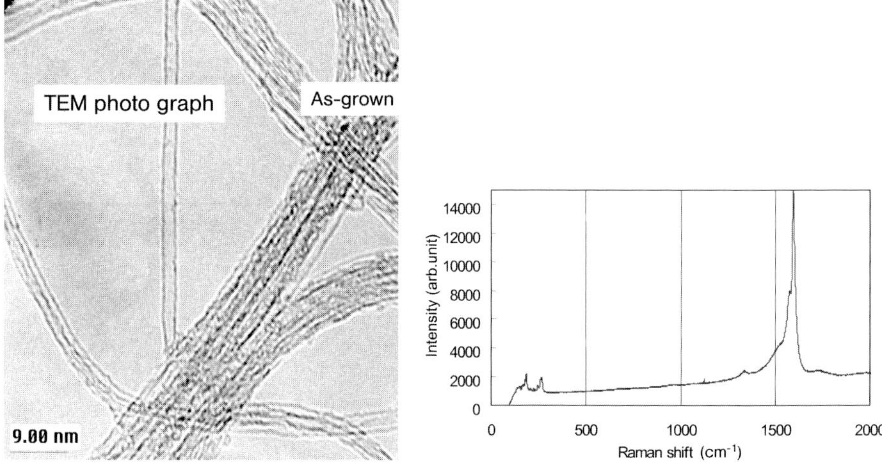

Fig. 2 Clean single-walled carbon nanotubes prepared by catalytic chemical vapor deposition using alcohol: the ACCVD process.

and the Raman spectrum clearly shows the radial breathing mode around $200\,\mathrm{cm}^{-1}$ and a G band around $1590\,\mathrm{cm}^{-1}$ with almost no D band at $1380\,\mathrm{cm}^{-1}$. These results indicate almost no amorphous carbon. The oxidation temperature in air is as high as $550\,^{\circ}\mathrm{C}$, indicating some defects in the graphene sheet. It is anticipated that these single-walled carbon nanotubes have high strength and good electrical conductivity.

5.3
Device Nanotech Research Institute, Inc. (DNRI)

5.3.1 R&D Projects at DNRI

Device Nanotech Research Institute Inc. (DNRI) is the third R&D company in the Mitsui XNRI group, established on 16 January 2003. This company is focusing on micro- and nanoelectromechanical systems (MEMS and NEMS) technology (Fig. 3). Micro- and nanofabrication technology is one of the most important technical issues for MEMS/NEMS device development. We have been researching the nanoimprinting fabrication technique, which is a very attractive procedure for simple and low-cost fabrication compared with conventional lithography, for making our MEMS/NEMS device structures. Four attractive promising areas – RF (radio frequency) MEMS, photonic MEMS, chemical MEMS, and bio-MEMS – are the main targets of our MEMS/NEMS devices made by the nanoimprinting fabrication process. The first prototype DNRI original MEMS/NEMS devices are photonic crystal devices for high-speed optical communication networks and microreactors for chemical reaction analysis or synthesis of organic solvent, which are in development.

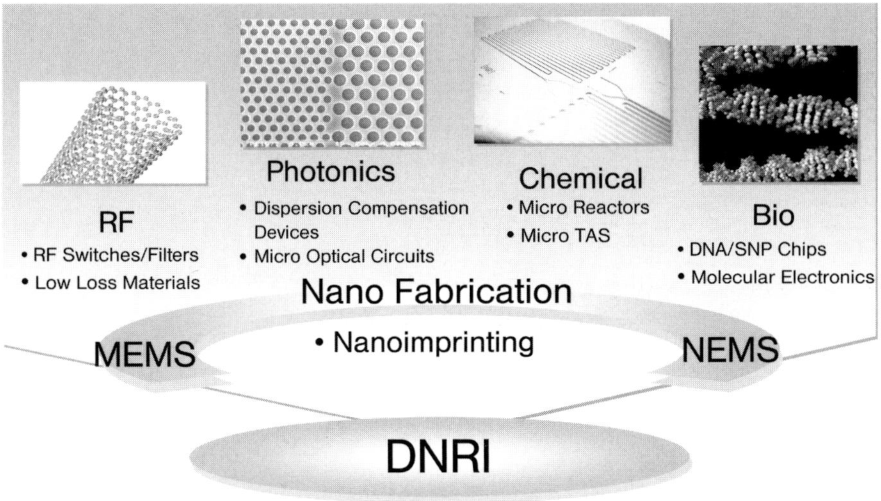

Fig. 3 Research and development projects at DNRI.

5.3.2 DNRI Nanoimprinting Technology Road Map

Commercially available nanoimprinting apparatus is already available from US and European companies and many results on nanoimprinting technology have been reported. However, many technical issues for real commercialization of nanoimprinting process still remain. The process throughput must be improved to apply nanoimprinting to mass production (Fig. 4). Also, a higher imprinting temperature, above 300 °C, is required to apply the process to polymer materials with high glass transition temperature (T_g), such as polycarbonate resins or PEEK polymers, and glasses. A third requirement is the need to be able to produce structures with high aspect ratios such as deep holes or tall pillars for photonic crystal devices and filter structures in microreactors. Furthermore, a multi-layer patterning technique is essential for VLSI fabrication. To satisfy these demanding requirements we are developing a new nanoimprinting system capable of higher throughput for 4–12 inch substrates, with a temperature range up to 550 °C, and higher aspect ratio structuring. The system will be released in the coming year.

5.3.3 Photonic Crystal Waveguides for Controlling Chromatic Dispersion

Controlling chromatic dispersion in optical fibers plays a crucial role in lightwave communications. Chromatic dispersion produces pulse broadening and distortion, and hence data transmission error such as cross-talk. Photonic crystal waveguides provide highly dispersive media because of wavelength-dependent multiple reflections in the periodic arrays of dielectric materials. Photonic crystal waveguides are thus useful for controlling chromatic dispersion. We simulated chromatic dispersion in silicon photonic crystal waveguides. The Si photonic crystal waveguides

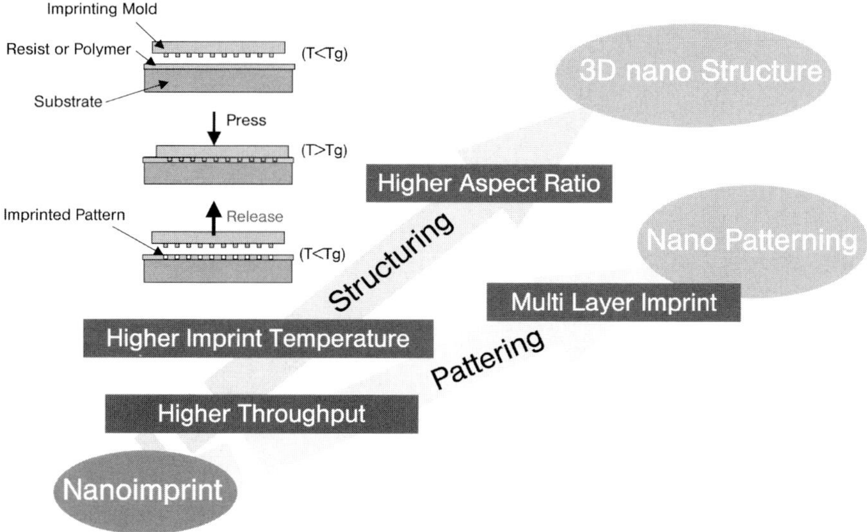

Fig. 4 DNRI nanoimprinting technology road map.

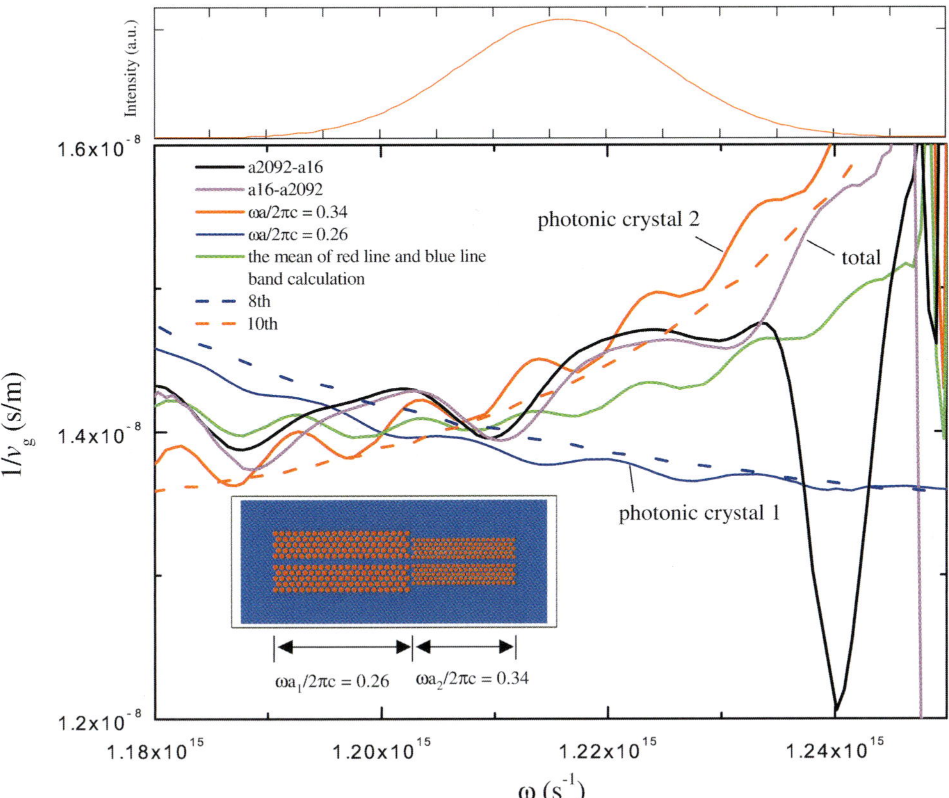

Fig. 5 Photonic crystal waveguides.

consisted of cylindrical air holes arranged in a form of two-dimensional triangular lattices in the Si crystal. The triangular photonic crystals were defined with a ratio $r/a = 0.4$. Here, r and a represent radius of an air hole and length of a unit cell of the triangular lattice, respectively.

In this chapter, we present an implementation of chromatic dispersion control using a novel tandem Si photonic crystal waveguide depicted in the inset of the Fig. 5. The tandem Si photonic crystal waveguide includes two parts in series, each of which is a photonic crystal waveguide based on a triangular photonic lattice. Both parts are identical, except for the factor $\omega a/2\pi c$ with angular frequency of light ω, and light velocity c. $\omega a/2\pi c$ is 0.26 in the first part (photonic crystal 1) of the photonic crystal waveguide, while it is 0.34 in the second part (photonic crystal 2). A finite domain time-difference simulation generated characteristics of chromatic dispersion in the photonic crystal waveguide as plotted in the bottom graph of the figure. The numerical characteristics were obtained using an optical pulse as input. The spectrum of the optical pulse is shown in the top graph of the figure.

The first part has negative chromatic dispersion, namely, increase in group velocity v_g with increasing frequency (decreasing wavelength). The second part, on the other hand, shows the opposite sign of chromatic dispersion. When both parts are combined, the whole waveguide provides a nearly-flat group velocity over the entire spectrum of the optical pulse. This means that the total chromatic dispersion is nearly zero. Photonic crystal waveguides, therefore, realize compact optical devices for controlling chromatic dispersion.

5.4
Ecology Nanotech Research Institute (ENRI): Metallofullerene

Fullerene is known as a typical nanotechnology material, but we have found that metallofullerenes (MF) have their own distinctive natures very different from fullerenes that do not contain metals. Fig. 6 shows one schematic example of lanthanum fullerene, and the structure was simulated for two lanthanum atoms in a C_{80} fullerene cage. In another case we found one atom in a C_{82} fullerene. Both are isolated as stable compounds and characterization shows that lanthanum is located near the cage, not at its center.

In the synthesis of MF, we have been looking for higher yielding soot and its beneficiation. Finally we have developed rational processes in collaborative work with Prof. Shinohara at Nagoya University and Prof. Akasaka at Tsukuba University (Fig. 7). Our novel isolation technique can improve the yield efficiency 100 times more than the conventional method, which is designed to produce MF at 120 g per year as the capacity.

Our expected production volumes will unfold new research applications. One new result is the property found by Aida et al., who reported a novel material for a supramolecular magnet, which shows coupled spin rate of $S = 3/2$ (Fig. 8) (N. Toyama, T. Kato, T. Akasaka, K. Tashiro, T. Aida, Research Activities II, Department of Molecular Structure, Tsukuba University, **2003**, p. 54). This material may ultimately be applicable to small memory devices for medical use.

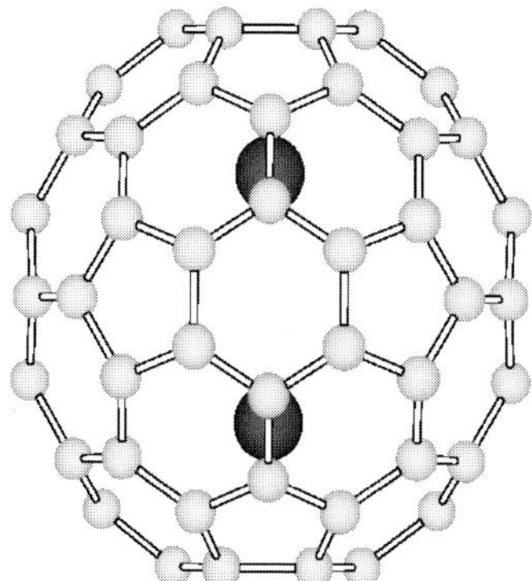

Fig. 6 $La_2@C_{80}$, block circles are La.

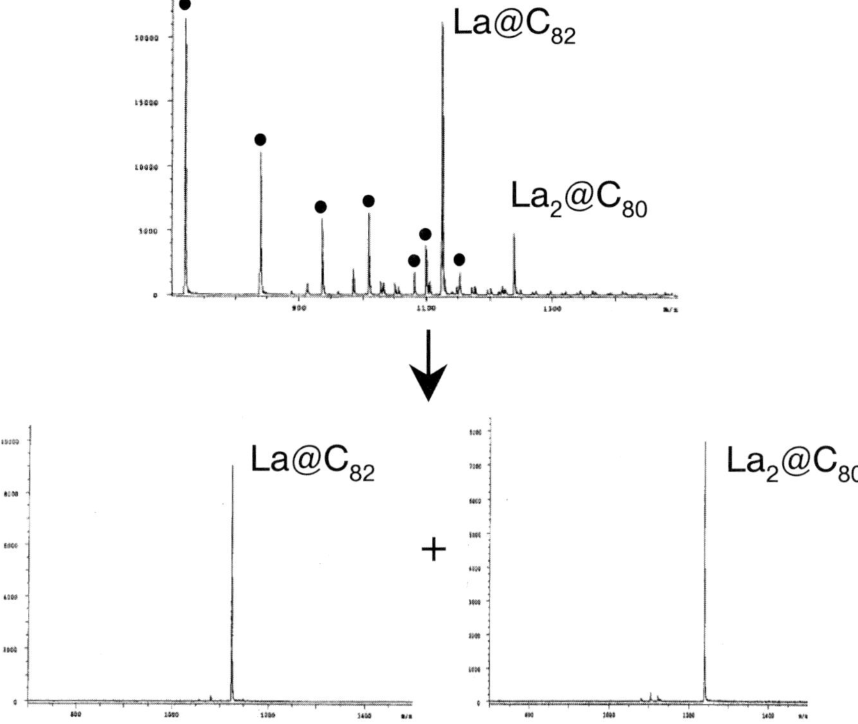

Fig. 7   TOF-MS spectra of soot and pure MFS isolated

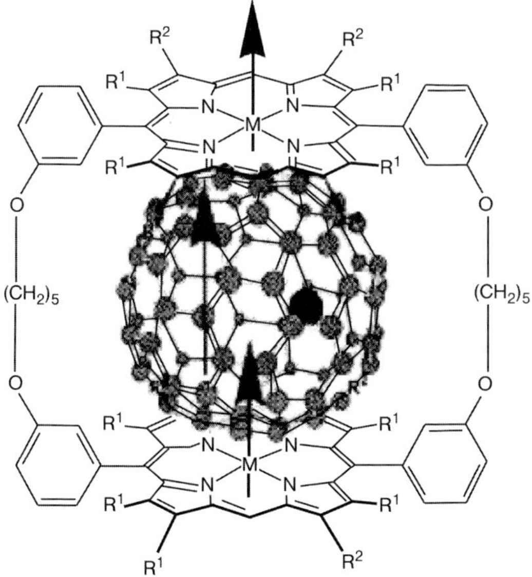

Fig. 8   Metalloporphyrin-MF-supramolecule with $S=3/2$.

5.5
INRI, Inc. (for Intellectual Property): Approach and Strategy

XNRI needs an intellectual property (IP) service to ensure commercialization of research efforts (Figs. 9 and 10). Through collaboration with universities, there are increasing requests for an IP service to assist with technology transfer from idea to product. The IP service is needed at all stages: novelty research of IP, development of IP strategy, writing patents, IP auditing and administration, and licensing or auction-like sales. The Japanese government has advocated strengthening IP services by subsidizing a huge budget for academic work and the Patent Office. We realized that there was no organization adept at total IP strategy in Japan, so we decided to establish INRI as the first IP strategic company to support the interface between laboratory and business.

6
Conclusions

A lack of quick thinking and failure to make quick decisions is not a major problem, but inaction or lack of vision resulting from inadequate information can be calamitous. An interdisciplinary approach is favored in which many different technical and commercial experts are combined into a project team. We need to nurture our new business model and discover how best to achieve the desired outcomes. Basic laboratory research is very valuable as the starting material, but total business design is important for achieving successful transfer to profitable products. We are concentrating on enhancing the capacity of all individuals in our organization, and developing effective collaboration without any wasted effort. In this complex nanotechnology world we are allocating resources to worthwhile projects.

Scientific research and development from XNRI, Nanotech Park, and collaborations elsewhere in the world will be exploited through optimal business design to maximize the value of licensing, new ventures, and worldwide manufacturing. In this way Mitsui will use global networks to achieve success. For this reason I have attended the Namix conference.

Once our new business model is acknowledged to be successful, I believe it will galvanize others to pursue many projects. Already we are being approached by various labs including some local government entities. We hope that our example will be a potent engine for the upturn of our dull economy, and also for solving many of the social problems of the 21st century world, such as green issues.

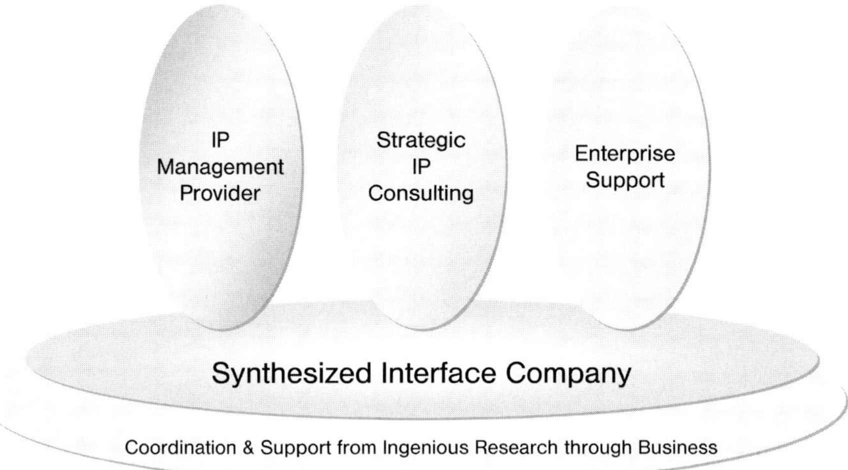

Fig. 9 Vision of INRI Inc.

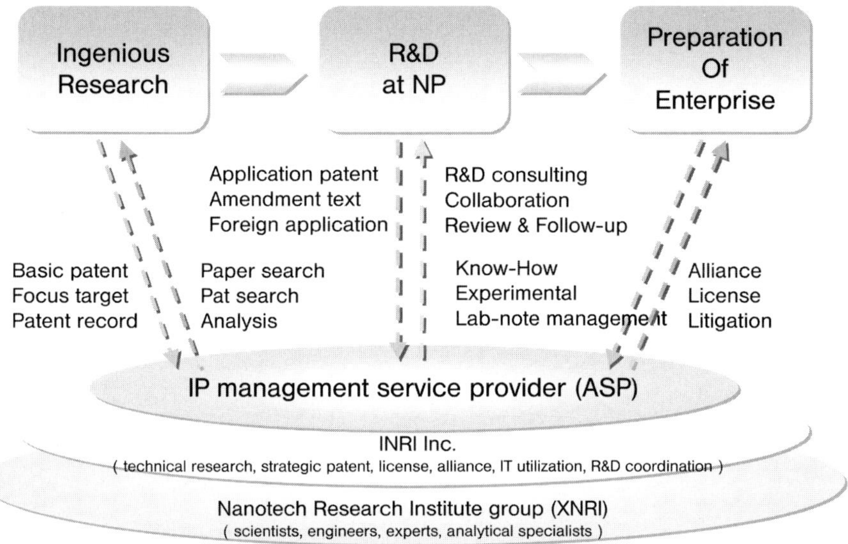

Fig. 10 Interface between research and business.

Nanomaterials and Smart Medical Devices

Ottilia Saxl

Until relatively recent times, medical knowledge was limited, as was the possibility of remedial effect. The doctor had little to sell: death was early, inevitable, and inexpensive. J. K. Galbraith

Replacing a defective component of the body or substituting for a deteriorated organ function by a natural or man-made counterpart is one of the major contributions of modern science to clinical medicine. P. Galletti

1
Introduction

Most of us have the expectation that medical science will "fix" organs or limbs whose use has been diminished or lost through age, accident, or genetic causes. Thirty years ago the "Six Million Dollar Man" was Hollywood fiction, but it was becoming ever more likely even then that medical science could rebuild a damaged person; as the replacement of heart, lungs, kidney, and liver were being demonstrated with increasing success. Now, with an even more profound understanding of how life works at the fundamental level, advances in medicine are dramatically increasing. Knowledge, combined with the demands of maintaining a good quality of life from a longer-living population, while also addressing the increase in lifestyle diseases, are providing opportunities for breakthroughs in medical care, fuelled by huge financial rewards.

According to statistics issued by the Department of Health and Human Services in the USA, *National health expenditures are projected to reach $3.1 trillion in 2012, growing at a average annual rate of 7.3 percent between 2002–2012. As a share of GDP, health spending in the US is projected to reach 17.7 percent by 2012, up from its 2001 level of 14.1 percent.* According to another statistic, posted on the liver4you website in May 2003: *more than 80,000 Americans are on the waiting list for kidney and liver transplants today.* There is a huge healthcare market out there, which nano- and microtechnologies are set to influence dramatically, and this chapter discusses some early advances that are already bringing about important changes.

The Nano–Micro Interface: Bridging the Micro and Nano Worlds
Edited by Hans-Jörg Fecht and Matthias Werner
Copyright © 2004 WILEY-VCH Verlag GmbH & Co. KGaA, Weinheim
ISBN: 3-527-30978-0

2
Why are we Seeing Advances Now?

The integration of nano- and microtechnology with biology is producing many medical breakthroughs in diagnostics, therapeutics, and bioengineering. A relatively simple example is the ability to create nanoparticles with electronic, optical, magnetic, or structural attributes. These nanoparticles can be linked chemically to biological molecules that can make them suitable for many uses, for example, as contrast agents for in-vivo optical and magnetic resonance imaging, as smart carriers for drug delivery applications, or as structural scaffolds for tissue engineering.

Nano- and microtechnology is also leading to a new generation of medical implants. For example, the combination of nanocomposite materials, biocompatible nanocoatings, ultraprecision machining with nanoscale tolerances, and manufacturing techniques derived from the electronics industry, enables novel cochlear and retinal implants to be developed. For example, a "learning" retinal implant has been developed by IIP GmbH that employs several nano- and microtechnologies. These include:

- a camera-on-chip to obtain images,
- a retina encoder to translate video images to nerve-compatible signals,
- a wireless signal for energy transmission into the eye, and
- a stimulator to electrically stimulate retinal cells.

The astonishing implications are that sufferers from retinitis pigmentosa and macular degeneration, previously classed as "blind", with this implant would have the possibility to be re-classified as sighted.

Opportunities also exist in applying nano- and microtechnologies to drug delivery. Due to advances in biotechnology, there has been considerable growth in the number of new protein drugs (about 15% per annum) on the market. Although only about 5% of drugs are presently protein based (generating about US$ 23 billion per annum), it is predicted that by 2020 protein drugs will comprise an astonishing 50% of all drugs. Protein drugs cannot be taken orally as they are acted upon by digestive enzymes. Injection directly into the blood stream is the obvious route, but not one that appeals to patients. Most people want an alternative, so the drug companies and others have developed other, preferable ways of administering these drugs. These include patches, implants, skin gels, coated pills, and nasal sprays. These delivery techniques themselves are not insignificant, enjoying a market of over US$ 20 billion/year.

One route to effective delivery of protein drugs to the bloodstream has been found to be via the lungs. This requires that the drug is administered in particulate form, but the particles can not be so large that they clog the lungs, or so small that they are exhaled. A novel means of creating suitably sized carrier particles, and coating them with drugs has been developed by a company called XstalBio (www.xstabio.com). Using the XstalBio technique, water-soluble microcrystals, which can be made of salts, sugars, or amino acids, can be prepared and coated with the protein drug. The beauty of the technique is that not only can the size of the particles can be controlled for optimum delivery, but also their composition, and that of the coating.

Another advance in the utilization of nano- and microtechnology has been in the treatment of several kinds of tumors. The problem of dealing with advanced tumors, often resistant to most other therapies, has been addressed through a technique called magnetic hyperthermia. Special iron-oxide nanoparticles with surfaces that attach specifically to cancer cells have been produced by a company in Germany called MFH, a spin-out of the Charité organization within the Humboldt University in Berlin. These particles are easily introduced into the tumor, and when an alternating external magnetic field is applied, the particles become selectively heated, killing the attached cancer cells. There are many surprising advantages associated with this treatment: the particles are tumor-specific, the effect is immediate, and there are no toxic side effects. This technique is now moving into clinical trials. The expected market for the magnetic field applicators is around US$ 300 million and, for the nanoparticles themselves, around US$ 1.5–2 billion.

To turn to controlled release, individual "demand" or need for therapeutic drugs, including replacement hormones, varies on a daily basis both in terms of timing and dosage. Three pills at equally spaced intervals each day is not the optimum administration regime, especially for hormones such as insulin; as well as many drugs. It has been an ongoing problem to address these important variations in requirements, which can change dramatically according to the time of day, exercise levels, timing of meals and so on. The more desirable and sensible ideal is to treat hormone deficiencies by mirroring the body's own mechanisms. Nano- and microtechnologies are providing the solution. As a first step in offering timed delivery, implantable chips with wells etched into the silicon structure are being developed in the USA. Drugs are stored in the wells and sealed by a thin layer of gold. An electronic signal can be applied that "breaks" the gold seal and releases a "dose" of the medication. This capability could lead towards implanted closed loop system with nanosensors that can determine the medication needs, triggering the appropriate delivery of hormone or drug only as and when required, and at appropriate levels.

This idea of "appropriate level" of medication is also critical, as individuals vary in their responses; too little drug and it has no effect; too much and the effect is toxic. The quantity that constitutes what exactly is a therapeutic level varies from individual to individual, and gauging this is extremely important. Even more worrying to the pharmaceutical companies is the number of deaths due to adverse reaction to relatively "common" prescription drugs, such as codeine and warfarin (marketed as Coumadin), and some drugs used to treat leukemia. A major study showed that prescription drugs cause up to 137 000 deaths a year in US hospitals, making adverse reactions to drugs the fourth leading cause of death.

It has been demonstrated that this response is directly linked to an individual's genetic make-up. Already several companies and medical laboratories are developing rapid genetic tests to determine who is likely to respond poorly, or not at all, to standard doses of common medications. For example, Roche is investing US$ 70 million on gene chips developed by the Affymetrix company that will check patients for common variations in two genes that play crucial roles in the breakdown of almost half of the prescription drugs now on the market. These tests represent the beginning of a fundamental change where the choice of drug and the amount prescribed will be related to the genetics of the patient.

One of the key advances underpinning medical diagnostic and analytical techniques, and leading to the gene chip technology outlined above, is the development of "lab-on-a-chip". This has been made possible by using manufacturing techniques developed by the electronics industry. Lab-on-a-chip technology hinges on the fact that the smaller the sample, the faster the reaction. Many medical analyses for individuals can now be performed on palm-sized instruments using tiny, self-calibrating and disposable sample holders. Results can be almost instantaneous, and processed on site or transmitted to processing centers for diagnosis. This technique has application not only in speeding up diagnosis for an individual but also for fast screening of disease in a population, as well as fast screening of new drug candidates.

Lab-on-a-chip technology (or the miniaturization of analytical and diagnostic techniques) is fundamental to utilizing our growing knowledge of genomics and proteomics. From this, new and better medical therapies can be developed based on understanding the *causes* of disease, as well as why we respond so violently to some drugs. In essence, lab-on-a-chip technology is the ultimate in miniaturization techniques, which allows us to analyze (bio)chemical materials at very high throughputs, making it possible to perform thousands of experiments in parallel.

Another step along the road to the Star Trek hospital is the development of sophisticated remote medical sensing and monitoring systems based on nano- and microtechnologies on-chip. The Laboratoire d'Electronique et de Technologies de l'Information (LETI), at CEA-Grenoble, in conjunction with Tronics and Absys has developed a remote-powered medical microsystem based on two tiny silicon chips. This particular system is capable of continuously monitoring critical pressures (blood, lungs) of patients on respirators. Located in the windpipe, and remote-powered via an antenna in a patch placed on the patient's chest, the system can transmit a pressure reading 200 times per second, enabling the respirator airflow to be adjusted according to the patient's needs. This integration of functions has many additional applications in terms of medical care for rural communities, the potential for patients to be treated at home, or monitored by a central computer in a hospital.

A company based near CEA-LETI in France is ST Microelectronics. This company has developed a new silicon-chip-based biometric recognition system. This is the TouchChip: a fingerprint sensor that combines image processing and algorithm processing in a single chip. This technology is being incorporated as a security device in laptops. What makes it even more interesting is the application of this chip-based technology into the skin care market. Some clever modifications have resulted in the TouchChip technology being applied to provide a detailed analysis of human skin condition, quickly and easily, for cosmetic applications.

Advances on this technique are being developed by L'Oréal in conjunction with other laboratories in order to detect the orientation of lines in the forearm (an indication of aging), quantify the actual hydration of the skin; and measure skin texture. This new SkinChip is capable of generating a "capacitance" map that can easily be converted into a hydration or dryness map. Assessing how dry or well-hydrated the skin is will allow L'Oréal to design beauty-care products better suited to age-related changes of skin.

So what does the future hold in terms of the prevention, control, and diagnosis of disease? If we think about some of the technologies discussed above that are in or close to market, it is not difficult to make some predictions. We are beginning to understand why some drugs are toxic to some individuals (and test for this), why the optimum dose varies from person to person, how to screen individuals for infectious or genetically transmitted diseases, how to use silicon chip technology for rapid sensing and diagnosis, and how to monitor and control therapies remotely.

What we then have is a likely future scenario. Individuals will be enabled to test themselves using non-invasive techniques, perhaps by placing a finger on a sensor that detect symptoms from sweat or pulse rates. Data can be transmitted to a central medical service via home computer, and either the patient will receive the diagnosis directly, or it will be sent to his or her primary care physician who will explain the appropriate course of treatment. This treatment will be tailored to the person's own genetic makeup. In many instances visiting a surgery for minor complaints will become unnecessary, reducing the spread of infection. Doctors may only be required to patch accidental damage to tissues (mild cuts, burns).

As discussed earlier, nano- and microtechnologies are also leading to the reality of the six-billion dollar man. Retinal and cochlear implants that can actually provide sight to the blind and quality hearing to the deaf are on the way, as are nerve regeneration techniques for those paralyzed through accident. Nanotechnology is also leading to novel materials that can be used as "scaffolds" around which organs such as liver and kidneys can be grown from a patient's own cells; and the combination of nano-, microelectronics, and artificial intelligence are already creating robotic surgeons that can successfully operate on moving organs such as the heart or lungs.

3
Conclusion

In conclusion, our increased understanding of living processes is providing us at last with a powerful and real means to create therapies that really work. It will enable those of us who can afford to pay for treatment to look forward to a happy and fit ripe old age. The interesting thing about nano- and microtechnologies is that they offer the potential to make new therapies available cheaply and it is to be hoped that the companies commercializing these technologies will make them available to all, regardless of wealth, and sooner rather than later.

II
Fundamentals and Technology

Bridging Dimensional and Microstructural Scaling Effects

Uwe Erb, Cedric Cheung, Mohammadreza Baghbanan, and Gino Palumbo

1
Introduction

Controlling and manipulating matter down to the length scale of one nanometer, which is the size domain of atoms and molecules, has been the dream of scientists for many years. With the achievements over the past few decades in increasing the resolving power of the characterization techniques used to look at atoms and molecules (e.g., microscopy, spectroscopy, diffraction) and the development of new tools to actually pick up and manipulate individual atoms and molecules (atomic force microscopy) or very small objects (nanotweezers), this dream has now become a reality. In the so-called bottom-up approach, nanostructures can now be created atom-by-atom or by assembly of components consisting of a few thousand atoms or molecules. Processes such as self-organization, self-assembly, or templating can be used to build a multitude of nanoarchitectures. For the top-down approach, many challenges to satisfy the demands for ever-increasing density of integration of many different components have already been met. While the large-scale implementation of many of these applications is still several years or perhaps decades away, many people agree that this technology has a tremendous long-term potential to completely revolutionize society, resulting in (among other things) a deeper understanding of nature, better quality of life, and increased productivity.

This chapter deals with two key areas in nanotechnology: *nanocrystalline materials* and *nano/microsystem technology*. Both areas are presently of great interest but have their origins in different research communities. Nanomaterials were mainly developed by materials scientists, while microsystems were the domain of electrical, electronic, and mechanical engineers. Unfortunately, however, so far the interaction between these two communities has been relatively poor and they do not seem to benefit as much as they should from each other and the rapid advances in their respective fields.

The Nano–Micro Interface: Bridging the Micro and Nano Worlds
Edited by Hans-Jörg Fecht and Matthias Werner
Copyright © 2004 WILEY-VCH Verlag GmbH & Co. KGaA, Weinheim
ISBN: 3-527-30978-0

2
Nanocrystalline Materials

One of the "oldest" areas of nanotechnology deals with nanostructured materials (first introduced in 1981 by H. Gleiter [1]) in which many properties of materials can be tremendously improved by grain- or crystal size reduction to less than 100 nm. The main structural characteristic of fully dense nanomaterials is the enhanced volume fraction of atoms located at the interfaces or grain boundaries between differently oriented crystals. As shown in Fig. 1 (adapted from [2]) for conventional polycrystalline materials with grain sizes larger than 1 µm (10^3 nm) this volume fraction is much less than 1%. However, when the grain size is reduced to values less than 100 nm (which is usually considered the upper grain-size limit for nanocrystalline materials) this volume fraction increases to significant values. For example, at 10 nm the volume fraction is about 30% and for grain sizes less than 5 nm the number of atoms associated with grain boundaries is actually higher than the number of atoms located in perfect crystal positions. It is therefore not surprising that nanocrystalline materials exhibit considerable changes in many of their physical, chemical, and mechanical properties.

Over the past two decades, close to 200 different synthesis routes for the manufacture of nanocrystalline materials have been developed (e.g., [3–6]). Top-down approaches that use conventional bulk starting materials include solid-state processing (e.g., mechanical attrition, severe plastic deformation, crystallization of amorphous precursors) and liquid-phase processing (e.g., rapid solidification, atomization). Bottom-up methods produce materials, often under conditions far from equilibrium, in an atom-by-atom fashion such as in vapor-phase processing (e.g., physical and chemical vapor deposition, inert gas condensation, sputtering),

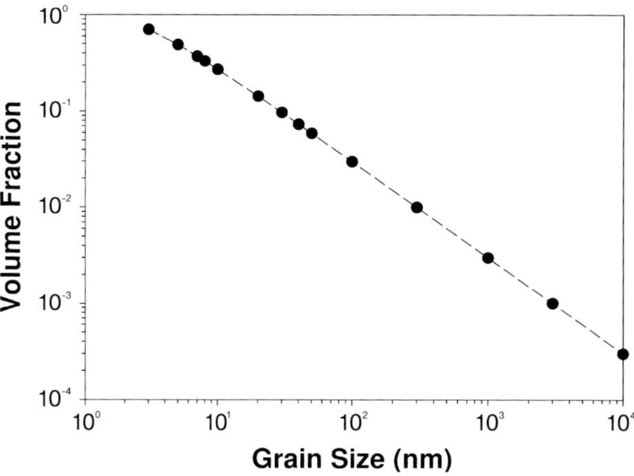

Fig. 1 Effect of grain size on the volume fraction of the interface component assuming a grain boundary thickness of 1 nm (adapted from [2]).

chemical synthesis (e.g., sol-gel, precipitation), and electrochemical synthesis (e.g., electroplating, electroless plating). Many of these techniques produce nanostructured powders, which must be further processed in secondary processing steps (e.g., hot isostatic pressing, sintering, thermal spraying) to produce useful structural materials. In this area of nanotechnology there is no immediate need for further synthesis methods. What is really needed is for materials engineers to focus on applications of the many nanomaterials already available today.

3
Nano/Microsystem Technology

Nano/microsystems can be described as intelligent miniaturized systems that combine sensing and/or actuating functions with processing functions. Such systems are typically multifunctional, combining two or more electrical, mechanical, optical, chemical, biological, magnetic, or other properties (such as in MEMS, NEMS).

Microsystems can be constructed by various micromachining methods on a single substrate in the form of a monolithic system (e.g., an all-Si device), or from parts using different manufacturing technologies on different substrates and connected together in hybrid systems. Although monolithic systems are typically more compact, any microsystem requiring multifunctional properties and materials (e.g., insulator, semiconductor, conductor, ferromagnet) can only be produced in hybrid form.

Advances in materials science and technology have played key roles in the evolution of microsystems. However, as pointed out recently by Spearing [7], the effects of length scale on the design and fabrication of miniaturized material systems, which are crucial for microsystem performance, are not completely understood today. Spearing [7] considered at least three scaling effects: quasi-fundamental scaling, extrinsic scaling, and mechanism-dependent scaling. Examples of quasi-fundamental scaling include the cube (e.g., volume, mass, magnetic force) versus square (e.g., area, power, electrostatic force, thermal loss) dependencies. Extrinsic scaling addresses issues such as tolerances, shape factors, and other real estate considerations never encountered in macroscopic design. The third scaling factor, mechanism-dependent scaling, deals with changes in properties when the characteristic length scale of a device approaches that of an intrinsic property change of the material, for example as a result of grain-size reduction.

This chapter addresses the latter of these scaling issues, namely the mechanism-dependent scaling of properties of microsystems based on the crystallinity and crystallographic structure–property relationships of typical *metallic materials* presently applied in (or considered for) the design of hybrid nano/microengineered systems. In most cases, these metallic microcomponents are fabricated by electrodeposition from aqueous or organic solutions (see Tab. 1) [8–18] but often do not perform as expected because many people in the nano/microsystem community have not yet appreciated the importance of mechanism-dependent scaling.

Tab. 1 Typical microcomponents produced by micromolding and electrodeposition.

Component	Material	Reference
Helical microcoils for microinductors and magnets	Au, Ag, Cu	8
Microrivets	Ni	9
Ferromagnetic microactuators, microgears	NiFe	10
HARM's for heat transfer, catalysis, acoustics	Ni	11
Multicoil helical springs, inkjet printer nozzles	Ni	12, 13
Microelectrodes for monitoring on-chip fluid chemistry	Au	14
Fuel atomizers for gas turbine engines	Ni	15
Wobble motor with integrated synchronous control	Ni	16
Gripper for micromanipulation tasks	Ni	17
Acceleration switches for automotive applications	Ni	18

4
Present Gap Between Nanomaterials and Nano/Microsystem Technology

As long as nano/microsystems are based on the monolithic approach, such as micromachining of single crystal Si, the main scaling effects of concern are quasi-fundamental and extrinsic scaling. However, with the increasing need to incorporate metals in hybrid systems, which can only be produced in polycrystalline form, structure-intrinsic factors have also become a major concern.

To date the most widely used techniques to produce nano/microsystem structures are all based on electrodeposition. Such techniques include LIGA (German acronym for Lithographie [lithography], Galvanoformung [electroforming] and Abformung [molding]), HARMS (high aspect-ratio moldings), DEM (deepetching, electroforming, and microreplication), and electroless plating. While the development of suitable molding technology based on lithography and the associated miniaturization of components has been remarkable, the crucial electrodeposition step itself has remained a simple extension of traditional electroplating technology. As a result, the internal structure of the components is the same as that observed in conventional electrodeposits. Essentially, with increasing deposit thickness a transition is observed from a very fine grained structure in the initial deposit layer to a highly textured, elongated (columnar) structure with grain widths of a few micrometers and lengths of tens of micrometers. A schematic diagram illustrating in cross section the microstructure produced by conventional plating is shown in Fig. 2a and an example of an actual microstructure is shown in Fig. 2b for the case of Ni produced from a Watts-type electrolyte [19]. In such a plated structure the thickness of the initial fine-grained layer close to the substrate is very strongly dependent upon plating variables (temperature, bath composition, pH). Fig. 2c, for example, shows the thickness of this fine grained layer as a function of bath pH [19]. The crystallographic texture of the columnar grains

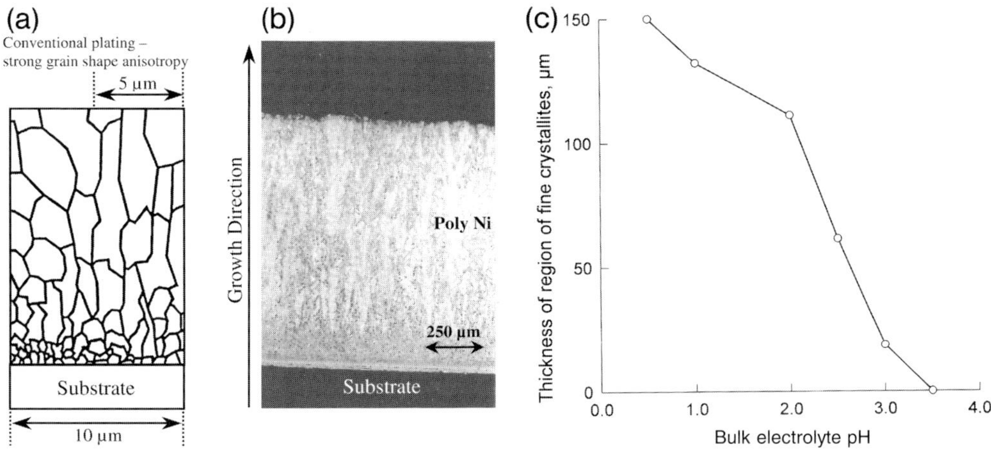

Fig. 2 Microcomponent produced by conventional electrodeposition:
(a) schematic cross section; (b) actual cross section of a Ni deposit;
(c) thickness of region of fine crystallites close to the substrate
as a function of bulk electrolyte pH.

growing out from the initial fine grained structure also depends strongly on plating variables, particularly cathode potential, bath pH, and the type of bath additives used (e.g., [19]). As a result of this structure, MEMS components show considerable variations in many properties, in particular mechanical properties. Attempts to achieve a more uniform structure by post deposition annealing have had some success but usually at the expense of strength. Tab. 2 summarizes the results of mechanical testing performed on LIGA components produced at different current densities as well as the effect of annealing [20]. Large, and often unpredictable, variations both in yield strength and Young's modulus are observed.

Tab. 2 Summary of mechanical properties of some LIGA components [20].

Sample	0.2% proof stress in tension [MPa]	Elastic modulus [GPa]
LIGA Ni, 50 mA/cm² as-deposited	277 ± 7.6	160 ± 1.0
LIGA Ni, 20 mA/cm² as-deposited	441 ± 27	156 ± 9.3
LIGA Ni, 50 mA/cm² annealed (600 °C/1 hr)	55	92
LIGA Cu, 30 mA/cm² as-deposited	239 ± 5.1	113 ± 21
LIGA Cu, 30 mA/cm² annealed (500 °C/1 hr)	122.5	76

The second problem in the present approach is that the grain size of components in the columnar region is simply too large. With decreasing overall width of the microcomponent, a critical level will be reached at which there are only a few grains over the entire cross section of the component, as indicated in Fig. 2a, for example, for a 5 µm wide component. With a further decrease in width, the component can even approach the bamboo structure with a significant loss in strength. Thus the microstructure of electrodeposited microcomponents presently in use is entirely incommensurate with property expectations both across the width and the thickness of the component on the basis of average grain size, grain-size gradients, and grain shape variations.

5

Bridging Dimensional and Microstructural Scaling Effects

Our approach to overcoming these problems is based on the premise that any further size reduction of metallic nano/microcomponents can only lead to satisfactory system response when the grain size in the material is simultaneously reduced to overcome the limitations of present micro/nanosystems. A microstructure that could entirely eliminate the grain-size gradient and grain shape limitations is shown schematically in Fig. 3. This component shows an equiaxed grain structure with a grain size much smaller than the external dimensions of the component. The question to be addressed is, of course, to what extent grain-size reduction is required to obtain a component with the desired property/perform-

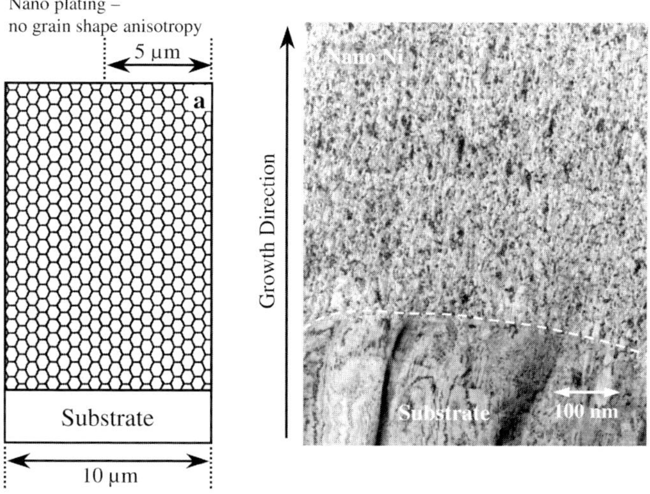

Fig. 3 Microcomponent produced by nanoelectrodeposition:
(a) schematic cross section; (b) actual cross section of nano Ni on
polycrystalline phosphor bronze substrate.

Tab. 3 Expected improvements of performance indicators for Ni by grain size reduction (normalized with respect to 10 μm grain size).

Performance indicator	*Grain size*			
	10 μm	*1 μm*	*100 nm*	*10 nm*
Specific strength (maximal σ_y/ρ)	1	2.2	3.4	4.6
Elastic energy storage (maximal σ^2/E)	1	4.8	11.7	21.5
Thermal shock resistance (maximal $\sigma_y/E \cdot a$)	1	2.2	3.5	4.9
Wear resistance (maximal $1/W$)	1	1.6	3.7	52

σ_y: yield strength; E: Young's modulus; ρ: density; a: thermal expansion coefficient; W: wear rate.

ance characteristics. For strength reasons it is suggested that the grain size should be at least one order of magnitude smaller than the smallest component dimension. For microcomponents with submicrometer external dimensions, this requirement dictates that the grain size should be in the nanometer range. For design considerations, it is therefore important to know the grain-size dependence of various properties over a wide grain-size range. Unfortunately, such data are presently not available for most metallic materials with external component dimensions in the micrometer or submicrometer range. However, considerable information is available on grain-size dependence of various properties from the study of bulk nanocrystalline materials [21, 22]. These data can be used at least as a starting point for designing the correct scale-adjusted microstructure.

Figs. 4 and 5 (adapted from [21]) show several important properties for bulk nanostructured nickel with either strong grain-size dependence (Fig. 4) or weak grain-size dependence (Fig. 5). It should be noted that these properties were measured on freestanding, porosity-free foil samples with thicknesses in the 100–200 μm range and grain sizes of about 10 μm down to 10 nm, that is, covering the grain-size range of importance for the design of nano/microcomponents.

Using the property changes with grain size shown in Figs. 4 and 5 and an Ashby type approach [23], performance indicators for some of the important properties of micro/nanocomponents can be readily assessed. Tab. 3 shows that remarkable improvements can be achieved in specific strength, elastic-energy storage, thermal-shock resistance, and wear resistance for Ni based solely on grain-size reduction from 10 μm to 10 nm. Of course, for any specific application other property changes with grain size must be considered in order to arrive at the optimum grain size for a particular component. For example, for some applications the increase in electrical resistivity at grain sizes less than 50 nm (Fig. 4d) could be either of concern (if high conductivity is required) or beneficial (for reduced eddy current losses). The important point to make here is that with a good understanding of the effect of grain size on all properties of concern for the development of a nano/microcomponent, grain size becomes an important design parameter in efforts to achieve optimum and reliable system performance.

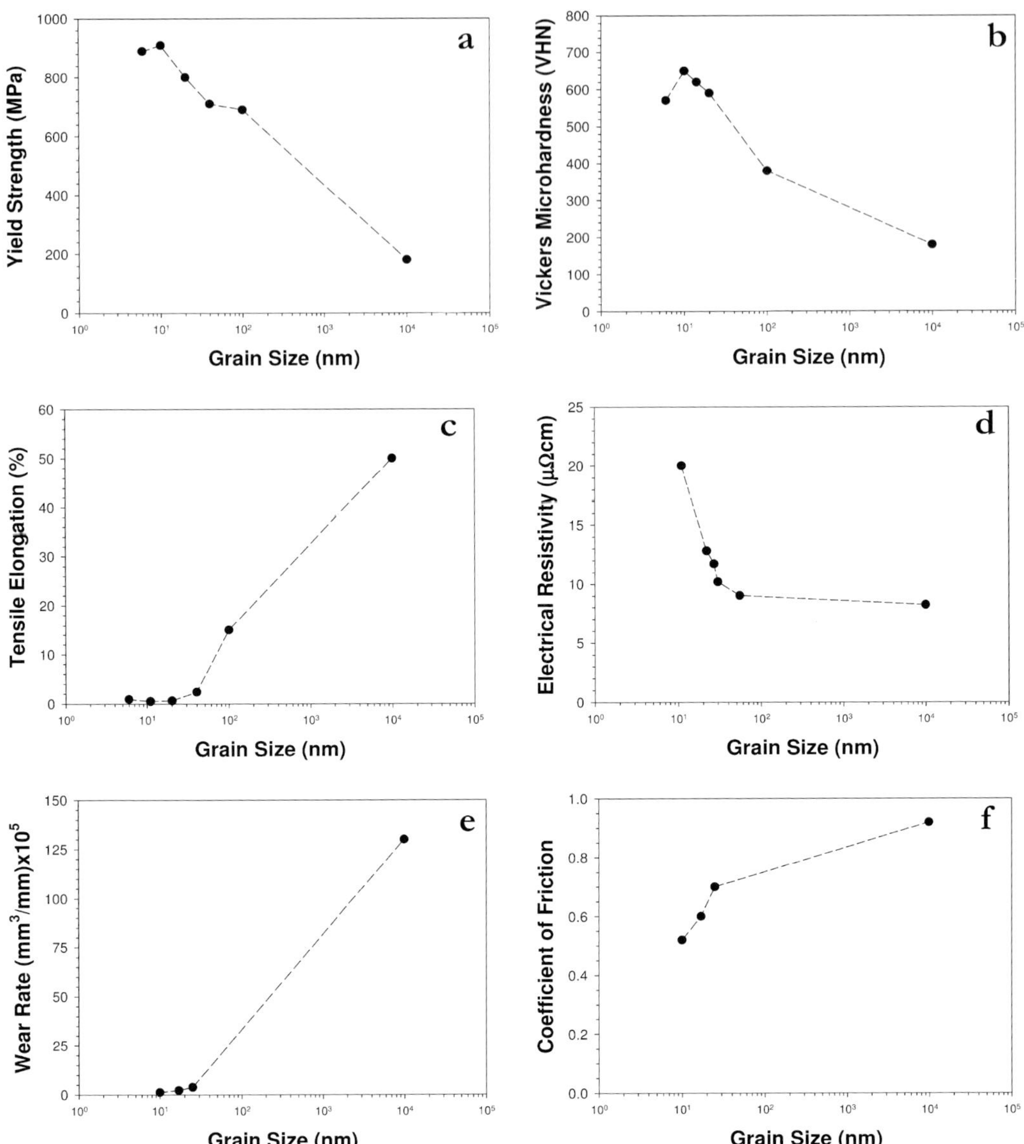

Fig. 4 Grain-size dependence of various properties for Ni:
(a) yield strength; (b) Vickers microhardness; (c) tensile elongation; (d) electrical resistivity; (e) wear rate; (f) coefficient of friction (adapted from [21]).

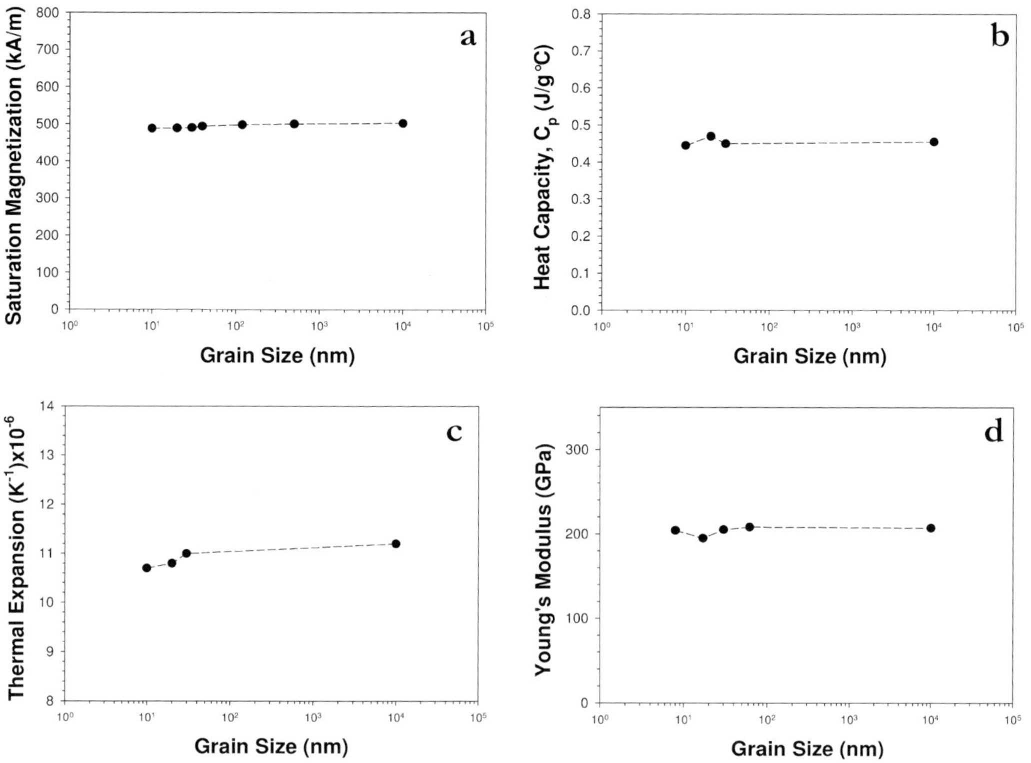

Fig. 5 Properties of Ni with weak grain-size dependence:
(a) saturation magnetization; (b) heat capacity; (c) thermal expansion
coefficient; (d) Young's modulus (adapted from [21]).

6
Conclusions

The application of conventional electrodeposition technology to the manufacture
of micro/nanosystem components results in microstructures that are incommensurate with the dimensional constraints imposed by the external component size.
Considerable improvements in component performance and reliability can be
achieved by appropriate microstructural design involving overall grain-size reduction and elimination of the transition from the fine grain to columnar structure.

Acknowledgments

Financial support from the Natural Sciences and Engineering Research Council of Canada is gratefully acknowledged.

References

1 H. Gleiter, in *Deformation of Polycrystals: Proc. 2nd Risø Int. Symp. on Metallurgy and Materials Science*, Risø National Laboratory, Roskilde, Denmark (1981) 4.

2 G. Palumbo, S.J. Thorpe, and K.T. Aust, *Scripta Metall. Mater.* 1990, 24, 1347.

3 Proc. 2nd Int. Conf. on Nanostructured Materials, *Nanostr. Mater.* 1995, 6, 1–1026.

4 Proc. 3rd Int. Conf. on Nanostructured Materials, *Nanostr. Mater.* 1997, 9, 1–771.

5 Proc. 4th Int. Conf. on Nanostructured Materials, *Nanostr. Mater.* 1999, 12, 1–1188.

6 Proc. 5th Int. Conf. on Nanostructured Materials, *Scripta Mater.* 2001, 44, 1161.

7 S.M. Spearing, *Acta Mater.* 2000, 48, 179.

8 J.A. Rogers, R.J. Jackman, and G.M. Whitesides, *J. Microelectromech. Syst.* 1997, 6, 184.

9 B. Shivkumar and C.J. Kim, *J. Microelectromech. Syst.* 1997, 6, 217.

10 J.W. Judy and R.S. Muller, *J. Microelectromech. Syst.* 1997, 6, 249.

11 C. Marques, Y.M. Desta, J. Rogers, M.C. Murphy, and K. Kelly, *J. Microelectromech. Syst.* 1997, 6, 329.

12 J.D. Madden and I.W. Hunter, *J. Microelectromech. Syst.* 1996, 5, 24.

13 J.D. Lee, J.B. Yoon, J.K. Kim, H.J. Chung, C.S. Lee, H.D. Lee, H.J. Lee, C.K. Kim, and C.H. Han, *J. Microelectromech. Syst.* 1999, 8, 229.

14 H.E. Ayliffe, A.B. Frazier, and R.D. Rabbit, *J. Microelectromech. Syst.* 1999, 8, 50.

15 N. Rajan, M. Mehregany, C.A. Zorman, S. Stefanescu, and T.P. Kicher, *J. Microelectromech. Syst.* 1999, 8, 251.

16 V.D. Samper, A.J. Sangster, R.L. Reuben, and U. Wallrabe, *J. Microelectromech. Syst.* 1999, 8, 214.

17 M.C. Carozza, A. Menciassi, G. Tiezzi, and P. Dario, *J. Microelectromech. Syst.* 1998, 8, 141.

18 T. Tonnesen, O. Ludke, J. Noetzel, J. Binder, and G. Mader, *J. Micromech. Microeng.* 1997, 7, 237.

19 K.S. Gill, PhD Thesis, Queen's University, Kingston, Ontario, Canada 1990.

20 T.E. Buchheit, T.R. Christenson, D.T. Schmale, and D.A. Lavan, *Mater. Res. Soc. Symp.* 1999, 546, 121.

21 U. Erb, K.T. Aust, G. Palumbo, J.L. McCrea, and F. Gonzalez, in: *Processing and Fabrication of Advanced Materials IX*, T.S. Srivatsan et al. (eds.), ASM International, Materials Park, OH 2001, 253.

22 U. Erb, K.T. Aust, and G. Palumbo, in: *Nanostructured Materials*, C.C. Koch (ed.), Noyes Publications/William Andrew Publications, Norwich, NY 2002, 179.

23 M.F. Ashby, *Materials Selection in Mechanical Design*, Pergamon Press, Oxford 1992.

Bridging the Gap Between Nanometer and Meter

Matthias Meyer, Jürgen Koglin, and Thomas Fries

1
Introduction

Some of the goals of nanotechnology are: unlimited optimization of product properties in the field of power engineering (fuel cells, batteries, solar cells), environmental technology (resource cycles, waste management, cleaning), information technology (high-density storage, powerful processors), and in the sector of health. Those are the identified future markets that offer good opportunities for exploiting nanotechnology. For developing this technology we need a completely new set of metrology devices.

To fill the gap between the milli- and micrometer technology and the nano- and subnanometer technology, it is now possible for the surface measuring instrument MicroGlider® [1] to be optionally equipped with AFM. The AFM is fixed to the instrument in addition to the standard optical topography sensor. If necessary, a spot will be selected within the available overview measurement to determine the measuring range of the AFM. The AFM is able to investigate structures down to the molecular range. Alternatively, it is also possible for all high-resolution AFM measurements to be automatically carried out in measurement range of the optical topography sensor to investigate larger profiles and surfaces. The measuring instrument enables the combination of maximum measuring ranges of 100 mm, 350 mm, or 600 mm with resolutions down to the sub-nanometer range in one single instrument. A variety of AFM modes is included, as is the new AFAM (atomic force acoustic microscopy) mode.

2
Motivation

The focus of this chapter is on metrology in the nanometer range and its coupling to larger orders of magnitude. Of course, measurement in the nanoscale range is the domain of AFM [2]. But to fill the gap between the milli- and micrometer technology and the nano- and subnanometer technology it actually needs more. There is an open question about the interface between the AFM range and

The Nano–Micro Interface: Bridging the Micro and Nano Worlds
Edited by Hans-Jörg Fecht and Matthias Werner
Copyright © 2004 WILEY-VCH Verlag GmbH & Co. KGaA, Weinheim
ISBN: 3-527-30978-0

the "real", everyday world. Quantitative measurements have to be done with the same traceability, with the same ease of use, and with the same grade of standardization as in the milli- and micrometer range. Nanometrology has to be defined now and must be able to perform under industrial conditions.

The production process is the driving force. The investigation of structure and roughness of surfaces is gaining more and more technological relevance. The demands on components and finished surfaces have been increasing substantially during recent years. Quality assurance and global competition also play their own part in aggravating the situation. Nanotechnology is making its way as a completely new industry, having all the needs of "main stream industry", but also the special problem of the size of the products.

This is an essential part of the game. The relationship between the significant dimension that you want to produce, and the precision you can get, may be called the signal-to-noise ratio of your production. In classical precision mechanics the device size is about 1 m and the precision is up to 1 μm. This means six orders of magnitude for the signal-to-noise ratio of such a production procedure. Producing a micro- or nanodevice, say at 0.05 mm scale, would require a precision of 50 pm, one tenth of the radius of a gold atom. It is just impossible to produce this accuracy.

To establish real products, a perfect supplier industry is needed. To get a perfect supplier industry standards for dimensions, enabling metrology, and of course instruments to measure those dimensions are needed [3]. With the ongoing rapid development a quantitative, high-resolution analysis of geometric surface parameters is becoming necessary. Topography, roughness, and contour have to be measured precisely and non-destructively.

The existing DIN/ISO standards do not meet these needs. The existing DIN/ISO procedures do not fit the relevant dimensions. There is no clear definition of 3D-roughness values yet. There is a need for new characteristic data sets specially defined for nanotechnology applications.

Conversely, typical instruments such as tactile profilers, confocal microscopes, or 3D-interferometers do not fulfil the combination of high resolution with significant measuring ranges in 3D and non-destructive, metrological measurement.

The related problems deal with:
- low resolution, destructive measuring, bad aspect ratio, restriction to 2D-evaluation in the case of tactile profilometers,
- bad performance at edges, small range of view, missing metrology in the case of 3D-interferometers,
- poor height and x, y measurement range in the case of confocal microscopes.

To resume, there is need for:
- 2D as well as 3D measurement,
- non-destructive measurement,
- high-resolution measurement,
- high aspect ratio,
- metrological measurement in all three directions,

- complete statistical (roughness) evaluation in 3D and 2D,
- quantitative evaluation of 2D structures, such as profile, contour, and area,
- quantitative evaluation of 3D structures, such as volume and shape, and
- new software features to evaluate the data with respect to nanotechnology needs.

3
Bridging the Gap

There is a long list of publications about AFM dating back to 1986 and now as well as the academic papers there are also a lot of articles about industrial applications. We give a few key references here [2–5]. Given the high performance of AFM, and having solved all the known technical problems, AFM is still restricted to research labs and a few high tech companies because the gap in dimensions between state-of-the-art industrial metrology and the nano range can not be bridged by AFM alone.

The MicroGlider® (Fig. 1) combines two sensors in one machine, which makes it possible to investigate large ranges at lower resolution and smaller ranges at very high resolution. The instrument is equipped with a chromatic optical sensor (CWL) and an atomic force microscope (AFM), both mounted permanently. Without changing the measuring device or sample mounting, complete component parts for industrial purposes (e.g., process or quality control) can be measured using a chromatic optical sensor, and the AFM can be used to investigate the surface with nanometer resolution at selected sites.

The optical sensor allows rapid and accurate topographical measurements on sample sizes from 200 μm × 200 μm up to 600 mm × 600 mm using a precision air-bearing table with magnetic linear drives as a scanning device. The x,y resolu-

Fig. 1 MicroGlider 350 mm with air-bearing scanner, sensor set-up with camera, chromatic sensor, and AFM.

tion of this sensor is 1–2 μm. The z range may be chosen from 300 μm up to 50 mm, giving a maximum resolution of 3 nm. The sensor includes no moving parts resulting in some remarkable properties: a very high maximum scanning speed of up to 100 mm/s; no artifacts at steps, because the whole z range is always active; and no wear within the sensor, meaning stable calibration data.

For additional ease of use, the optical sensor is complemented by a positioning camera, which makes it possible to define the scanning area. The optical sensor is mounted on a motorized linear axis to facilitate an automatic approach into the measurement range of the sensor.

The AFM is available with scanning ranges between 20 μm × 20 μm and 80 μm × 80 μm. The z range is 2–6 μm. The AFM resolution is better than 1 nm in all three axes. It is mounted on a separate axis to allow completely independent approach and withdrawal of the AFM head. The sample is conveniently positioned under the sensor head with the air-bearing table, using either the positioning camera or a previously recorded topographical image with the optical sensor.

The AFM may be equipped with various measuring modes such as magnetic force, lateral force, force modulation, phase shift, liquid compatibility, and atomic force acoustical microscopy (AFAM). This mode allows for investigating mechanical properties and elastic modulus of surfaces (see below for details).

During AFM measurements, however, the air-bearing table is left depressurized and firmly seated on the granite base to minimize any vibrations. Safety interlocks are provided to prevent damage caused by xy-table movement to the AFM head while it is lowered onto the surface. As an additional degree of freedom, both sensors may be mounted on a rigid manually adjustable slide to cover a large range in specimen thickness (Fig. 2).

Fig. 2 Arrangement of chromatic sensor and AFM for adapting to large sample heights.

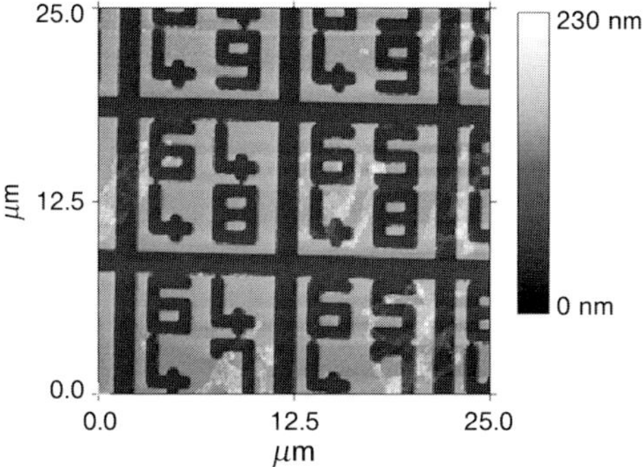

Fig. 3 Standard for calibrating the offset of camera, chromatic sensor, and AFM with respect to each other.

Owing to the scanning range being between a few nanometers and nearly 1 m, the calibration of the MicroGlider is an important feature. Both sensors have to be calibrated separately. To calibrate the AFM the known AFM standards (Figs. 3 and 4), which are structures etched in silicon and chrome on glass, are used [6, 7]. Those standards include a coordinate-system-like grid with 10 μm spacing used to calibrate the AFM xy-scanner and to determine the offset between AFM scanning area and the spot of the optical sensor.

The chromatic sensor is calibrated by means of an interferometer. The calibration is performed once by the manufacturer and can be checked using a specified gauge block or a height calibration standard.

The xy-table of the MicroGlider is controlled by a closed-loop feedback system using Heidenhain glass rulers. The original height deviation of the table for the whole range is better than 1 μm. This is a static deviation that is not due to any noise. So, by measuring a reference plane with the instrument on an optical flat and by deducting this "zero plane" a reproducibility in height for the whole range such as 350 mm × 350 mm of better than 100 nm is achieved. This high-precision table makes it possible to perform metrological measurements for the whole range of movement. To investigate surfaces with the AFM and with the optical sensor at the same surface position the offset between both sensors is calibrated. Therefore measurements on an AFM calibration standard are taken with both

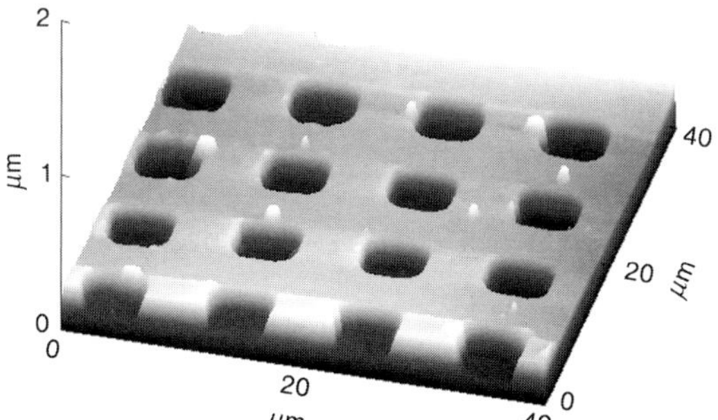

Fig. 4 Calibration standard for micro- and nanometer measurements.

sensors separately. From these data the distance between the sensors can be determined.

An important part of modern metrology instruments is the software. First there is the data acquisition software. This has to be comfortable in use, and there must be options for calibration routines, automation of measuring cycles, and the possibility to run the instrument with only "a few clicks" for use in the production process. The last point is a real problem, as most AFMs still need highly skilled people to operate them, which is an obstacle to acceptance of the technology in industry, but the data acquisition software is of a high standard in most commercially available instruments.

Second, the evaluation software in most commercial instruments is not good. However, there is help. Some specific evaluation software packages available on the market are able to read the data from all kinds of instruments, giving the chance to add on high-performance data treatment not only to AFMs. The instrument discussed here uses the Mark III software package [8] established for working with different instruments in the industrial area as well as in research.

4
Examples of Measurement

4.1
Structured Si Wafer

The first example of an application of the described instrument shows a typical use covering the whole metrology range from 300 mm down to some nanometers, really bridging the gap. Also different parameters, such as topography, warpage and bow, profiles and roughness are calculated for this sample. The sample is a 300 mm silicon wafer that is already structured to produce a storage device. The first question arising for this sample is bow and warpage. In the measurement of the whole wafer these data are easily calculated from the topographical information shown in Fig. 5a. In this specific case the bow measures between 10–50 µm across the complete wafer, depending on the direction of the profile.

To get further information about the wafer and the respective structures on the wafer, the following measurements are done with rising resolution. All measurements are done in the same machine without dismounting the sample but using optical as well as AFM methods.

Fig. 5b and c gives the arrangement and the orientation of the structured areas on the wafer. Fig. 5d–e represents the characterization of shape and 3D dimensions of the single structures on this surface. Finally the AFM measurements in Fig. 5f and g reveal the fine structure on the pits and the respective roughness data of these devices. By covering all dimensions of metrology for this product it is possible to control and verify all relevant steps of the production process.

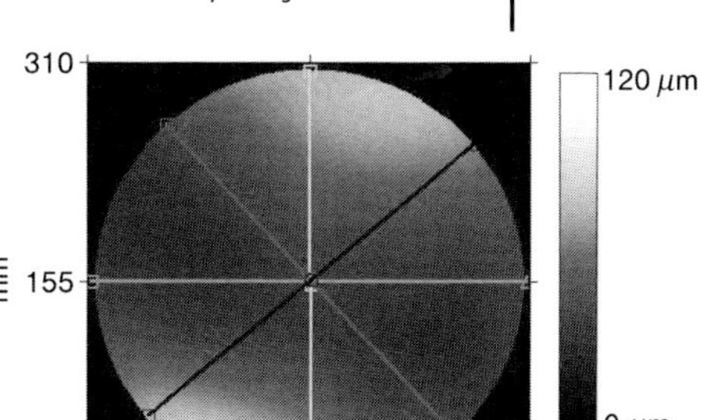

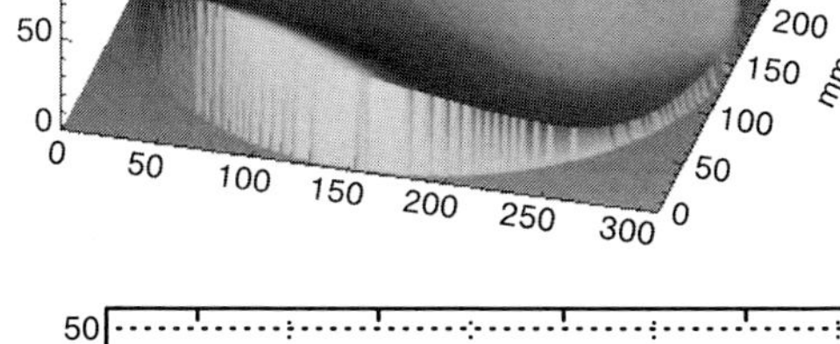

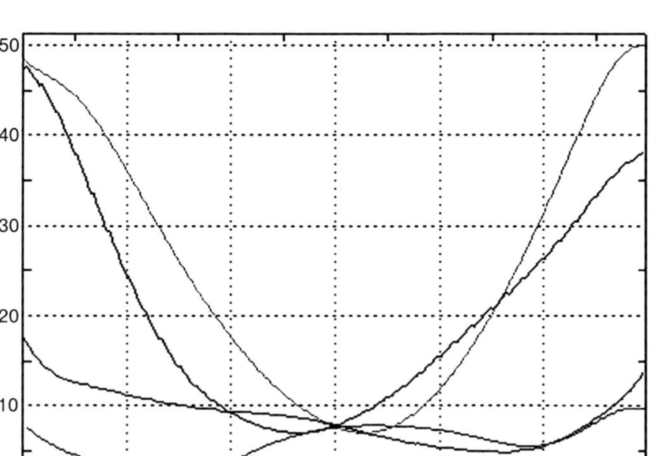

(a)

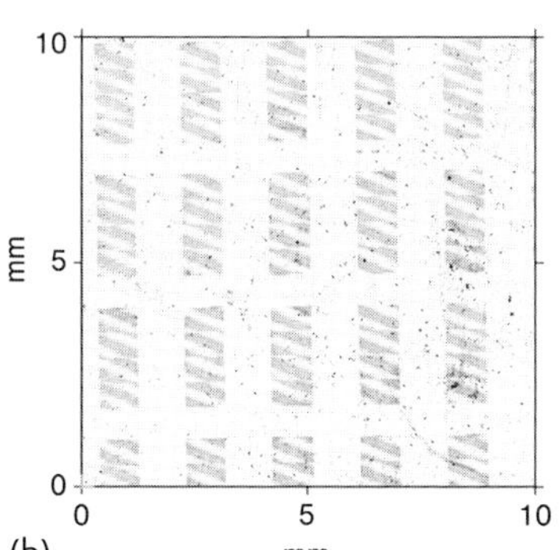

(b)

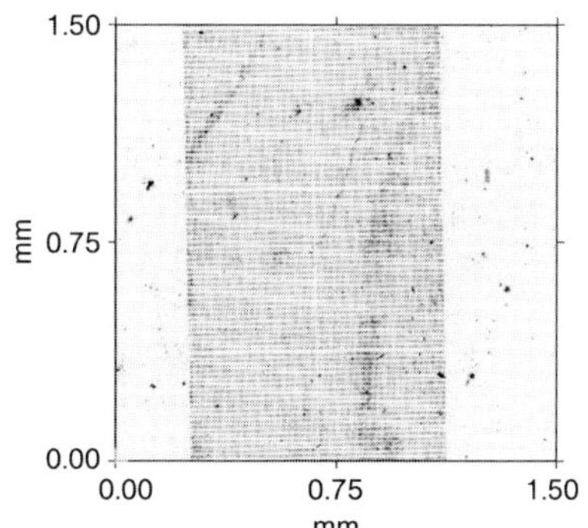

(c)

Fig. 5 (a) Warpage of a 300 mm wafer in 3D, top view, and profiles;
(b) and (c) arrangement and orientation of structures on wafer.

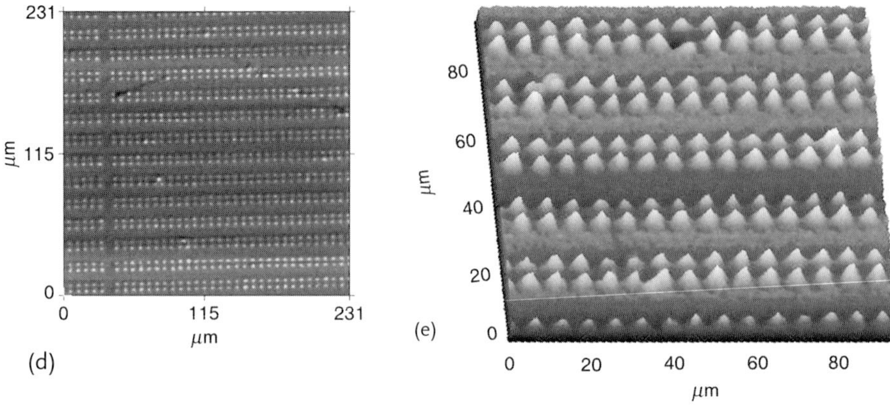

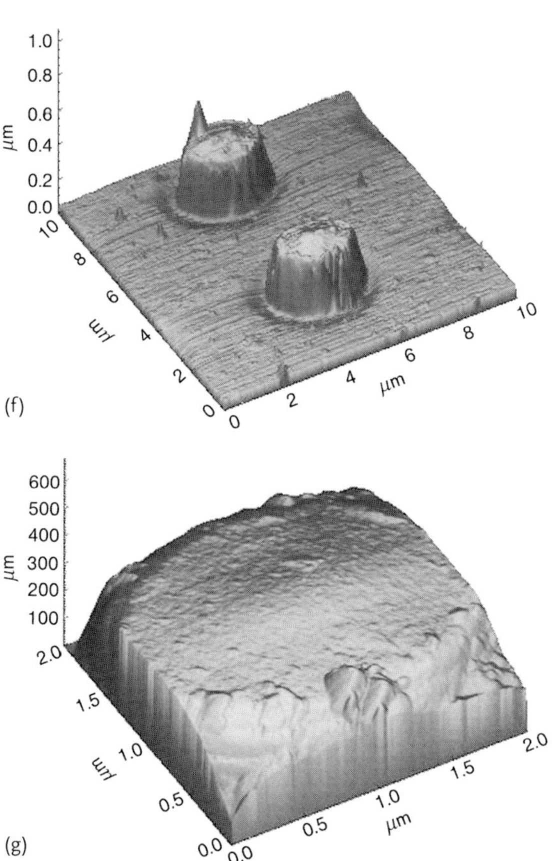

Fig. 5 (d) and (e) functional features on the structured wafer;
(f) and (g) structural details measured with AFM.

4.2
Bump Measurements

The second example is taken from a microelectronics application. A device of 25 mm × 25 mm equipped with a complete array of solder bumps is under investigation. With samples of this size the system described may be loaded with a whole set of samples. The measuring process and evaluation may then be performed completely automatically.

Fig. 6a shows the topography of the whole device, giving details about the single bumps. Details on top of the different solder bumps are shown in Fig. 6b, giving a higher resolution measurement. It is possible to take profiles in any direction out of the 3D measurement or just take any new profiles just by measuring single lines with the instrument. A height determination of the bumps is shown in Fig. 6c, also demonstrating the ability to characterize step heights in similar applications.

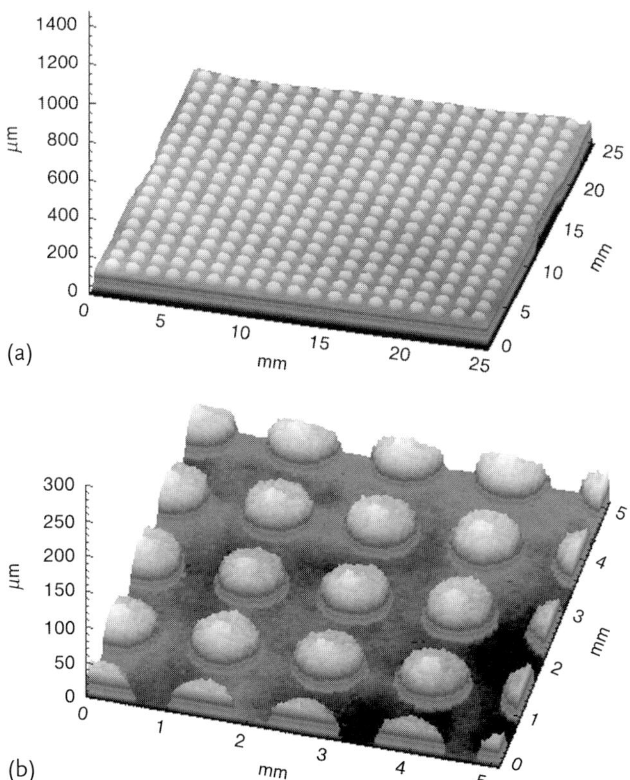

Fig. 6   (a) and (b) Array of solder bumps in 3D topography.

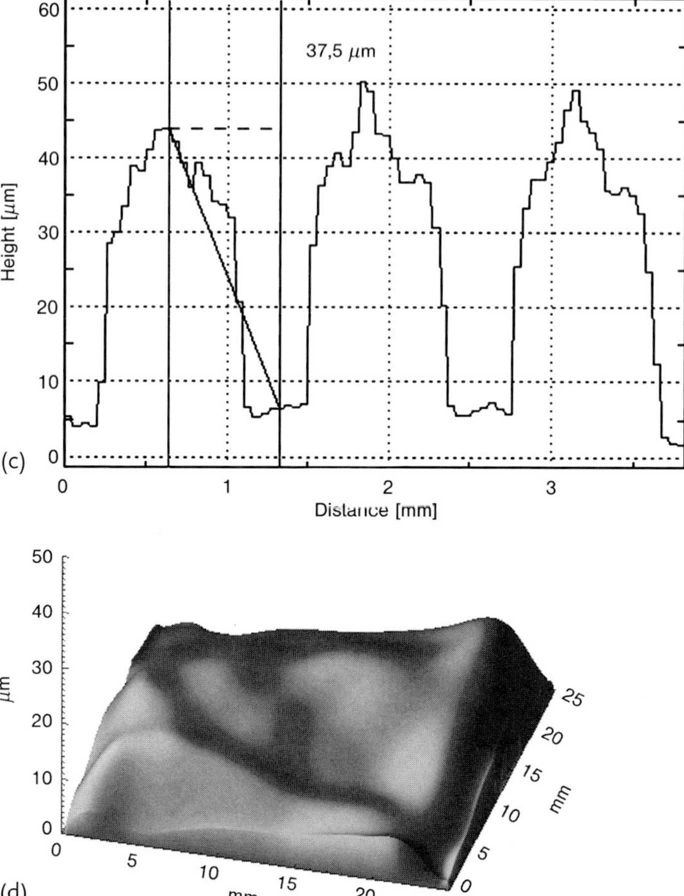

Fig. 6 (c) Profile of bumps for height and width evaluation;
(d) warp of solder bump device by neglecting bump structures.

For this device again warpage or planarity are key features. If the tolerances for the height deviation across the substrate is too big, the solder contacts will fail after being processed because not every pin will contact the substrate. Fig. 6d shows the data for planarity on this surface that extracted from the measured data. For this evaluation the solder bumps are neglected in the software evaluation and only the original substrate surface is taken into account. Some of the main efforts of using the instrument for this application are the high speed, the automation, the wide range of sample size, and achievable resolution.

4.3
LED Housing

Another field of application is electronics and packaging. In this example the housing of an LED device is investigated. There are several dimensions of surface data relevant for controlling the production process. First there is the complete housing as shown in the 3D measurement (Fig. 7a). This has to fit to the encapsulation. Also the three contact pins have to be fixed properly within the insulating glue that can be recognized between the pins and the housing.

But besides this macroscopic view, the pin surfaces also have to be investigated (Fig. 7b). The focus in this case is in planarity and roughness. Both are essential for providing a good connectivity and fitting while welding the wires of the LED to the housing. The three pins are not all of the same material, due to the corresponding wires of the LED. So, every single pin surface has to be characterized in detail. This is done by 3D topography, planarity, and the complete set of statistical roughness data (Fig. 7c and d).

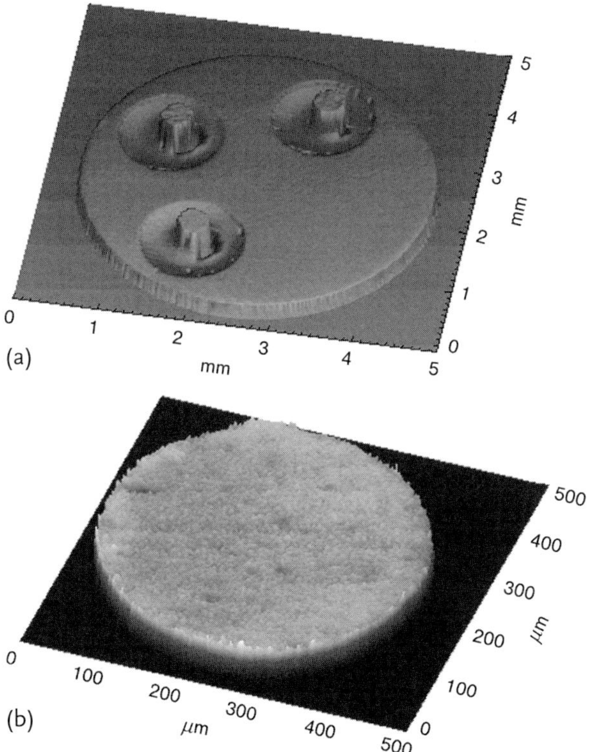

Fig. 7 (a) Complete topography of LED housing; (b) and (c) topography and planarity of a single pin; (d) roughness evaluation of a single pin.

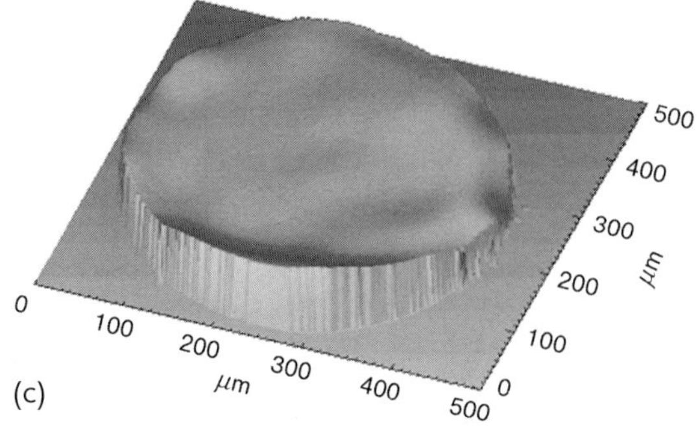

Ra	0.356 µm
Rq	0.468 µm
Rz(ISO)	1.550 µm
Rz(DIN)	1.706 µm
Rmax	2.377 µm
Rp	1.278 µm
Rv	1.459 µm
Rt	2.737 µm
Rsk	-0.300
Rku	3.604
RPc	24.162/cm
Rk	0.934 µm
Rpk	0.442 µm
Rvk	0.688 µm
Mr1	11.607 %
Mr2	84.821 %

(d)

Fig. 7 (c) and (d)

4.4
Film Thickness Measurement

This example deals with the measurement of thin transparent films of hybrid polymers on wafers. The film thickness is to be measured. If the film has been partially removed, the film thickness could be determined from a profile across the wafer surface and film upper surface, where the thickness would be the height difference. Unfortunately, it is often not allowed to remove parts of the film.

Conventional contact stylus measuring systems are unsuitable for this application, as they use mechanical contact techniques and scratch the soft surfaces being measured. Non-contact, optical measuring systems using confocal, autofocus, or triangulation sensors fail on thin, transparent films, as the reflected light from the upper and lower surfaces of the film cannot be individually evaluated.

The system described here uses an interferometric film thickness sensor. The sensor measures the light reflected from each of the film surfaces and determines for each wavelength the interference of the two light beams. The thickness can be measured in a range up to 200 µm with a resolution of 10 nm.

In the MicroGlider, interferometric measurement enables highly resolved 3D mapping of a layer thickness. The instrument further may be upgraded with a two-sensor twin holder, carrying the interferometric sensor and the chromatic distance sensor. This powerful system rapidly measures both film thickness and surface topography with the highest local thickness resolution.

The 3D view in Fig. 8a shows a section of the measured polymer layer. The film thickness, and not the height of the surface, is shown in the z direction. The quadratic structures show clearly where the film has been removed. The measured film thickness is evaluated by the Mark III analysis software.

The film thickness variation along a profile is shown in Fig. 8c. This is taken out of the 3D measurement shown again in Fig. 8b, representing a mapping of the film thickness across the surface of the device.

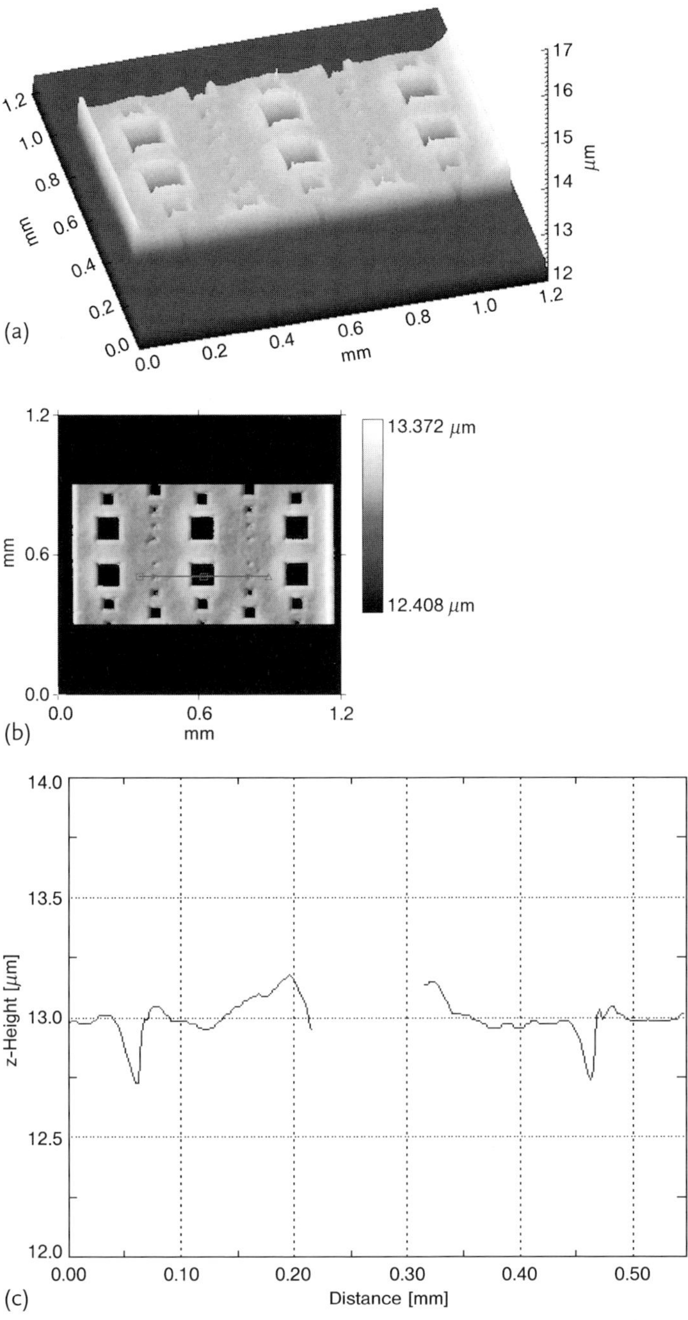

Fig. 8 (a) Mapping of film thickness on a wafer; (b) and (c) top view of mapping and film thickness profile across the surface.

4.5
Elastic Properties with AFAM

Atomic force acoustical microscopy, AFAM, is a new variation of widely spread scanning probe microscopy [9]. The instrument is an option available as addition to the AFM series of MicroGlider. The principle of this mode of operation is easy to understand: the sample under investigation is stimulated by ultrasonic waves. While scanning with an AFM this leads to interactions with cantilever frequencies. The response function of the cantilever is analyzed.

For proper results, frequencies up to 3 MHz are used. It is possible to register 3D images with this mode or to take spectra at specific places, or even do a mapping of frequency spectra. The information from these measurements covers elastic properties of surfaces, phase boundaries, and properties and characterization of surface hardness. By including the software and hardware into the standard AFM and optical profiler series, this new approach to mechanical characterization of surfaces is open for industrial use.

AFM makes it possible to measure surface topography quantitatively with best resolution. But the quantitative evaluation of surface properties, such as elastic modulus still remains a challenge. Via the evaluation of these properties small defects, such as particles, materials inhomogenities, doping profiles or others, will be identified.

The physical principle behind AFAM is ultrasonic contact resonance spectrocopy [9]. This makes it possible to determine elastic properties of a surface with a resolution down to the nanometer. The principle is to evaluate the shift between the free and the contact resonances of an AFM cantilever, determine the local contact stiffness, and finally calculate the local indentation modulus. The AFAM function is added onto an existing AFM: the sample is stimulated by an ultrasonic transducer.

The FRT AFM uses an optical interferometer as a position control for the cantilever. The signal from the interferometer is decoupled from the feedback loop into a second circuit allowing for complete frequency analysis. To show the performance of the AFAM arrangement some measurements are presented. The sample under investigation is a polymer with embedded ceramic particles. Different types of cantilevers (contact/non-contact) and different frequencies were used.

Fig. 9 a shows the topography of the sample and an AFAM measurement with a non-contact cantilever. The contact frequencies of the two materials are shown for two different modes in Fig. 9 b. The spectra show specific differences between the two materials. In the AFAM image the topography of the AFM image is clearly identified, and additional features are revealed. The next step in evaluation is the quantification of the elastic properties of the materials, such as the elastic modulus.

The results for four different frequencies are shown in Fig. 10. In this case a contact cantilever with diamond coating was used. By variation of the applied frequencies different information and resolution of the respective features is achieved. This gives an idea of the possibilities that are open by changing the measuring parameters.

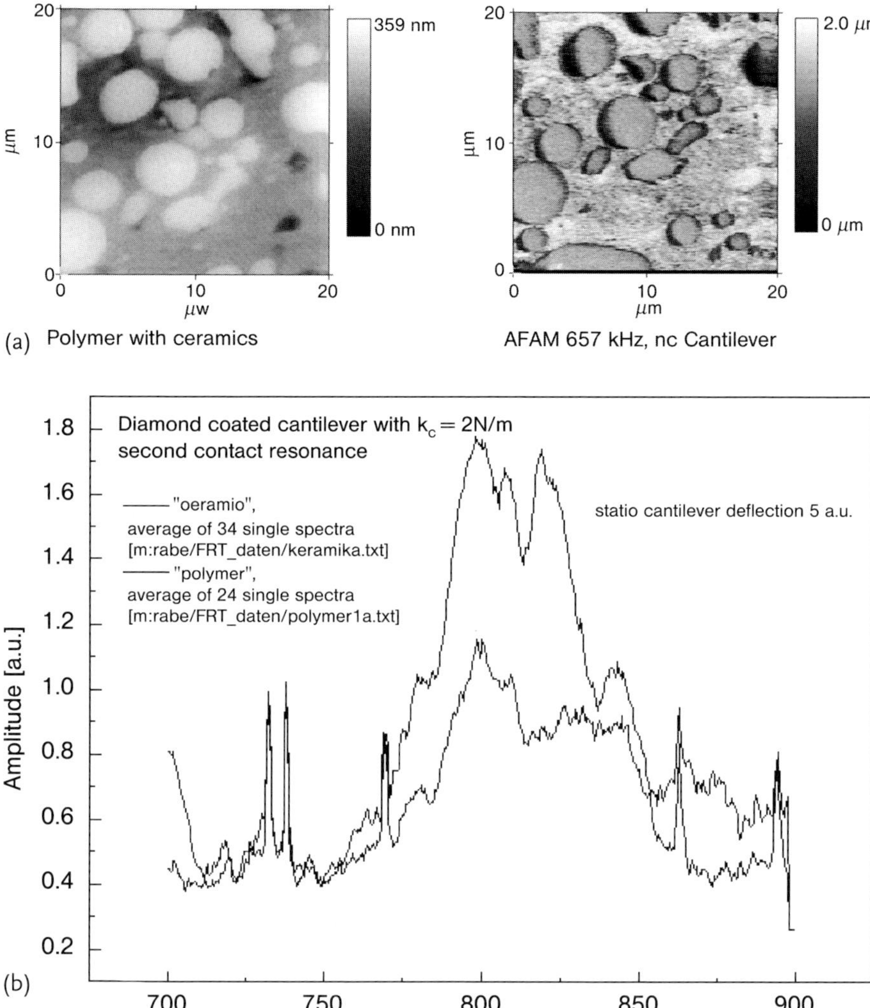

Fig. 9 (a) Topography and AFAM measurement of ceramic on polymer; (b) frequency spectra of the different materials.

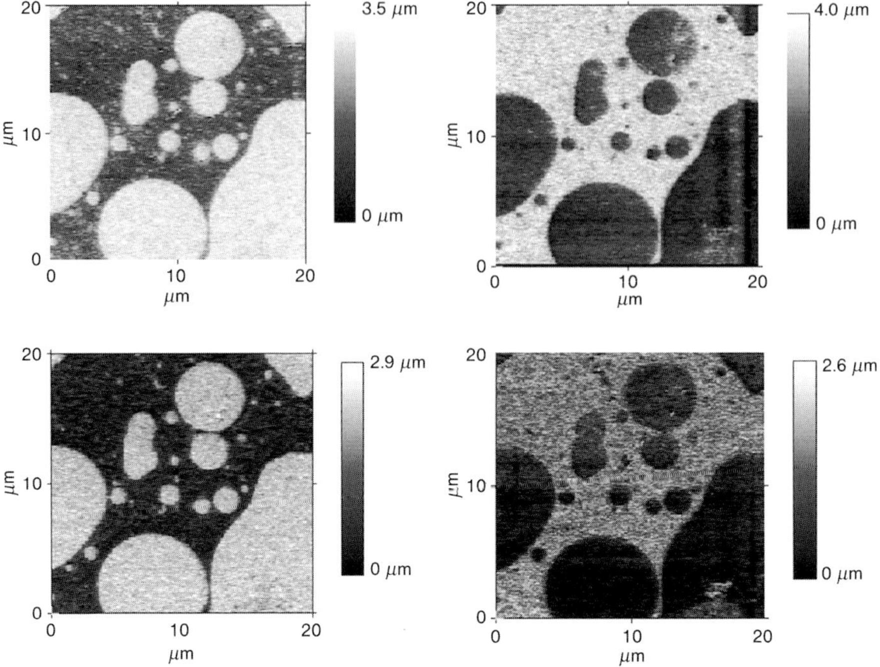

AFAM, 365 kHz, 415 kHz, 809 kHz, 838 kHz, diam.

Fig. 10 AFAM measurements at four different ultrasonic frequencies.

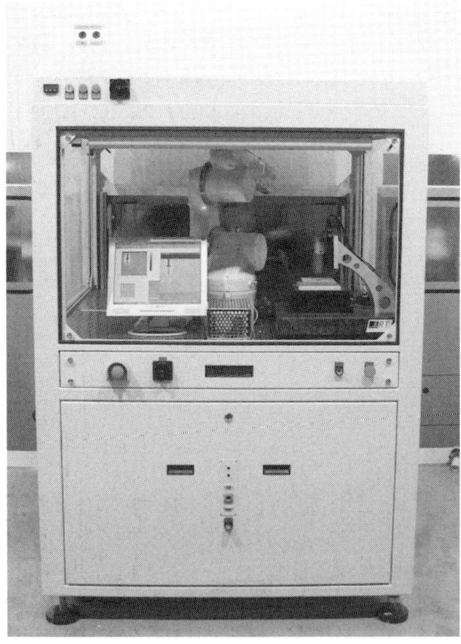

Fig. 11 For use in production the measuring instruments are equipped with industry compatible housing and robotics for automation may be integrated.

Acknowledgments

This work was partially supported by the German Ministry of Education and Research (BMBF).

References

1 MicroGlider surface measuring instrument, Fries Research & Technology GmbH, Bergisch Gladbach, Germany, www.frt-gmbh.com

2 G. BINNIG, C. F. QUATE, and CH. GERBER, *Phys. Rev. Lett.* **1986**, *56*, 930.

3 *Proc. 5th Seminar on Quantitative Microscopy and 1st Seminar on Nanoscale Calibration Standards and Methods, NANOSCALE 2001*, Bergisch Gladbach, Germany, K. KASCHE, W. MIRANDÉ and G. WILKENING, PTB Bericht, ISBN 3-89701-840-3.

4 *Scanning Probe Microscopy*, Cambridge University Press **1994**, ISBN 0-521-42847-5.

5 TH. FRIES, in: *Proc. Prüfverfahren für dünne Schichten*, **1995**, 41 VDI-TZ, Germany; and TH. FRIES, in: *Proc. Micromat 2000*, **2000**, 987, ISBN 3-932434-15-3.

6 L. KOENDERS and G. WILKENING, in: *Proc. 5th Seminar on Quantitative Microscopy and 1st Seminar on Nanoscale Calibration Standards and Methods, NANOSCALE 2001*, Bergisch Gladbach, Germany, K. HASCHE, W. MIRANDÉ and G. WILKENING, PTB Bericht, ISBN 3-89701-840-3.

7 TH. FRIES, in: *Proc. SINCS 2002*, Shanghai International Nanotechnology Cooperation Symposium, Shanghai, China.

8 Mark III software package, Fries Research & Technology GmbH, Bergisch Gladbach, Germany, www.frt-gmbh.com

9 U. RABE, K. JANSER, and W. ARNOLD, in: *Proc. Int. Symp. Acoustical Imaging 22*, Plenum Press, New York **1996**.

Nanometer-Scale View of the Electrified Interface: Scanning Probe Microscopy Study

Peter Müller, Laura Rossi, and Santos F. Alvarado

1
Introduction

The continuing research interest in shrinking the dimensions of electronic devices from the micrometer to the nanometer scale is of great importance. This trend is driven by economic reasons and by present technological needs. Shrinking the dimensions of electrified interfacial systems from the micrometer- to the nanometer-scale regime often leads to the appearance of effects in the electronic properties that are not easily scalable. For example, the morphology of nanometer-scale thin films on a defined substrate is an important parameter as it strongly influences the transport properties of charge carriers in thin films. In addition, at organic–metal or organic–organic interfaces, the morphological structure and the way the interfacing molecules interact chemically are determining factors for the electronic properties, and thus crucial for the characteristic energies needed to inject charge carriers.

As an application [1], organic light-emitting devices (OLEDs) consist of a stack of very precise nanometer-thick organic layers, in which each has a specified function [2]. An additional challenge of nanometer-scale structuring is posed by nanometer-sized transistors [3]. Emerging technologies such as spintronics [4] and quantum molecular electronics [5, 6] are based on nanometer-scale physical phenomena. The interest in organic electronic devices is reflected by the number of experimental and theoretical investigations aimed at elucidating the underlying physical processes [7–13]. Particular attention is paid to charge-carrier injection from a metal electrode into an organic thin film or from one organic thin film into another, and charge-carrier transport within the organic thin films, all of which are crucial parameters for device operation.

Scanning tunneling microscopy (STM) imaging has been shown to be a powerful tool for investigating these processes locally with a resolution on the nanometer scale. Moreover, characterization by STM distance versus potential (z–V) spectroscopy can be used to determine accurate values for the electronic properties, especially of organic materials. For an overview of STM-based techniques in this field, the reader is referred elsewhere [14–16].

The Nano–Micro Interface: Bridging the Micro and Nano Worlds
Edited by Hans-Jörg Fecht and Matthias Werner
Copyright © 2004 WILEY-VCH Verlag GmbH & Co. KGaA, Weinheim
ISBN: 3-527-30978-0

The barriers for the injection of charge carriers from a metal electrode into an organic thin film are determined by the work function of the metal electrode, and the energy levels of the highest occupied (HOMO) and the lowest unoccupied molecular orbital (LUMO) of the organic film. Typically, the position of the HOMO level (E_H) with respect to the Fermi level of the metal electrode (E_F) is set equal to the difference in the ionization potentials (IP) of the materials and the work function of the metal (Φ_w), usually obtained by means of ultraviolet (UPS) or X-ray photon spectroscopy (XPS), or by cyclic voltametry (CV) experiments.

$$E_H = IP_0 - \Phi_w \tag{1}$$

This is the so-called common vacuum level (CVL) approximation [17] (see Fig. 1a). The position of the LUMO level (E_L) relative to E_F is then calculated by adding the optical band gap (E_a) to E_H

$$E_L = E_H + E_a \tag{2}$$

where E_a is taken from the optical absorption spectrum.

Eq. (1), however, neglects the influence of interfacial effects such as dipole layers and image forces, and the calculation of E_L, done using Eq. (2), does not take the exciton binding energy (E_b) [18, 20] into account. In organic materials, E_b is important because it exceeds the thermal energy (kT) at room temperature

(a)

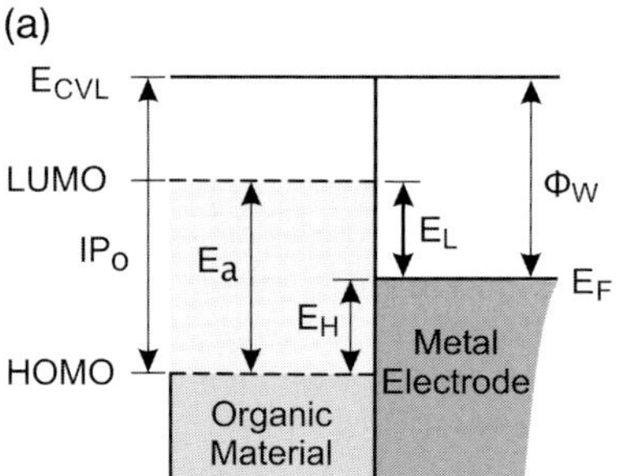

(b)

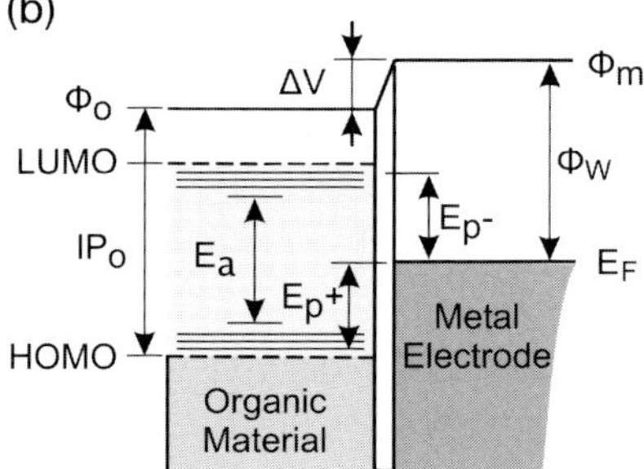

Fig. 1 (a) Diagram of the common vacuum level (CVL) model, where E_{CVL} is the absolute vacuum level, Φ_w the work function of the electrode, E_F the Fermi level of the electrode, LUMO the level of the lowest unoccupied and HOMO the level of the highest occupied molecular orbital, E_L (E_H) the injection energy for electrons (holes) relative to E_F, IP_0 the ionization potential of the organic thin film, and E_a the optical-absorption band gap. (b) Diagram of the extended CVL model. A potential shift ΔV is introduced, where Φ_m and Φ_0 are the vacuum levels of the electrode and the organic thin film, respectively. This allows the inclusion of the effects of image forces or interfacial dipole layers in the model. The lines below the LUMO and above the HOMO symbolize the polaronic states of the organic thin film, where E_{p-} and E_{p+} are the energies, relative to E_F of the metal electrode, needed to inject charge carriers into these polaronic states. Because of the exciton binding energy and interfacial effects, such as the injection of charge carriers into optically inaccessible states, the optically measured E_a may differ from the charge injection gap.

(RT). The CVL approximation also disregards the molecular levels of optically inaccessible electronic states at the interface. Evidence of significant deviations of the CVL model for the organic–metal interface has been widely reported in studies based on techniques such as UPS and XPS [9, 21–25], internal photoemission [19, 26], and STM [27, 28]. Fig. 1 b depicts an extended model that takes all the deviations into account.

2
STM z–V Spectroscopy

In the following, we discuss how STM $z–V$ spectroscopy can be used as a direct probe of the injection energies, E_{p-} and E_{p+}, for electrons and holes into polaronic states of an organic material (Fig. 1 b). In contrast to conventional STM current–voltage ($I–V$) spectroscopy [29–31], where the feedback loop remains open during operation and measures the energy dependence of the density of states (DOS) [32], STM $z–V$ spectroscopy probes the DOS via the voltage-dependent tip displacement at constant tunneling current. Fig. 2 shows the three phases of a $z–V$ spectrum of a thin film, labeled A, B, and C. An example of a complete $z–V$ spectrum is depicted in Fig. 3.

Under standard STM conditions, the tunneling resistance (R_t) is much larger than the resistance of the organic layers (R_{org}) which can be neglected (Fig. 2, Phase A). Thus R_t, in Ω, is given by

$$R_t = V_t/I_t \tag{3}$$

where V_t is the tunneling bias voltage applied and I_t the tunneling current. Charge-carrier injection occurs through the vacuum barrier into the organic material. This means that V_t drops almost completely at the vacuum tunneling barrier. The charge-injection energy, E_t in eV, can be tuned by adjusting the magnitude and polarity of V_t

$$|E_t| = |eV_t| - |E_{pi}| \tag{4}$$

for $|eV_t| > |E_{pi}|$, where e is the electron charge and E_{pi} is the injection barrier for electrons ($i = -$) or holes ($i = +$) into corresponding electronic states. The polarity of V_t determines the type of polarons to inject (electrons or holes). Reducing the bias voltage under constant-current conditions first forces the tip into close proximity of and ultimately into direct physical contact with the organic thin film ($R_{org} \geq R_t$), causing the vacuum barrier to collapse.

When contact mode has been attained, the Fermi level of the tip, E_{Ft}, is typically located at an energy level within the forbidden energy gap of the organic material (Fig. 2, Phase B). The high electrical field across the organic layer leads to band bending and to the formation of a Schottky-like diode. The gradient of the electric field strongly depends on the curvature of the injecting electrodes. Thus,

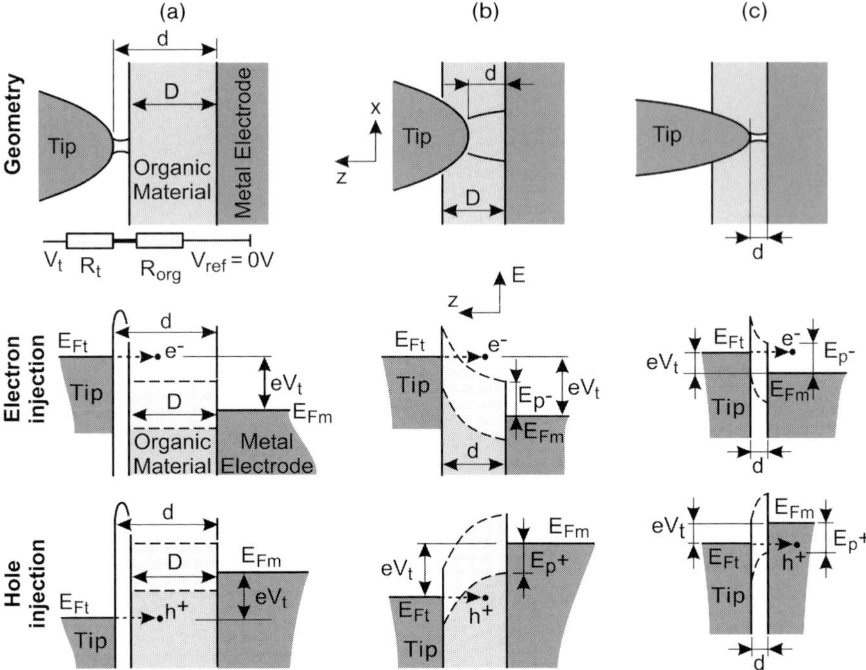

Fig. 2 The three phases of a z–V spectroscopy measurement. In Phase A, the tip is tunneling in standard STM mode on the organic thin film through a vacuum barrier. The resistor chain shown below the geometrical configuration depicts the relevant resistances between tip and electrode that have to be taken into account. In Phase B, the STM tip first enters into contact with and then penetrates the organic thin film. In Phase C, the tip reaches the electrode and, in the case of a metal, enters the well-known logarithmic z–V curve for metals. D denotes the thickness of the organic thin film, d the tip-to-electrode distance, eV_t the tip potential, e^- and h^+ denote the electron and hole, E_{Fm} and E_{Ft} the Fermi level of the electrode and tip, E_{p-} and E_{p+} the polaronic injection energies for electrons and holes relative to E_{Fm}, z and x are geometrical dimensions describing the tip movement, and E is the electric field.

the highest electric-field drop appears at the apex of the STM tip, which under ideal conditions will be the predominant site for charge-carrier injection into the organic thin film. For example, with a typical STM tip-apex radius of $R = 50$ nm, one can obtain continuous injection current densities in the range of $j = 10–10^4$ A/cm^2 at a V_t of a few volts, which is orders of magnitude higher than current densities for the planar interface between the organic material and the metal electrode. This is in agreement with the qualitative statement that in an organic thin film the electric field gradient as a function of distance is linear for an atomically flat electrode, whereas it is quadratically dependent on the inverse of the radius for an STM tip apex. Consequently, the current passing through the STM-tip–thin-film-substrate system is expected to be dominated by a unipolar injection of charge carriers from the STM tip. Note that step sites of rough substrates may act

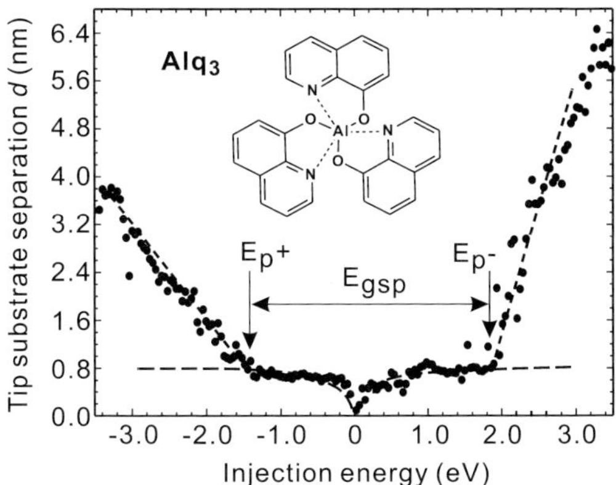

Fig. 3 z–V spectroscopy curve of Alq$_3$ on an atomically flat Au(111) single-crystal electrode. The three phases of a z–V spectroscopy measurement are clearly visible for electron injection (right-hand side). At energies higher than 3.0 eV, tunneling through a vacuum barrier is observed. Between 3.0 and 1.8 eV, the tip penetrates the Alq$_3$ thin film. At the point labeled E_{p-}, the tip reaches the DOS of the gold substrate and starts to follow a logarithmic curve that is characteristic of a z–V spectrum of a metal (long-dashed line). The same procedure can be applied for the left-hand side, which shows the characteristics for the injection of holes. Note that as a direct consequence of Eq. (4), the injection energies for electrons are positive ($V_t < 0$ V), whereas those for holes are negative ($V_t > 0$ V). Adapted from [14].

as injection points, which can lead to bipolar charge-carrier injection. The injection process can either be thermionic emission or tunneling [33, 34], depending on the barrier heights, the field strength at the interface, and temperature. For barrier heights of approximately 1 eV and average fields in the range 0.1–1 V/nm, mainly tunneling injection [35] has to be expected, whereas at barrier heights well below 1 eV thermionic emission might become significant. A more detailed description of tip geometries and their local current densities can be found elsewhere [14, 36, 37]. Regarding tip penetration, if the magnitude of V_t is decreased for constant I_t, the STM feedback loop causes a decrease of the tip–sample distance d to compensate the decrease in R_t. In other words, the tip motion compensates for the reduction of the electric field associated with the change in V_t to maintain a constant tunneling current across the barrier. Because the tip contact area increases with decreasing d, the electric field required for injection and transport should be reduced when the tip moves into the sample. This effect, however, is relatively small owing to the exponential dependence of the injection probability on the electric field and the geometrical dependencies of the tip discussed. Therefore, the effective injection area can be treated as constant throughout Phase B, except at the very early stage, when the vacuum barrier collapses and the tip enters into contact with the thin film.

Injection of charge carriers into the organic material is possible until $|V_t|$ has been reduced to the value at which the Fermi level of the tip enters the forbidden energy gap at the interface with the metal electrode (Fig. 2, Phase C). The point at which this transition occurs determines the position of the lowest electron polaron state (E_{p-}) for an applied $V_t < 0$ V or of the highest polaron state (E_{p+}) for $V_t > 0$ V. In this phase, a characteristic logarithmic z–V curve is observed when either tunneling directly into the metal electrode or non-resonantly through a monomolecular organic layer in contact with the electrode. The transition between Phases B and C, which is typically marked by a sharp decrease in the steepness of the z–V curve, is a direct measure of the potential barrier to inject either electrons or holes from a buried electrode into a thin film.

Note that the z–V technique works only for fluids and soft materials that yield to the pressure exerted by the STM tip.

From the polaronic quantities E_{p-} and E_{p+} measured, the single-particle energy gap (E_{gsp} in eV) can be obtained by taking [19]

$$E_{\mathrm{gsp}} = E_{p-} - E_{p+} \tag{5}$$

The combination of an electron polaron ($p-$) and a hole polaron ($p+$) results in the formation of an exciton. The exciton binding energy (in eV) has been defined [18] as

$$E_b = E_{\mathrm{gsp}} - E_a \tag{6}$$

where E_a is the energy required to create a molecular excitation as determined from optical absorption spectra. Equations (5) and (6) show that supplying the exciton with the energy E_b creates a pair of oppositely charged polarons. In an organic material, for example, the radiative decay of such a singlet exciton results in the emission of a photon.

The local mobilities of the electrons (μ^-) and holes (μ^+) of relevance in charge-carrier transport can be qualitatively estimated by comparing the slopes ($\mathrm{d}z/\mathrm{d}V$) of the z–V curves for Phase B in Fig. 2. For a given I_t and a thin film of thickness D, one can see that, to sustain the current flow through the injection area, a lower electric field is required for a high-conductance than for a low-conductance material. Thus, in contact mode (Phase B), the average slopes of the z–V curves are directly affected by the local conductivity of the sample. This implies that in point-contact mode, z–V curves contain information about the transport properties of organic materials. Experiments indicate that for a given sample conductance the rate of the tip penetration is proportional to R_t (short-dashed line in Fig. 3).

3
Experimental Details

The measurements were carried out at RT under ultrahigh vacuum (UHV) conditions, typically at a base pressure in the 10^{-10} mbar range. An adjacent treatment chamber is equipped with the instrumentation used for substrate cleaning and preparation. The detailed description of the experimental setup can be found elsewhere [15, 38]. To acquire the STM z–V spectra, commercial ion-milled platinum-iridium tips were used. The organic thin films of tris-(8-hydroxyquinoline)aluminum (Alq$_3$) and copper phthalocyanine (CuPc) were prepared *in situ* by thermal evaporation on atomically flat Au(111) and Ag(111) single-crystal substrates at RT [14]. The as-grown organic thin films are typically 3–5 nm thick. For further details regarding sample preparation and the morphologies observed using STM, the reader is referred elsewhere [14, 27, 39].

3.1
Alq$_3$ Thin Films on Au(111)

STM images taken at different locations and on various samples reveal that the surface of very thin (a few nanometer thick) films exhibits unaligned nanocrystallites [14]. Fig. 3 shows a typical z–V spectroscopy curve collected on Alq$_3$ on Au(111). For this system we statistically obtained $E_{p-}=1.15\pm0.18$ eV for electron injection and $E_{p+}=-1.81\pm0.25$ eV for hole injection, relative to the Au(111) Fermi level.

In comparing the barrier-height values determined by z–V spectroscopy with those predicted by the CVL approximation, bearing in mind the measurement uncertainties, only differences of more than 0.3–0.4 eV are significant. The scatter of IP values reported in the literature is in this range. Despite this deviation, the CVL model predicts higher energy levels for Alq$_3$ on a Au(111) substrate than the result obtained by z–V spectroscopy. Using Eq. (1) and Eq. (2), the barrier height for electron injection (Fig. 1a) can be estimated to be $E_L=1.9$–2.5 eV, for $\Phi_{Au(111)}=5.31$ eV [26], the $IP_{Alq3}=5.57$–6.0 eV [21, 22, 40–42], and the optical band gap measured on an Alq$_3$ thin film $E_a=2.75$ eV [43]. The direct measurement by z–V spectroscopy (Fig. 1b), however, yields $E_{p-}=1.15$ eV. The single-particle energy gap obtained from Eq. (5) is $E_{gsp}=2.96\pm0.13$ eV, which is clearly higher than E_a derived from the optical-absorption band gap [44]. Using Eq. (6), an exciton binding energy of $\sim 210\pm130$ meV can be obtained.

It is also found that for an Alq$_3$ thin film the electron injection barrier is approximately 0.2 eV higher on Au(111) than on Ag(111) [14]. The magnitude of this shift is much smaller than the value of 0.57 eV calculated in the framework of the CVL model by taking the difference between the work functions of these two metal surfaces ($\Phi_{Ag(111)}=4.74$ eV) [45].

All these deviations from the pure CVL model indicate the effect of image forces or the formation of a dipole layer at the interface owing to the transfer of negative charges from Alq$_3$ to the metal substrate. The value for E_{p+} confirms the results of UPS measurements on the Alq$_3$/Au(111) interface [9, 21, 22].

A statistical analysis of measurements taken on Au(111) and Ag(111) yields a distribution of injection thresholds for electrons that exhibits a peak at $E_{p-}=1.15$ and 1.7 eV, respectively. This indicates that two different types of surface configurations exist between the organic material and the metal electrode. The E_{gsp} values for the two configurations are 2.96 and 3.04 eV. This behavior can either be explained by the different morphologies of the Alq$_3$ thin films [46], the different electronic properties of the two existing Alq$_3$ isomers, the meridianal and the facial configurations [47], or by different orientations of the Alq$_3$ on the substrate [14, 27, 48].

The slope dz/dV of the z–V curve of Phase B acquired with negative tip bias (E_{p-}) is much steeper than that of the curve taken with positive tip bias (E_{p+}). As discussed previously [14], this confirms that Alq$_3$ preferentially transports electrons and that μ^- is much higher than μ^+ [49].

3.2
CuPc Thin Films on Au(111)

STM images of CuPc thin films grown on a Au(111) substrate at RT are featureless, indicating that these films are completely disordered. After annealing at 600 K for 1 h, however, the organic thin film was found to crystallize, exhibiting various polymorphic phases (Fig. 4). The crystallites are about 30–40 monomolecular layers thick, several 100 nm in diameter, and strongly resemble the bulk α and β phases [39]. Three examples are shown in Fig. 5.

On the disordered phase, the injection energies measured for electrons and holes are $E_{p-}=0.55\pm0.15$ and $E_{p+}=-0.55\pm15$ eV, respectively, relative to the E_F of the Au(111) metal electrode, yielding $E_{gsp}=1.1\pm0.2$ eV [39]. In a statistical analysis of the polymorph phases α, β_1 and β_2 (Fig. 5), the following injection energies

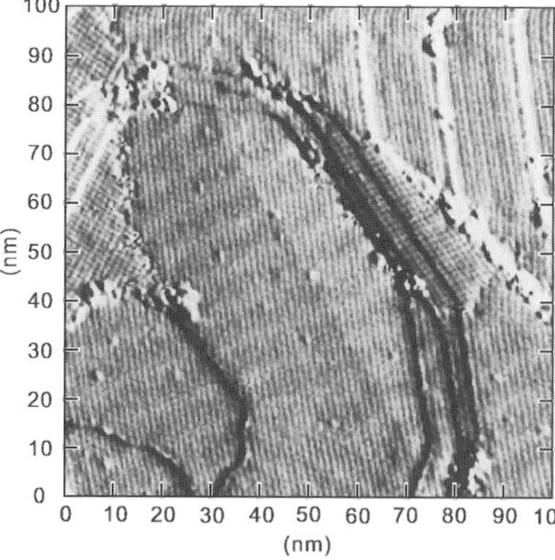

Fig. 4 STM image of polymorph grains of CuPc on a Au(111) substrate annealed at 600 K for 1 h. The STM settings were $I_t=1.8$ pA and $V_t=3.3$ V.

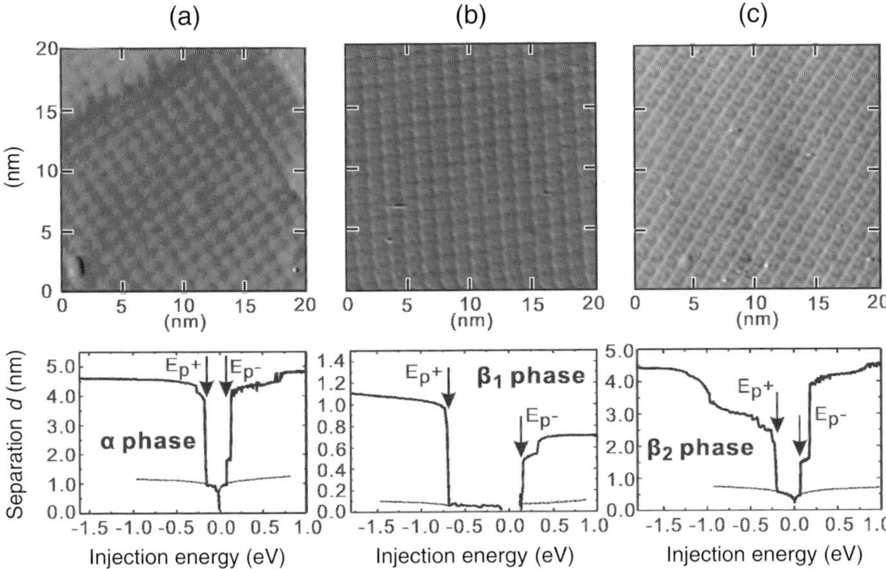

Fig. 5 Three morphological phases of CuPc grains grown on Au(111) and their z–V spectra. The exact values of E_{p-} and E_{p+} have to be obtained by statistical analysis. (a) shows the α-phase (quasi-tetragonal), (b) the β_1 phase (monoclinic), and (c) the β_2 phase (monoclinic). The STM settings were: (a) $I_t = 20$ pA, $V_t = 3.5$ V; (b) $I_t = 60$ pA, $V_t = -1$ V, and (c) $I_t = 3$ pA, $V_t = -1$ V. (b) adapted from elsewhere [39].

were measured: $E_{p-} = 0.08 \pm 0.04$, 0.21 ± 0.07, and 0.1 ± 0.05 eV, and $E_{p+} = -0.1 \pm 0.04$, -0.62 ± 0.17, and -0.2 ± 0.05 eV, relatively to the E_F of the Au(111) substrate. The results suggest very small single-particle energy gaps of $E_{gsp} = 0.18 \pm 0.06$, 0.83 ± 0.19, and 0.3 ± 0.07 eV, respectively.

Considering the ionization potentials reported for CuPc thin films, $IP_{CuPc} = 4.7$–5.3 eV [40, 42, 50, 51], an $E_H = 0$ to 0.6 eV relative to the E_F of the Au(111) substrate would be expected when applying the CVL approximation. Theoretically, both the HOMO and the LUMO level are equal to or higher than the E_F of the Au(111) substrate. The results show, however, that in all cases E_H is negative; therefore, the HOMO level clearly lies below the Fermi level of the Au(111) substrate, which indicates the presence of a dipole layer, of a net charge at the interface, or the effect of image forces.

The slopes of the z–V curves acquired on polymorphic thin films are approximately one order of magnitude steeper than that of the disordered phase, indicating much higher μ^- and μ^+ values for polycrystalline molecular arrangements [39].

In general, the polymorphic phases of CuPc thin films seem to exhibit a much smaller E_{gsp} than the disordered samples do [39]. This clearly reflects the influence of the molecular packing on the energy levels. In both cases, that is, in disordered and polymorphic films, E_{gsp} appears to be smaller than the optical-absorption band gap $E_a = 1.6$–1.7 eV, as reported in the literature [52, 53]. This suggests that charge carriers are injected from the Au(111) substrate into optically inactive states of the CuPc thin film. Note, however, that this interpretation assumes unipolar injection from the tip. For a detailed theoretical discussion of the effect of optically inactive states, the reader is referred elsewhere [16, 39].

4
Concluding Remarks

Scanning probe microscopy is a powerful tool for characterizing surface morphologies with nanometer-scale spatial resolution. Applying z–V spectroscopy, the electronic bulk and interfacial properties of organic–substrate and organic–organic as well as of solution–substrate interfaces can be analyzed. An important capability of this technique is that it allows the direct measurement of the molecular-level alignments of both occupied and empty states. In addition, it is also possible to probe charge-carrier injection into electronic states that optically are inaccessible.

From the injection barriers (E_{p-} and E_{p+}), the single-particle energy gap (E_{gsp}) and, together with the optical-absorption band gap (E_a), the exciton binding energy (E_b) can be determined. Furthermore, a qualitative measure of the local electron and hole mobilities can be derived from the slope dz/dV of the z–V curves.

Measurements at the Alq$_3$/Au(111) interface reveal that, because of the formation of a dipole layer or effects of image forces, a potential shift is induced that influences the injection energy for charge carriers. The CuPc/Au(111) system turned out to be an ideal case for investigating the influence of the morphology, which can be disordered or polymorph. Charge-carrier transport clearly increases with the molecular-packaging density, whereas E_{gsp} is reduced. In all measurements on the CuPc/Au(111) system, a smaller energy for the single-particle gap than for the optical-absorption band gap was observed. That indicates charge injection into optically inaccessible polaronic states.

Because the injection barriers for electron- and hole-charge carriers obtained from STM z–V spectroscopy are quantitatively very accurate, they form an ideal basis for future theoretical investigations and modeling [54].

As mentioned above, measuring electronic properties by means of STM z–V spectroscopy is applied to thin films consisting of soft materials or to solution-based systems. Investigations of more rigid material systems, however, would be of great interest, in particular in the field of semiconductors.

Acknowledgments

Some of this work was performed within the Training and Mobility of Researchers Network EUROLED, and supported by grants from the Swiss Federal Office for Education and Science and by the European Commission. The authors thank W. Riess for his support and fruitful discussions as well as the IBM Zurich Lab's Engineering and Publications groups for their help.

References

1 T. Tsujimura, Y. Kobayashi, K. Mura-
yama, A. Tanaka, M. Morooka,
E. Fukumoto, H. Fujimoto, J. Sekine,
K. Kanoh, K. Takeda, K. Miwa,
M. Asano, N. Ikeda, S. Kohara, S. Ono,
C.-T. Chung, R.-M. Chen, J.-W. Chung,
C.-W. Huang, H.-R. Guo, C.-C. Yang,
C.-C. Hsu, H.-J. Huang, W. Riess,
H. Riel, S. Karg, T. Beierlein,
D. Gundlach, S. F. Alvarado, C. Rost,
P. Müller, F. Libsch, M. Maestro,
R. Polastre, A. Lien, J. Sanford,
R. Kaufmann, in: *Society for Information
Display Symposium (SID 03), Baltimore,
MD. Technical Digest,* **2003**, Paper 4.1,
p. 6.

2 W. Riess, H. Riel, T. Beierlein,
W. Brütting, P. Müller, P. F. Seidler,
IBM J. Res. Develop. **2001**, 45, 770.

3 B. Doris, M. Ieong, T. Kanarsky,
Y. Zhang, R. A. Roy, O. Dokumaci,
Z. Ren, F.-F. Jamin, L. Shi, W. Natzle,
H.-J. Huang, J. Mezzapelle, A. Mocuta,
S. Womack, M. Gibelyuk, E. C. Jones,
R. J. Miller, H. S. Philip, W. Haensch,
in: *IEEE International Electron Devices
Meeting (IEDM 02), San Francisco, CA,
Technical Digest,* **2002**, Paper 10.6.1,
p. 267.

4 S. Das Sarmas, *Am. Sci.* **2001**, 89, 516.

5 R. Martel, T. Schmidt, H. R. Shea,
T. Hertel, Ph. Avouris, *Appl. Phys. Lett.*
1998, 73, 2447.

6 L. M. K. Vandersypen, M. Steffen,
G. Breyta, C. S. Yannoni, R. Cleve,
I. L. Chuang, *Phys. Rev. Lett.* **2000**, 85,
5452.

7 H. Ishii, K. Sugiyama, E. Ito, K. Seki,
Adv. Mater. **1999**, 11, 605.

8 A. Rajagopal, C. I. Wu, A. Kahn,
J. Appl. Phys. **1998**, 83, 2649.

9 S. T. Lee, X. Y. Hou, M. G. Mason,
C. W. Tang, *Appl. Phys. Lett.* **1998**, 72,
1593.

10 W. Brütting, S. Berleb, A. Mückel,
Org. Electron. **2001**, 2, 1.

11 H. Bässler, *Phys. Status Solidi B* **1993**,
175, 15.

12 N. Graham, R. Friend, in: *Solid State
Physics, Advances in Research and Applica-
tions* (Eds: H. Ehrenreich, F. Spaepen),
Academic Press, San Diego, CA, **1995**,
Vol. 49, p. 2.

13 L. Yan, Y. Gao, *Thin Solid Films* **2002**,
417, 101.

14 S. F. Alvarado, L. Rossi, P. Müller,
P. F. Seidler, *IBM J. Res. Develop.* **2001**,
45, 89.

15 P. Müller, S. F. Alvarado, L. Rossi,
W. Riess, *Mater. Phys. Mech.* **2001**, 4, 76.

16 S. F. Alvarado, in *Conjugated Polymer
and Molecular Interfaces* (Eds.: W. R.
Salaneck, K. Seki, A. Kahn, J.-J. Piraux),
Marcel Dekker Inc., New York, **2001**,
p. 473.

17 W. Schottky, *Z. Phys.* **1942**, 118, 539.

18 E. M. Conwell, *Synth. Met.* **1996**, 83, 101.

19 I. H. Campbell, T. W. Hagler, D. L.
Smith, J. P. Ferraris, *Phys. Rev. Lett.*
1996, 76, 1900.

20 D. Woehrle, L. Kreienhoop,
D. Schlettwein, in: *Phthalocyanines
Properties and Applications* (Eds.: C. C.
Leznoff, A. B. P. Lever), VCH Publishers,
New York, **1996**, Vol. 4, p. 223.

21 K. Sugiyama, D. Yoshimura, T. Miya-
mae, T. Miyazaki, H. Ishii, Y. Ouchi,
K. Seki, *J. Appl. Phys.* **1998**, 83, 4928.

22 A. Rajagopal, C. I. Wu, A. Kahn,
J. Appl. Phys. **1998**, 83, 2649.

23 T. Mori, H. Fujikawa, S. Tokito,
Y. Taga, *Appl. Phys. Lett.* **1998**, 73, 2763.

24 R. Schlaf, B. A. Parkinson, P. A. Lee,
K. W. Nebesny, A. R. Armstrong, *J. Phys.
Chem. B* **1999**, 103, 2984.

25 S. C. Veenstra, U. Stalmach, V. V.
Krasnikov, G. Hadziioannou,
H. T. Jonkman, A. Heres, G. Sawatsky,
Appl. Phys. Lett. **2000**, 76, 2253.

26 G. L. J. Rikken, Y. A. R. R. Kessener,
D. Braun, E. G. J. Staring, A. Demandt,
Synth. Met. **1994**, 67, 115.

27 S. F. Alvarado, L. Libioulle, P. F.
Seidler, *Synth. Met.* **1997**, 91, 69.

28 S. F. Alvarado, P. F. Seidler, D. G. Lid-
zey, D. D. C. Bradley, *Phys. Rev. Lett.*
1998, 81, 1082.

29 N. D. Lang, *Phys. Rev. Lett.* **1987**, 58, 45.

30 S. De Cheveigne, J. Klein, A. Leger, in:
Tunneling Spectroscopy (Ed. P. K. Hans-
ma), Plenum Press, New York, **1982**,
p. 109.

31 R. Wiesendanger, *Scanning Probe Microscopy and Spectroscopy, Methods and Applications*, Cambridge University Press, Cambridge, UK, **1994**.

32 R. Feenstra, *Phys. Rev. B* **1994**, *50*, 4561.

33 E. M. Conwell, E. W. Wu, *Appl. Phys. Lett.* **1997**, *70*, 1867.

34 M. N. Bussac, D. Michoud, L. Zuppiroli, *Phys. Rev. Lett.* **1998**, *81*, 1678.

35 A. van der Ziel, in: *Solid State Physical Electronics*, 3rd edn (Ed.: N. Holonyak), Prentice-Hall, Inc., Englewood Cliffs, NJ, **1976**, pp. 152-183.

36 M. A. Lampert, P. Mark, *Current Injection in Solids* (Eds. H. G. Booker, N. DeClaris), Academic Press, New York, **1970**.

37 M. K. Miller, A. Cerezo, M. G. Hetherington, G. D. W. Smith, in: *Atom Probe Field Ion Microscopy*, Oxford University Press, Oxford, UK, **1996**, p. 41.

38 S. F. Alvarado, W. Riess, M. Jandke, P. Strohriegl, *Org. Electron.* **2001**, *2*, 75.

39 S. F. Alvarado, L. Rossi, P. Müller, W. Riess, *Synth. Met.* **2001**, *122*, 73.

40 C. Hosokawa, H. Higashi, H. Nakamura, T. Kusumoto, *Appl. Phys. Lett.* **1995**, *67*, 3853.

41 M. Propst, R. Haight, *Appl. Phys. Lett.* **1997**, *71*, 202.

42 S. T. Lee, Y. M. Wang, X. Y. Hou, C. W. Wang, *Appl. Phys. Lett.* **1999**, *74*, 670.

43 A. Aziz, K. L. Narasimhan, *Synth. Met.* **2002**, *131*, 71.

44 S. F. Alvarado, W. Riess, *Mat. Res. Soc. Symp. Proc.* **2001**, *665*, 121.

45 H. B. Michaelson, *J. Appl. Phys.* **1977**, *48*, 4729.

46. M. Birkmann, G. Gadret, M. Muccini, C. Taliani, N. Masciocchi, A. Sironi, *J. Am. Chem. Soc.* **2000**, *122*, 5147.

47 M. Amati, F. Leji, *J. Phys. Chem. A* **2003**, *107*, 2560.

48 A. Curioni, W. Andreoni, *IBM J. Res. Develop.* **2001**, *45*, 101.

49 T. Tsutsui, H. Tokuhisa, M. Era, in: *Polymer Photonic Devices* (Eds.: B. Kippelen, D. D. C. Bradley), *Proc. SPIE*, SPIE, Bellingham, WA, **1998**, Vol. 3281, p. 230.

50 T. Chassé, C.-I. Wu, I. G. Hill, A. Kahn, *J. Appl. Phys.* **1999**, *85*, 6589.

51 A. J. Ikushima, T. Kanno, S. Yoshida, A. Maeda, *Thin Solid Films* **1996**, *273*, 35.

52 E. A. Lucia, F. D. Verderame, *J. Chem. Phys.* **1968**, *48*, 2674.

53 B. R. Hollebone, M. J. Stillman, *J. Chem. Soc. Faraday Trans. II* **1978**, *74*, 2107.

54 M. Kemerink, S. F. Alvarado, P. Müller, P. M. Koenraad, H. W. M. Salemink, J. H. Wolter, R. A. J. Janssen, *Phys. Rev. B* **2004**, *70*, 45202.

New Technology for an Application-Specific Lab-on-a-Chip

Heiko Schäfer, Steffen Chemnitz, Konstantin Seibel, Volodymyr Koziy,
Alexander Fischer, Dietmar Ehrhardt, and Markus Böhm

1
Introduction

The miniaturization of chemical labs has numerous advantages, since the quantities of reagents can be reduced dramatically and chemical processes can be accelerated and highly parallelized. However, a growing number of microsystem technology applications, in particular in the field of microfluidics with its applications in life science, need novel fabrication methods for the control of the microfluidic system, to improve fluid handling and data acquisition. The novel approach presented here includes the vertical integration of an ASIC and a microfluidic system on only one chip, as described in Chemnitz et al. [1].

The microfluidic system is deposited in a CMOS-compatible post-process. As seen in Fig. 1, the application-specific lab-on-a-microchip (ALM) consists of four modules on top of each other. The ASIC acts simultaneously as substrate and contains custom specific circuitry for the control of the microfluidic system, such as

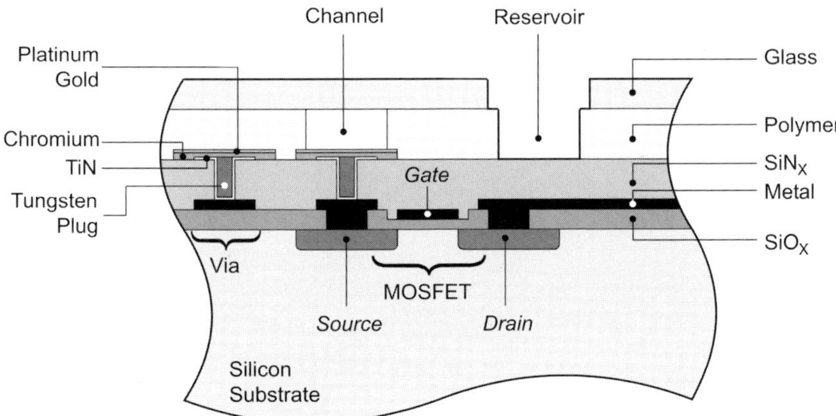

Fig. 1 Cross section of the application-specific lab-on-a-microchip (ALM) based on crystalline silicon.

The Nano–Micro Interface: Bridging the Micro and Nano Worlds
Edited by Hans-Jörg Fecht and Matthias Werner
Copyright © 2004 WILEY-VCH Verlag GmbH & Co. KGaA, Weinheim
ISBN: 3-527-30978-0

for liquid transportation, process compartments, and on-chip analysis tools. In the lab layer the chemical processes take place. It consists of a polymer in which channels and compartments are brought in by lithography, such as micropumps, mixers, transistor arrays, and detectors. On top of the polymer layer a Pyrex glass plate is placed to seal the microfluidic system. It also acts as a protective layer against environmental influences. The interface acts as contact between the ASIC and the lab layer

In commercially available ASICs, AlCuSi or AlCuGe are generally used for the top metal layer, often covered by titanium nitride (TiN). Since these materials cannot be used as electrical contacts to the microfluidic system, such as for electroosmotic pumps or electrophoresis systems, the AlCuSi/AlCuGe layers are etched off, so that the vias (tungsten plugs) in the highly planarized CMP (chemical mechanical polishing) treated intermetal dielectric SiN, become accessible. Usually the plugs are surrounded by an approximately 50 nm thick TiN layer, which improves the contact resistance. Instead of the standard metal, electrode materials suited for the chemical processes are deposited by PVD (physical vapor deposition), typically gold, platinum, palladium or chromium, as described below. First experiments were carried out with 80 nm thick chromium electrodes. The contact resistance between the ASIC metal and the chromium layer is in the range of a few Ohms ($R_{via} = 1\,\Omega$; $R_{TiN\text{-}Cr} = 8\,\Omega$ for a $10\,\mu m \times 10\,\mu m$ contact) for geometries shown in Fig. 2. The reproducibility of the contacts is excellent and was verified using test structures with up to 10 000 contacts connected in series without measured interruption.

At present, most of the microfluidic systems are fabricated by means of micromechanical procedures on glass or polymers. By combination of two microtechnologies, the "state of the art" thin-film-technology and micromechanical processes on glass, microfluidic systems will reach a dramatic increase of their capability. In fact, the increase in the production costs for microfluidic chips depends

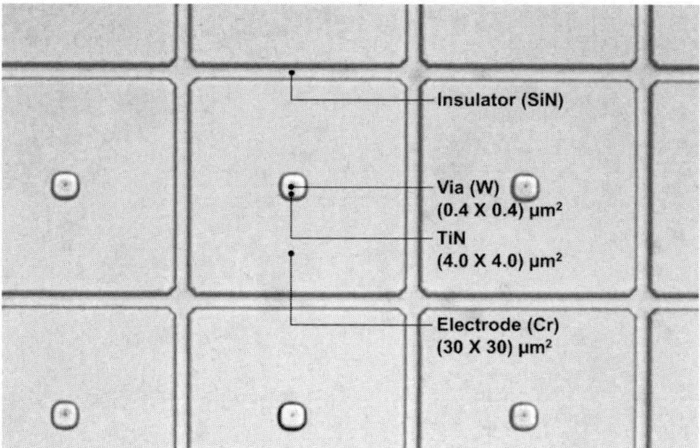

Fig. 2 Top view of the ASIC interface, fabricated in 0.35 µm process technology.

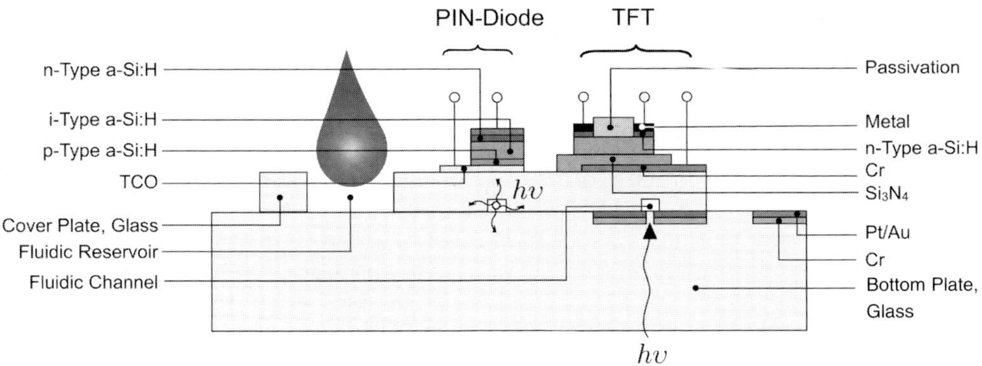

Fig. 3 Cross section of the application-specific lab-on-a-microchip (ALM) combined with active devices based on amorphous silicon.

on the cost-effective integration of these two microfabrication technologies. The costs for analysis are expected to decrease dramatically as the new technologies become available.

Amorphous hydrogenated silicon pin-diodes are widely used for large-area sensor arrays as well as for advanced image sensors to convert the incident radiation to a current. The advantage of a-Si:H consists in its better photoelectric properties compared with crystalline silicon diodes and the possibility of fabricating them at low temperatures.

Fig. 3 shows the cross section of the ALM combined with active devices based on amorphous silicon described in Schäfer et al. [2]. The device consists of two glass plates that are sandwiched together using oxygen plasma bonding. The thicker bottom plate (1.2 mm) contains the contacts to the microfluidic channels while the thinner top plate (0.15 mm) contains the microfluidic system. Alternatively, SU-8 (see below) may be used to form the microfluidic channels. It is bonded face down onto the bottom substrate and contains on its reverse side hydrogenated amorphous silicon (a-Si:H) based pin-diodes for optical detection and TFTs (thin film transistors) to transfer the electric charge caused by the incident radiation on the pin-diodes. Both the pin-diodes and the TFTs are manufactured by PECVD (plasma enhanced chemical vapor deposition) from silane, ammonia, and dopant gases at temperatures around 200 °C. Sputtered ZnO:Al is used for the pin-diodes as a semitransparent front contact: Cr as the rear contact. The transistors use Cr for the gate electrode and Al for the source and drain contacts.

2
Fabrication Technologies

On top of the electrodes a polymer layer is deposited in a spin-on process, which is baked at low temperatures, compatible to the underlying microchip. Post processing of the ASIC at temperatures of around 210 °C did not show any measurable influence on transistor leakage currents and threshold voltages. Cyclotene (Dow Chemicals Europe, Horgen, Switzerland) may be used because of its extremely high bulk resistivity of $10^{19}\,\Omega$/cm and a breakdown voltage greater than $10^6\,\mathrm{V\,cm^{-1}}$. Furthermore, Cyclotene exhibits an excellent transmission from ultraviolet to near infrared light, which is described elsewhere [3].

Alternatively, SU-8 (micro resist technologies, Berlin, Germany) may be used for the polymer. It is widely known in MEMS (microelectromechanical systems) applications, in particular as a master for hot embossing, as a resist for a couple of hundreds of microns thick structures, and because of its excellent aspect ratio [4–6].

In a spin-on process, the negative photoresist SU-8 is deposited on silicon wafers and patterned photolithographically and developed, resulting in a topology shown in Fig. 4b. Following hard baking, the surface properties of the polymer could be modified, which is important for a self-priming microfluidic system and which enhances the electroosmotic flow in a capillary [7]. The contact angle is reduced by plasma treatment for different times at 50 W, 100 mTorr and 80 sccm O_2 according to Fig. 4a. In an additional heating step, such as for 3 h at 200 °C, the hydrophilic property turns back towards higher contact angles, so that a precise adjustment is possible, controlled by the heating time.

To prevent the metallization layer from unwanted corrosion, two different technologies were developed to substitute the top metal layer by either a chromium–gold or a chromium–platinum layer. In the first process, according to the complete ALM back-end process flow in Fig. 5, an approximately 10 nm thick chromium layer and an approximately 50 nm thick gold layer is placed on an oxidized

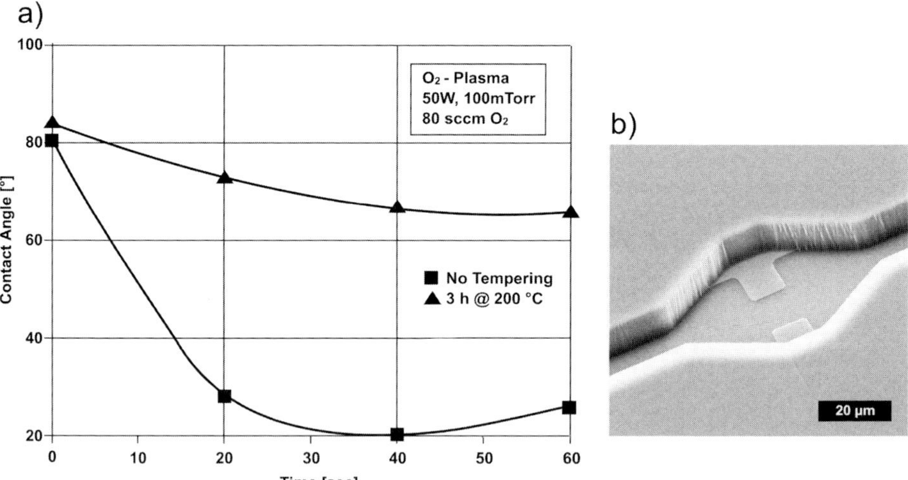

Fig. 4 (a) Contact angle of SU-8 after plasma treatment for different times, and additional heating; (b) SEM image of a conductive sensor, placed along an 15 µm deep and 15 µm wide SU-8 channel on 150 nm high-chromium electrodes on a thermally oxidized silicon wafer.

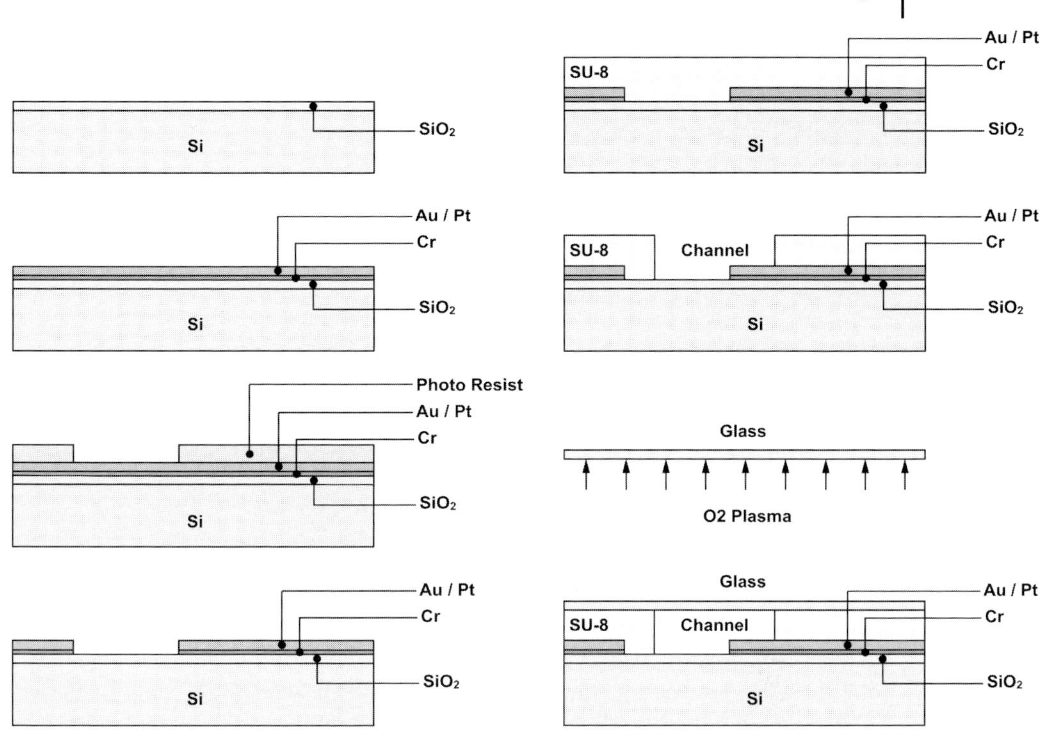

Fig. 5 Diagram of the complete ALM back-end process flow based on crystalline silicon.

silicon wafer by PVD. Both processes are initialized by a reversed sputtering step, which removes the inevitable thin film of water caused by handling the substrates at atmospheric pressures.

In the third step of Fig. 5 a photoresist is deposited and patterned, so that the unwanted metal becomes accessible. The gold layer is wet etched in a potassium iodide/iodine solution (2.67 g KI: 0.67 g I_2: 100 mL H_2O) for around 45 s. Directly after gold etching the now-accessible chromium is removed by wet etching, too. Finally, the photoresist is removed and the wafer is cleaned with acetone and propanol in an ultrasonic bath, followed by rinsing with de-ionized water.

A more sophisticated approach to fabricating metallization layers is shown in Fig. 6, where in only one lithographical step the chromium and the platinum layers are deposited, after HF-etching with the same photoresist pattern. The purpose of the top layer of platinum is to reduce the effects of metal corrosion directly in contact with the fluid at low frequencies and high voltages. Furthermore, platinum exhibits a linear temperature dependence of specific resistance, which could be useful, such as for temperature sensors and heaters. For creating embedded electrodes, the glass substrate is exposed to ammonium buffered HF-etchant for 30 s resulting in an approximately 60 nm deep topology.

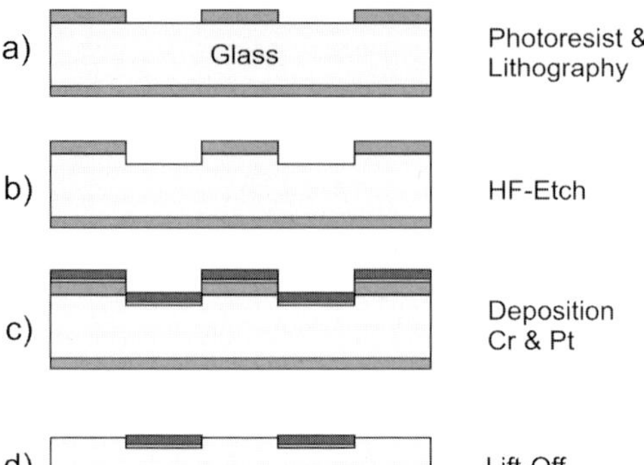

Fig. 6 Diagram of chromium and platinum deposition and lift-off process.

On top of the polymer layer a Pyrex glass plate is placed to seal the channels. It also acts as a protective layer against environmental influences. The 150 µm thick glass cap is exposed to an oxygen plasma, according to the last-but-one process step in Fig. 5. Directly after plasma treatment, the glass cap is placed with moderate pressure onto the substrate, which was preheated in an oven at 95 °C, to improve the bonding forces [8–10]. Instantly after placement, an irreversible sealing occurs; attempting to break the seal results in damage to the bulk glass or silicon wafer [11]. Finally, the microfluidic chip is placed into an oven for another 45 min at 95 °C. Fig. 18 shows the resulting wafer, containing capillary electrophoresis structures, micropumps, mixers, reservoirs, conductive sensors, and bondpads.

3
Experimental Results

The experimental results presented in this chapter focus on crucial properties of individual devices necessary to fabricate an integrated ALM.

3.1
Amorphous Silicon Pin-Diodes

At the top of the cover plate depicted in Fig. 3 diodes are deposited. The b/w photodiodes can be realized in the form of pin-layer sequences or Schottky devices, both of which have been successfully implemented in TFA (thin film on ASIC) sensors [12]. A pin-diode, depicted on the left of Fig. 7 in detail, consists of a light-absorbing intrinsic a-Si:H layer sandwiched between two heavily doped layers that provide the electric field necessary for the collection of photogenerated carriers in the i-layer. Optimization of the device performance resulted in a configuration in which the light enters through a wide bandgap a-SiC:H p-layer, which is produced by adding methane (CH_4) to the silane (SiH_4). The layer thicknesses of the optimized structure are 20 nm (p-layer), 1350 nm (i-layer), and 30 nm (n-layer). While sputtered Cr is used for the rear contact a sputtered ZnO:Al TCO (transparent conductive oxide) with a

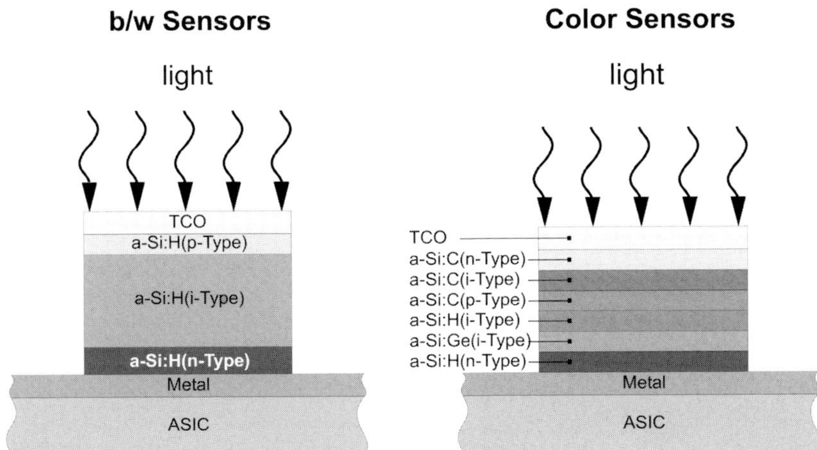

Fig. 7   Black/white and color sensors in a-Si:H technology.

thickness of 400 nm acts as front contact tacing the microfluidic channel. Its transmission is, as shown in Fig. 8, approximately 90% for visible and near IR light.

Amorphous hydrogenated silicon inherently exhibits poor charge transport characteristics compared to x-Si, which is a consequence of the low electron and hole mobilities. However, a-Si:H shows much better photoelectric properties than x-Si. The absorption coefficient is significantly higher for a-Si:H in the visible spectral range, which is relevant for many applications. The maximum is located at 550 nm, so the spectral response is similar to that of the human eye. The low dark current of a-Si:H pin-diodes allows the fabrication of highly sensitive detectors far superior to diodes made from x-Si. Although the transient behavior of a-Si:H photodiodes is inferior to that of crystalline silicon devices, frame rates far above video rate have been demonstrated by Böhm [13].

In order to detect radiation from, for example, laser stimulated molecules it is necessary to consider the dimension of the volume to be measured. If one molecule is present within a cube of $1\,\mu m^3$ and if such a molecule emits 10^6 photons per second if laser stimulated, a total of 10^9 photons are generated per second in a channel compartment of $10\,\mu m^3$. Assuming that 10% of these photons are collected in a detector and generate electron–hole pairs, photocurrents of 10^8 electrons per second are expected. If the molecules emit only 10^3 photons per second, such as without stimulation, and if the electrical and geometrical collection efficiency is only 1%, 10^4 electrons per second are expected. Thus the dark current of the photodiode should be on the order of a few electrons per second, which is a highly demanding application. Amorphous silicon pin-diodes appear to meet these requirements as described by Lulè et al. [14]. Present experimental results reveal dark currents of as low as 100 electrons per second per pixel of size $7.4\,\mu m \times 7.4\,\mu m$ at $20\,°C$ (Fig. 9). It is expected that an improvement of another order of magnitude can be achieved by further optimization [15]. Also, for many applications, molecule concentrations may be by a factor of up to 1000 higher.

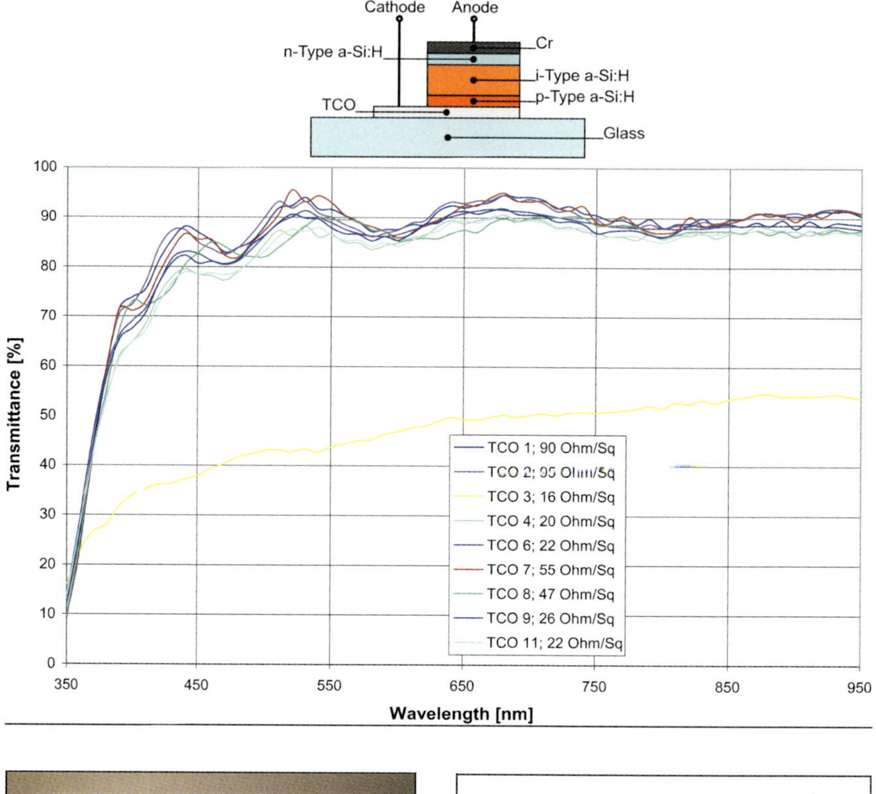

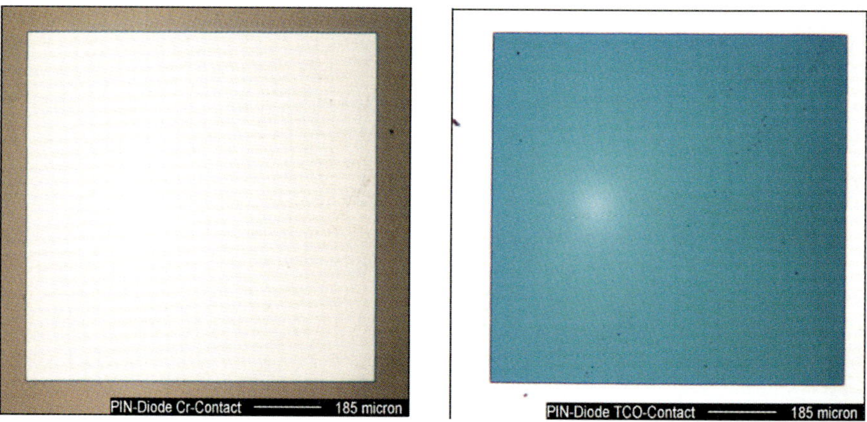

Fig. 8 Cross section and photomicrograph of an a-Si:H pin-diode with contacts fabricated by sputtering Cr and TCO.

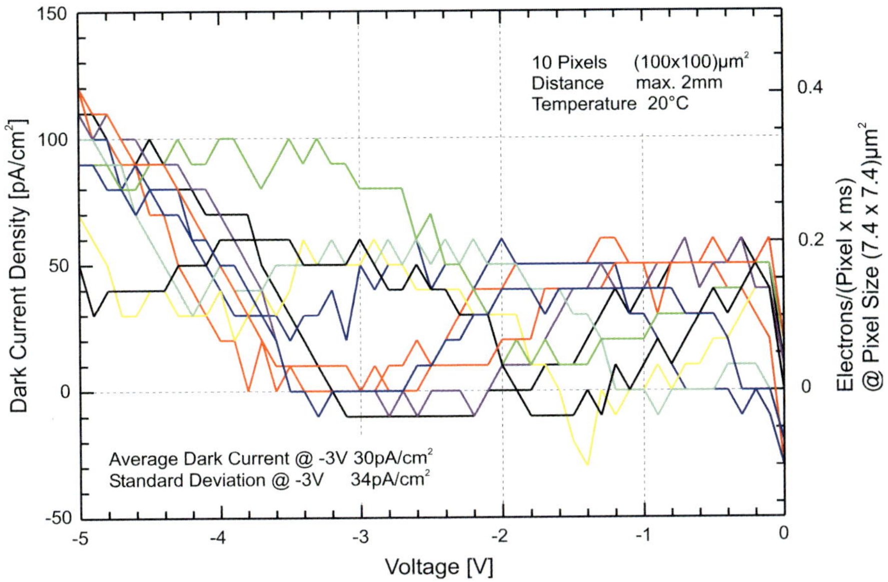

Fig. 9   Dark current density of an optimized pin-diode as a function of bias voltage.

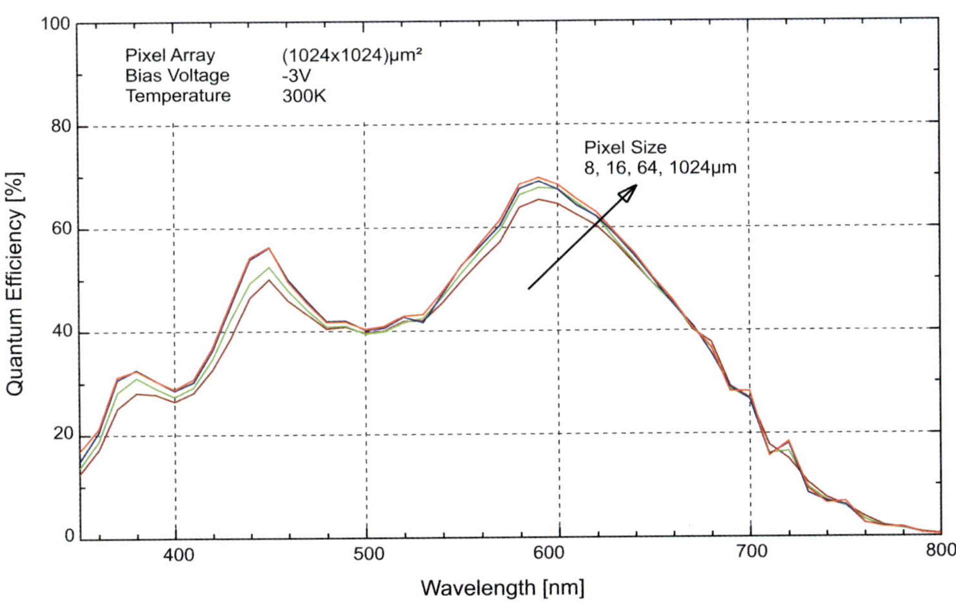

Fig. 10   Quantum efficiency of an optimized pin-diode as a function of wavelength.

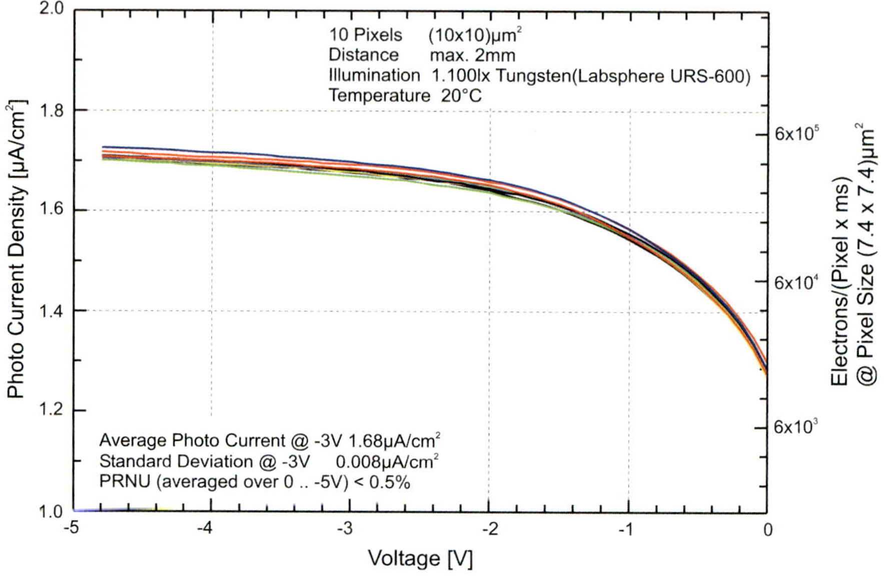

Fig. 11 Photocurrent density of an optimized pin-diode as a function of bias voltage.

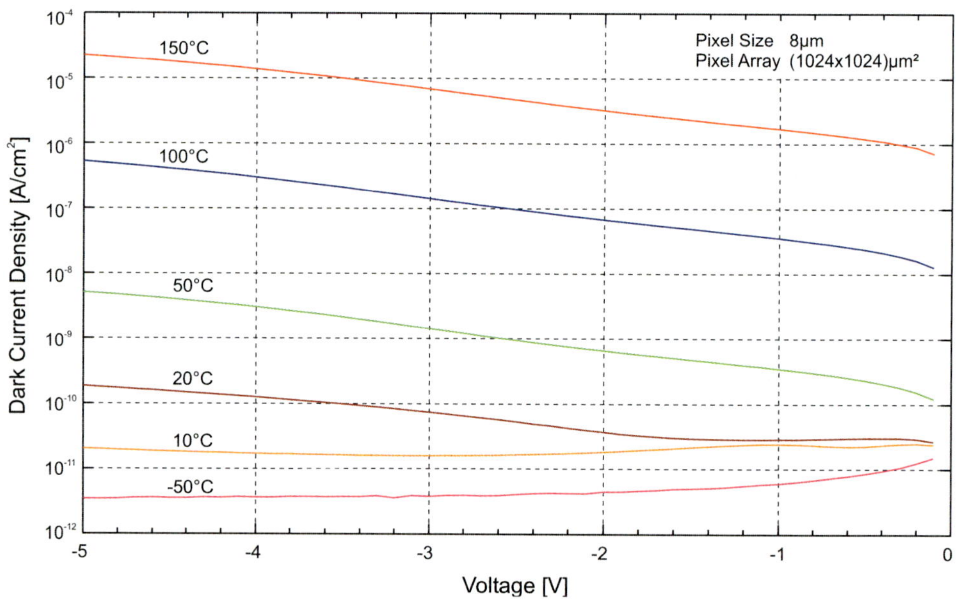

Fig. 12 Temperature dependence of the dark current density of an optimized pin-diode as a function of bias voltage.

Tab. 1   Process parameters for multispectral detectors.

Substrate	Corning glass coated with TCO
RF power supply	$0.02-0.04$ W/cm^2
Deposition facility	RF-PECVD system
Deposition rate	Approx. 3.5 Å/s
Top contact	Aluminium or chromium, 200 nm thick

Deposition parameters	*Substrate temperature (°C)*	*Reaction gas*
n-type layers	300	SiH$_4$ and PH$_3$
p-type layers	250 °C	SiH$_4$ and B$_2$H$_6$
i-type layers bandgap = 1.78 eV	300 °C	silane
i-type layers, carbonized bandgap = 1.78–1.95 eV	300 °C	silane and CH$_4$

Fig. 9 to Fig. 12 show measurements at an optimized pin-structure. The pin-diodes exhibit a quantum efficiency of approximately 70% at 580 nm and room temperature dark currents of less than 30 pA/cm^2, corresponding to above mentioned 100 electrons per second per pixel.

In addition to simple black/white detection, amorphous silicon multilayers, depicted on the right of Fig. 7, are capable of recognizing three or more colors. For this purpose, advantage is taken of the wavelength dependence of the absorption coefficient in amorphous silicon and the resulting carrier generation profile. In order to detect three colors in the same detector, the thin film system is subdivided into three i-layers with different electrical properties described elsewhere [16, 17]. Tab. 1 gives the process parameters used for fabrication of multispectral detectors. The pi^{3}n-structure allows to shift the main collection region of charge carriers from the blue sensitive region to the green and red sensitive region by increasing the reverse bias voltage. Fig. 13 depicts the photo- and dark current characteristics of the detector. The dynamic range, referred to 1000 lx illumination, is around 90 dB in the relevant voltage range from 0 to –5 V. Fig. 14 demonstrates the voltage controlled

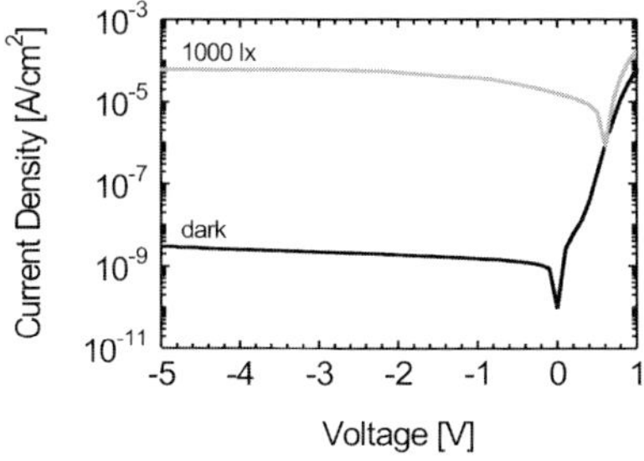

Fig. 13   Current density of an optimized pi^{3}n-diode as a function of bias voltage for darkness and illumination of 1000 lx.

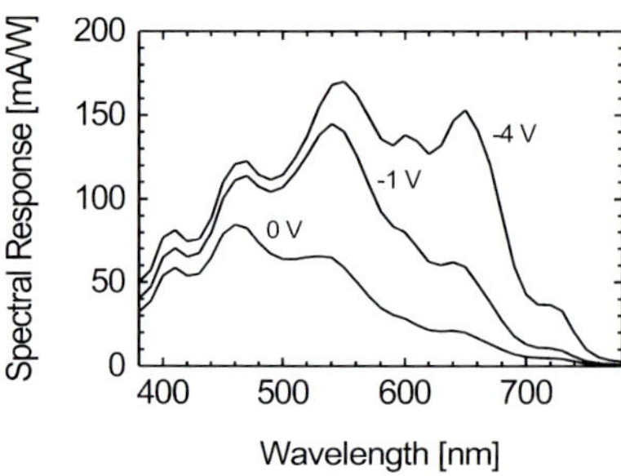

Fig. 14 Uncorrected spectral response of an optimized pi^{3}n-diode as a function of wavelength.

variation of the spectral response. The peak responsivity moves from the blue spectral range for 0 V to the green range for –1 V to the red range for –4 V. The three spectral responses are linearity independent, and a simple linear color correction has been employed for generating three slightly overlapping responses. The multispectral detector has been fabricated and tested successfully in the novel TFA imager CAESAR (color array with enhanced sensitivity and resolution) described in detail by Sommer et al. [18]. Multispectral diodes with bias controlled spectral response could be used in ALMs to allow on-chip microchromatography.

3.2
Amorphous Silicon Thin Film Transistors

The elementary active device for a readout circuit based on a glass substrate is a thin film transistor. Its basic structure, principle of operation, fabrication, and performance have been discussed in depth in numerous previous papers [19, 20]. Therefore it shall be summarized only briefly here.

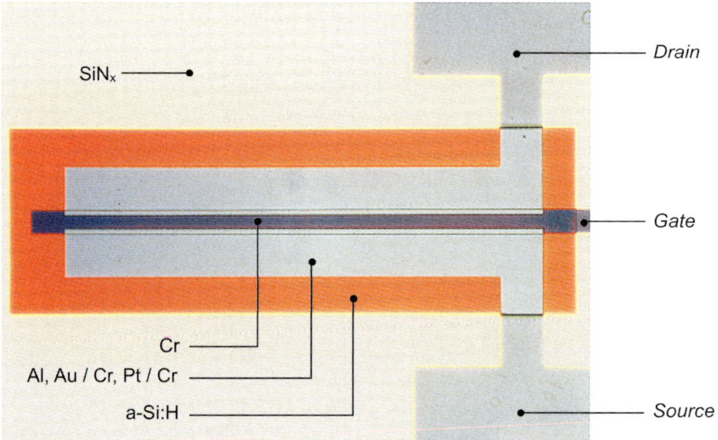

Fig. 15 Photomicrograph of an a-Si:H thin film transistor manufactured by PECVD; the channel length of the transistor is l=25.4 µm and the w/l ratio is 43.7.

Fig. 15 shows a photomicrograph of an a-Si:H thin film transistor manufactured by PECVD. The TFTs can be used as switches for the readout of an array of pin-diodes, for example, arranged along an electrophoretic separation channel. The transistors employ an inverted staggered architecture with silicon nitride as gate dielectric. In contrast to a crystalline silicon MOSFET, the gate is at the bottom and source and drain are on the top of the active layers. This configuration exhibits better interface properties and is easier to manufacture than the reversed structure.

At first, Cr is evaporated onto a glass substrate and patterned to define gate electrodes. Then the dielectric layer ($\sim$ 100–200 nm), consisting of silicon nitride, an intrinsic amorphous silicon layer ($\sim$ 100–200 nm), and a thin doped n-layer ($\sim$ 10 nm) are deposited sequentially without breaking vacuum. The a-Si:H layers are patterned selectively with respect to the underlying dielectric layer to define active transistor areas. Contact windows are opened in the silicon nitride to access the gate and to allow source/drain-gate interconnects. Subsequently, another metal layer (e.g., Cr or Al) is evaporated and patterned for the source and drain contacts.

The standard amorphous silicon TFT is an n-channel enhancement type device. The experimental transfer characteristics $I_D=f(V_{GS}\,[V_{DS}=\text{const}])$ and output characteristics $I_D=f(V_{DS}\,[V_{GS}=\text{const}])$ are shown in Fig. 16. In the off-state the very low conductivity of intrinsic a-Si:H, which has a bandgap of about 1.7 eV, inhibits current flow from source to drain. The leakage currents are in the range of pA. The conductivity can be increased by six to seven orders of magnitude for drain source voltages smaller than 10 V.

An array of electrically controlled microcompartments could be used scientifically to investigate thermochemical or electrolytic reactions. According to Fig. 17 the electrical controlled microcompartment consists of two glass plates with sput-

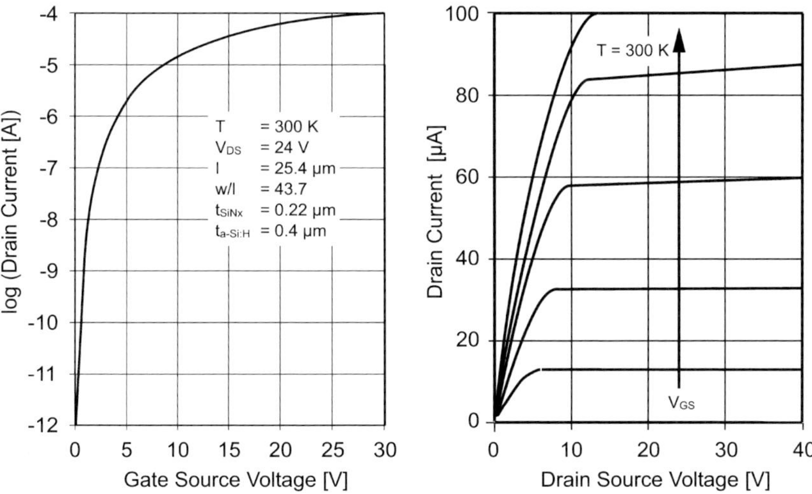

Fig. 16 Experimental transfer $I_D=f(V_{GS}\,[V_{DS}=\text{const}])$ and output characteristics $I_D=f(V_{DS}\,[V_{GS}=\text{const}])$ of an a-Si:H thin film transistor.

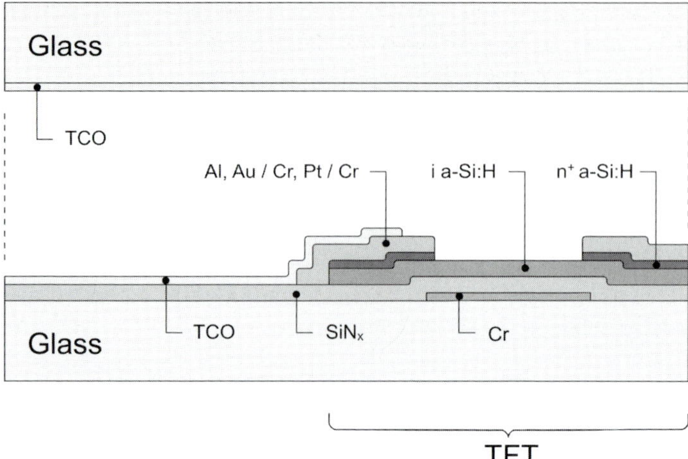

Fig. 17 Cross section of the device architecture of an electrical controlled microcompartment.

tered electrodes, such as Au or Pt, sandwiched together by using oxygen plasma bonding. The bottom plate acts as substrate and contains above an insulation layer from SiN_x a standard amorphous silicon TFT for readout contacted by sputtered ZnO:Al TCO. A distance of, for example, 10 µm between the two plates is adjusted by using glass spacers, as used in liquid crystalline displays. Sputtered TCO on the top plate acts as the backplate electrode.

3.3
Microfluidic Devices

After fabrication of the test system, described in Chemnitz et al. [21], the microfluidic properties of the test system, depicted in Fig. 18, have been investigated

Fig. 18 Photomicrograph of a fabricated 3 inch silicon wafer including test devices.

Fig. 19 Microscopic image of a circular reservoir and a channel, vertically sandwiched between the glass cover plate and a electrokinetic micropump with its typical interdigitated electrodes.

using a simple arrangement of two reservoirs, connected by a channel. As shown in Fig. 19, an electrokinetic micropump was integrated into the channel. It consists of asymmetric interdigitated pairs of electrodes as shown in detail in Fig. 20, comparable to the design described by Studer et al. [22]. The narrow electrode is 2 µm wide, followed by a gap of 3 µm to the other electrode, which is 5 µm wide. These electrode pairs have a periodicity of 23 µm.

An AC voltage applied to the electrodes leads to an electric field in the fluid, inducing a negatively charged double layer on the electrode at the positive voltage and a positively charged double layer on the other electrode, respectively. Half a period later, both the electric field and the surface charge change sign, such that the resulting fluid velocity is not reversed and a global flow is expected for an asymmetric electrode structure [23]. At high frequencies the double layer doesn't have time to build up during each half a period, while low frequencies lead to a decreased electrical field because of completed built-up of the surface charge and therefore a maximum fluid velocity is expected for a given frequency [24].

For investigation of the fluidic properties of the electrokinetic micropump the reservoirs were loaded with de-ionized water, which was doped with 1 µm fluorescent latex beads (F-8819, MoBiTec, Göttingen, Germany). Owing to the surface properties of the polymer layer and the reduced contact angle of the glass cap, the microfluidic system is self-priming by capillary forces. Following loading, the head pressure of each reservoir was balanced such that no significant hydrostatic flow occurred in the channel.

A rectangular voltage was applied to the electrodes of the electrokinetic micropump, which results in hydrodynamic pumping in the field-free regions of the same channel. The fluid transport was monitored using an optical microscope, equipped with a CCD video camera and a VCR. Particle velocity was determined by counting the video frames the particles need to pass a known distance within a field-free region of the channel. Particle velocities between 0 µm s^{-1} and 50 µm s^{-1} were achieved for voltages between 0 V and 8 V and frequencies between 2 kHz and 10 kHz, which is roughly comparable with those velocities presented by Brown et al. and Studer et al. [22, 25]. However, for optimization of the micropump's properties more detailed investigation is needed. For example, it is ex-

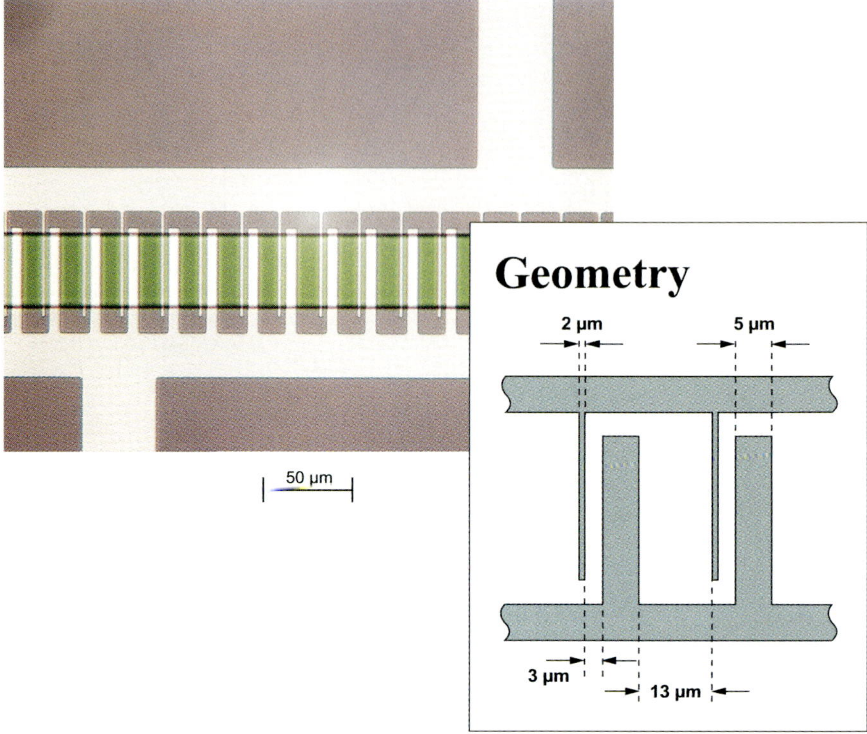

Fig. 20 Detail enlargement of the electrokinetic micropump with interdigitated electrodes.

pected that miniaturization would increase the fluid velocity because of the increased electrical field at unaltered voltage.

Other test devices have also been investigated. Fig. 21 demonstrates a photomicrograph and principles of a micromixer using Coanda effect in a newly modified Tesla structure for mixing purpose as described in detail by Hong et al. [26]. Fluids will tend to flow nearby an angle surface or a step angled surface by the Coanda effect, which is used to guide the fluid to collide. In the micromixer, one of the fluids is divided into two substreams and then merged again with the other fluid from the main channel of the micromixer. Then they are mixed with the other substream, producing a strong turbulence around the sub channel of the micromixer.

Electrophoretic separation is one of the most commonly used applications in micro total analysis systems. Fig. 22 shows a photomicrograph of a test device using the electrophoretic separation technique. Experiments to the behavior of the plug injection and non-dispersive angles are still under investigation.

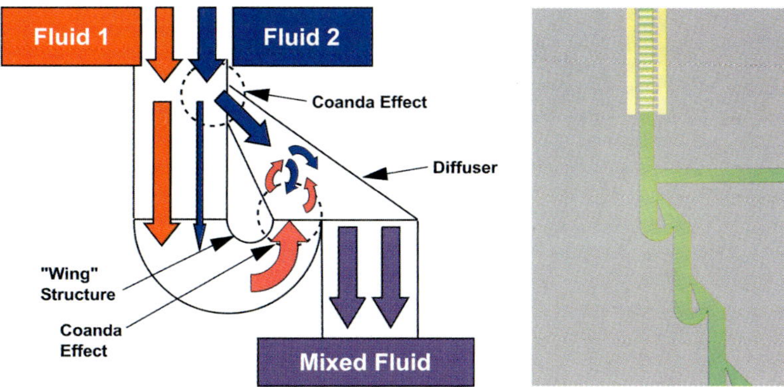

Fig. 21 Photomicrograph and principles of a micromixer using Coanda effect.

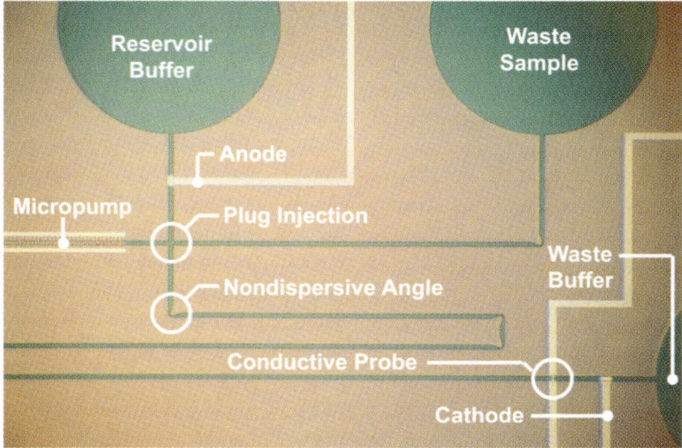

Fig. 22 Photomicrograph of a test device using the electrophoretic separation technique.

4
Conclusions

In conclusion, this chapter suggests a novel architecture for an ALM. The combination of two microtechnologies, namely the "state of the art" thin-film-technology and micromechanical processes, allows the fusion of chemical and electronic systems in a completely new type of lab chip based on micro- and nanochemistry and surface interaction. The microfluidic system and the microcompartments, which until now have been produced separately, such as on glass by micromechanical methods, are placed directly on top of an electronic integrated circuit. The microchip contains all necessary control- and analysis-hardware. The result is a compact multifunctional, three-dimensional lab-on-a-chip module on a silicon substrate including both microfluidics and electronics. Integration of active amorphous silicon devices completes the toolset to fuse electronic and chemical functions. The concept supports optical analysis with microcapillary combined with controlling and timing of a fluidic flow by integration of electrokinetic micropumps and micromixers. First experimental results demonstrate the feasibility of this new approach.

Acknowledgments

The authors are grateful to H.-J. Butt, K. Graf, and E. Bonaccurso (MPI for Polymer Research, Mainz, Germany) for helpful discussion and supporting measurements at the atomic force microscope. We also thank H.-J. Deiseroth and C. Reiner (Inorganic Chemistry, University of Siegen, Germany) for providing measurements at the scanning electron microscope. Further, we would like to acknowledge the creative input of A.J. Meixner (Physical Chemistry, University of Siegen, Germany).

References

1 S. CHEMNITZ, D. EHRHARDT, M. BÖHM, *Statusseminar Chiptechnologien*, Frankfurt a. M., **2002**, 70–72.

2 H. SCHÄFER, S. CHEMNITZ, S. SCHUMACHER, V. KOIZY, A. FISCHER, A.J. MEIXNER, D. EHRHARDT, M. BÖHM, in: *Proc. SPIE Conf.*, 19–21 May **2003**, Maspalomas, Gran Canaria, Canary Islands, Spain.

3 S. CHEMNITZ, D. EHRHARDT, M. BÖHM, in: *Proc. 6th Int. Conf. Microreaction Technology*, New Orleans, **2002**, 326–328.

4 J.A. VAN KAN, A.A. BETTIOL, B.S. WEE, T.C. SUM, S.M. TANG, F. WATT, *Sens. Actuators A* **2001**, *92*, 370–374.

5 L.J. GUERIN, M. BOSSEL, M. DEMIERRE, S. CALMES, P. RENAUD, *Transductors 97*, **1997**, Chicago, USA, 1419–1422.

6 H. LORENZ, M. DESPONT, N. FAHRNI, N. LaBIANCA, P. RENAULT, P. VETTIGER, *J. Micromech. Microeng.* **1997**, *7*, 121–124.

7 E. BONACCURSO, H.-J. BUTT, V. FRANZ, K. GRAF, M. KAPPL, S. LOI, B. NIESENHAUS, S. CHEMNITZ, M. BÖHM, B. PETROVA, U. JONAS, H.W. SPIESS, *J. Am. Chem. Soc.* **2002**, *18*, 8056–8061.

8 A. WEINERT, P. AMIRFEIZ, S. BENGSSTON, *Sen. Actuators A* **2002**, *92*, 214–222.

9 D.J. DUFFY, O.J. SCHUELLER, S.T. BRITTAIN, G.M. WHITESIDES, *J. Micromech. Microeng.* **1999**, *9*, 211–217.

10 Z. WU, N. XANTHOPOULOS, F. REYMOND, J.S. ROSSIER, H.H. GIRAULT, *Electrophoresis* **2002**, *23*, 782–790.

11 J.C. McDONALD, D.J. DUFFY, J.R. ANDERSON, D.T. CHIU, H. WU, O.J. SCHUELLER, G.M. WHITESIDES, *Electrophoresis* **2000**, *21*, 27–40.

12 B. SCHNEIDER, P. RIEVE, M. BÖHM, *Handbook of Computer Vision and Applications*, Academic Press, Boston, **1999**, 237–270.

13 M. BÖHM, *J. Non-Cryst. Solids* **2000**, *266–269*, 1145–1151.

14 T. LULÉ, S. BENTHIEN, H. KELLER, F. MÜTZE, P. RIEVE, K. SEIBEL, M. SOMMER, M. BÖHM, *IEEE Trans. Electron. Dev.* **2000**, *47*, 2110–2122.

15 F. BLECHER, B. SCHNEIDER, J. STERZEL, M. HILLEBRANDT, S. BENTHIEN, M. BÖHM, *J. Non-Cryst. Solids* **2000**, *266–269*, 1188–1192.

16 Q. ZHU, S. COORS, B. SCHNEIDER, P. RIEVE, M. BÖHM, *IEEE Trans. Electron. Dev.* **1998**, *45*, 1393–1398.

17 Q. Zhu, J. Sterzel, B. Schneider, S. Coors, M. Böhm, *J. Appl. Phys.* **1998**, *83*, 3906–3910.

18 M. Sommer, P. Rieve, M. Verhoeven, M. Böhm, B. Schneider, B. van Uffel, F. Librecht, in: *Proc. 1999 IEEE Workshop on Charge-Coupled Devices and Advanced Image Sensors*, Nagano, Japan, 10–12 June **1999**.

19 M. Böhm, S. Salamon, Z. Kiss, *Appl. Phys. A* **1988**, *45*, 53.

20 M. Böhm, Solid State Technology Series, *Disk & Display Technology*, **1988**, *21*, 129.

21 S. Chemnitz, H. Schäfer, S. Schumacher, V. Koizy, A. Fischer, A. J. Meixner, D. Ehrhardt, M. Böhm, in: *Proc. SPIE Conference*, 19–21 May **2003**, Maspalomas, Gran Canaria, Canary Islands, Spain.

22 V. Studer, A. Pepin, Y. Chen, A. Adjari, *Microelectronic Engineering* **2002**, *61–62*, 915–920.

23 A. Adjari, *Appl. Phys. A* **2002**, *75*, 271–274.

24 N. G. Green, A. Ramos, A. Gonzáles, H. Morgan, A. Castellanos, *Phys. Rev. E* **2000**, *61*, 4011–4018.

25 A. B. D. Brown, C. G. Smith, A. R. Rennie, *Phys. Rev. E* **2000**, *63*, 016305.

26 C. Hong, J.-W. Choi, C. H. Ahn, *Proc. μTAS* **2001**, 31.

Impact of Nanoscience on Heterogeneous Catalysis

Sharifah Bee Abd Hamid and Robert Schlögl

1
Introduction

Heterogeneous catalysis is the science and technology of changing the rates of chemical reactions. The technology is needed to produce all liquid fuels, polymers, functional materials, and artificial biological molecules. Catalysis is engaged with the making and breaking of specific chemical bonds. The specificity of the process in terms of chemo- and regioselectivity leading to economical and sustainable production processes is a sensitive function of the local electronic structure of the interaction zone of a few atoms between reactants and catalyst. The catalytic reaction depends, however, not only on the molecular properties of the system, but also in several domains of space and time on the properties of the catalyst–reactor system with respect to transport of energy and materials. The correlation between catalytic performance and molecular properties is thus weakened and can not be used as guide for catalyst development.

High-throughput experimentation (HTHE) and combinatorial methods have been advocated to remedy the lack of empirical optimization strategies. A knowledge-based strategy is required to define compositional libraries and experimental procedures with a reasonable success. Testing nanostructured materials by high throughput techniques is a highly promising possibility.

2
Nanotechnology in Catalysis

In heterogeneous catalysis, size effects are understood as the dependence of performance on the average geometrical size of the active material [1, 2]. These strong effects can be difficult to explain. The simple argument that highly dispersed active materials expose more active sites per unit weight is correct, but insufficient to account for all phenomena, as the properties of materials change with size when their dispersion (defined as the ratio between bulk and surface atoms) reaches finite values. This effect leads to nonlinear structure–activity relations.

The Nano–Micro Interface: Bridging the Micro and Nano Worlds
Edited by Hans-Jörg Fecht and Matthias Werner
Copyright © 2004 WILEY-VCH Verlag GmbH & Co. KGaA, Weinheim
ISBN: 3-527-30978-0

Chemical bonds are typically 0.2 nm long, whereas typical stable catalytic particles are 3–6 nm in size. The ratio between these two dimensions can be called the size parameter and it is found that a maximum of catalytic function occurs at values of the size parameter much larger than unity. It is thus justified to interrogate the impact of nanostructuring on catalysts [3]. This does not deny that catalysis relying on the deliberate synthesis of nanosized particles as active masses is probably the oldest and most successful application of nanotechnology.

The widespread attention that nanoscience has received in many areas of material and life sciences led to a loose definition of this field including molecular objects as well as small crystals of many thousands of identical atoms or molecules. The geometric size of a few nanometers [4] is thus not a strong definition. It is more useful to define nano-objects in catalysis as supramolecular ensembles of sizes that allow metastable bulk and surface configurations to be kinetically stabilized at conditions of investigation or application. Such objects are intermediates between crystals, aggregates, and molecules; they are defined by sharp boundaries, even when the objects are embedded in a matrix. These conditions are fulfilled for metallic elements with objects containing 2–5 atoms per edge of its polyhedron (often cubo-octahedra or icosahedra) or in semiconductors with 1–3 chemical bonds per edge. Fig. 1a illustrates the relation between particle site and number of atoms contained. Operating catalysts are made of particles with sizes between 2.5 and about 7 nm: in cases of encapsulated systems smaller particles can occur. Fig. 1b reveals that the ratio between surface and bulk atoms is for such particles still much larger than for bulk materials, pointing to a property of nanoparticles extremely important for catalysis: the fraction of coordinatively undersaturated (*cus*) atoms is large compared with bulk materials.

3

Electronic Structure and Catalysis

Nanoparticles are large enough to reveal the bulk electronic structure of metals or semiconductors. They are much larger than cluster objects, which are defined as entities at the transition from bulk to molecular properties (typically a few metal atoms for elements with highly delocalized valence electrons (alkali metals) and about 200 atoms for elements with strongly localized valence states (heavy main group metals) [5].

Nanostructures may be defined alternatively as objects exhibiting a significant fraction of their total number of atoms as *cus* with respect to the regular bulk structure. If 20% *cus* sites are accepted as the lower level of significance, then objects of about 8000 atoms are nano-objects falling for isotropic geometries in the same size range up to 6 nm (Fig. 1a) as defined above.

The *cus* sites exhibit local electronic structures decoupled from the band structure of the regular structure. Consequently, an energy gap opens (Kubo gap), the width of which depends on the number of valence electrons in the nano-object [6]. This size effect on the integral electronic structure is significant for small ob-

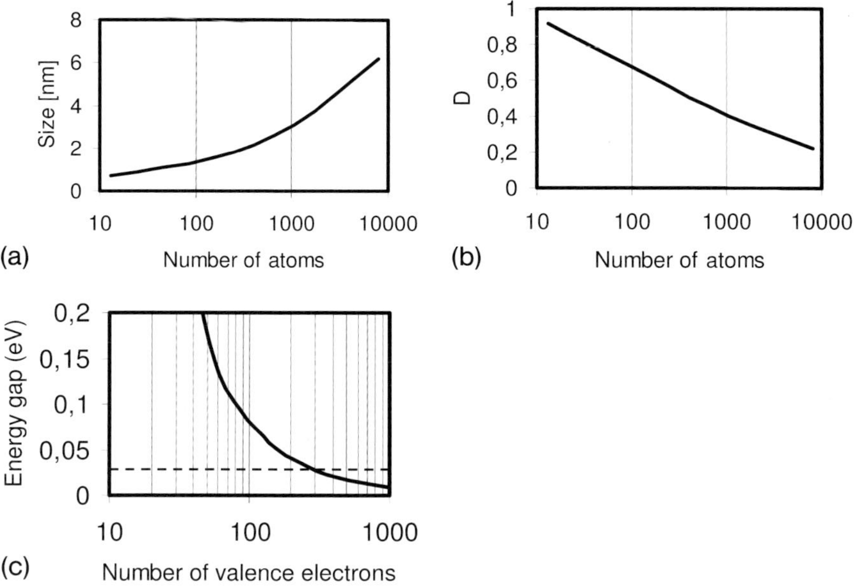

Fig. 1 (a) Relation between size and number of atoms for a typical metal with close-packed structure in cubo-octahedral shape. (b) Dispersion (ratio between bulk and surface atoms) for the same geometry. (c) Evolution of the Kubo gap as function of the number of valence electrons. The dashed line indicates the energy of room temperature excitation of electrons.

jects as can be seen from Fig. 1c. The band gap vanishes at 300 K (equivalent to about 25 meV) for objects with about 300 atoms. As catalytic reactions occur usually at elevated temperatures, the band gap effect is of significant influence only for small objects [7] of about 1 nm in size. Such objects are in non-encapsulated forms [8, 9] usually not very stable under reaction conditions and tend to sinter. Thus a ground state electronic structure modification well known to exist in small clusters [7, 10] is in most cases not responsible for a beneficial "nano-effect" in catalysis.

4
Geometric Structure and Catalysis

The relevance of the surface geometric structure is extensively known in heterogeneous catalysis and was originally designated by the term "structure sensitivity" [11, 12]. Typical stable metallic nano-objects that are active in catalysis are depicted in Fig. 2. Part (a) shows a Pt particle supported on silica [13], typical of a hydrogenation catalyst. It is clearly seen that the bulk is well-ordered metal (see also Fourier pattern) whereas the surface is blurred and facetted. Object (b) is a copper multiple-twinned particle (MTP) of icosahedral symmetry [14] (Fourier pat-

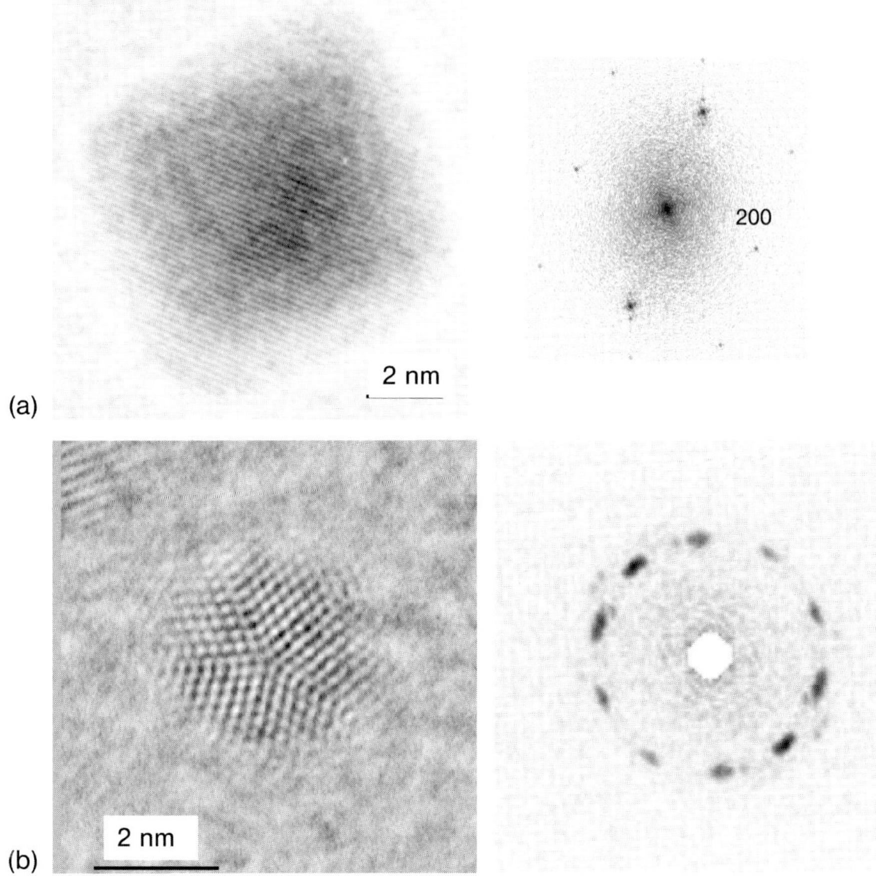

Fig. 2 (a) Pt nanoparticle deposited on a model silica support [13]. The internal structure is compatible with bulk Pt (see power spectrum at the right panel) but the surface of the 3-dimensional particle is rough and facetted. (b) Multiple-twinned Cu particle prepared by the gas aggregation technique [14]. The power spectrum reveals its perfect icosahedral symmetry by the satellite spot structure.

tern). It shows regular defect sites at the edges of the sub-units and exhibits a significant deviation from the bulk lattice constant [15, 16]. The sizes of both classes of objects exclude the operation of an integral ground state electronic effect. Their perimeters (surfaces) are non-uniform and exhibit irregularities and roughness. Such local geometric irregularities [17, 18] are well-known to exhibit strong variations in local electronic structure [19] and, for example, can easily revert the character of chemisorption from molecular to dissociative [20], a key feature in activating molecules for chemical reactions. An example of the control of chemical reactivity with the localization of *cus* sites was found in the interaction of methanol with Pd [21, 22] where edge atoms exhibit a different reactivity than terrace atoms

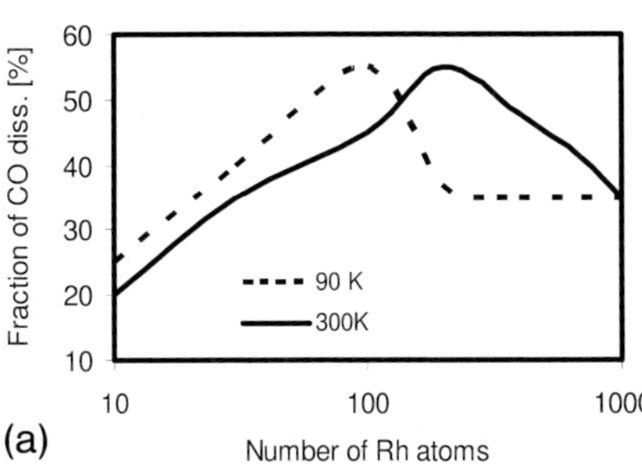

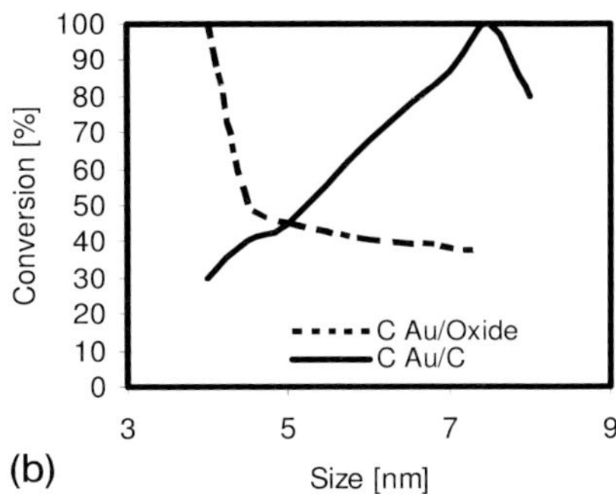

Fig. 3 Case studies of nano-effects in catalysts with supported metal nano-particles. (a) reveals that the deposition temperature of identically sized particles plays a decisive role in their function probed in chemisorption of CO [20]. (b) Structure–activity relationships for gold particles in glycol oxidation [24] for two families of catalysts on different supports. The strong effect of support properties on the relation is obvious.

albeit both types are *cus*. The examples indicate that fine details and a variation in the *local geometric structure* of metastable objects exhibiting non-equilibrium surface termination are a key element in the action of nano-objects in catalysis.

A nano-object is metastable and thus not defined in its properties uniquely by the combination of chemical identity and size; its synthetic pre-history and its interaction with the substrate are equally important, as illustrated in Fig. 3. The ratio between molecular and dissociative chemisorption of the CO probe molecule depends on the deposition temperature of *equally sized objects* [20] (example (a) in Fig. 3) and in example (b) of Fig. 3 the relation between geometric size and catalytic performance depends strongly on the chemical identity of the support of "inert" gold particles [23, 24].

5

Large Nano-Objects in Catalysis

The examples discussed so far are all isotropic and densely packed objects. Extended isolated objects with non-dense structures may carry a sizable amount of non-equilibrium surface features with the desired local chemical properties. Nano-structuring enhances the number density of such features and prevents the equilibration into a macroscopic solid (Ostwald ripening). Fig. 4a illustrates this with vanadium oxide nanorods [25] that were obtained with an extremely textured arrangement of the (001) atomic planes along the needle axis. The rods are accessible by seeding an oxide solution with carbon nanotubes of high hydrophilicity. The high magnification image reveals that the surface of the rods is covered with a monolayer of a one-dimensional ordered material most relevant for catalytic functions [25–29]. Bulk vanadium oxide crystallizing in platelets is free of such a

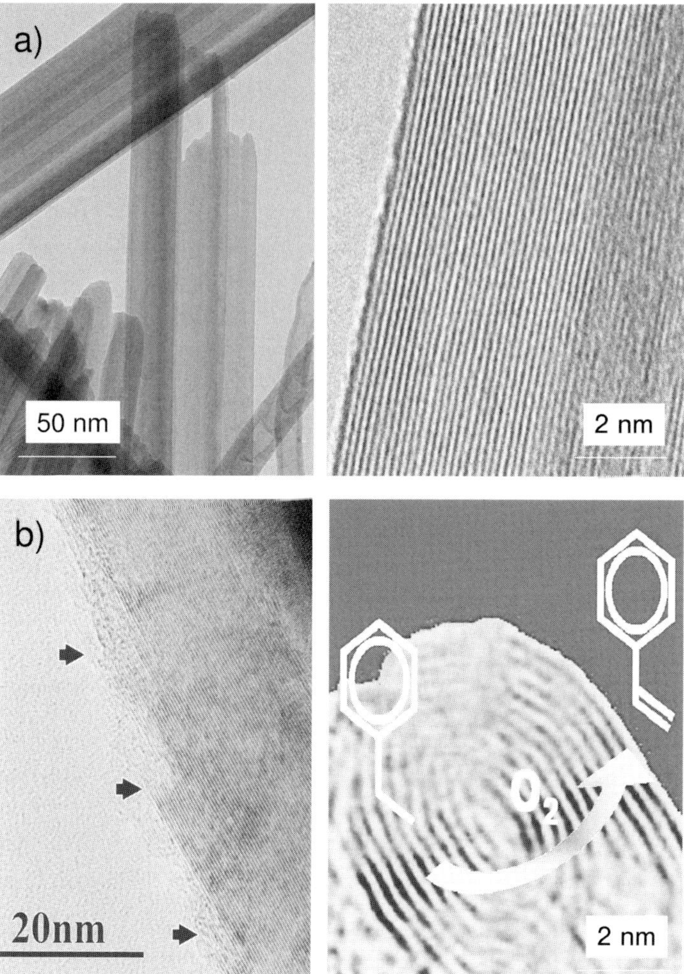

Fig. 4 Large nano-objects: (a) vanadium pentoxide nanorods grown from seeding a vanadate solution with carbon nanofilaments (bamboo type, see bottom right). The HRTEM image reveals the perfect internal ordering of the V_2O_5 structure and the amorphous termination layer. (b) A carbon nanofilament with a stepped outer surface (arrows) after use as catalyst in oxidative dehydrogenation of ethylbenzene to styrene. Some polystyrene covers as loose debris the surface of the still active catalyst. The onion-like carbon object exhibits a minimum of non-reactive stabilizing basal planes at the surface and a maximum catalytic performance in the reaction schematically shown is attained.

termination layer [30–32] even when grown in identical conditions. Fig. 4B shows a carbon nanofiber highly active in dehydrogenation of ethylbenzene to styrene [33]. The nanostructure allows for an angle between the graphene sheets and the needle axis giving rise to the tiled surface structure (arrows). The resulting carbon edges carry the active centers as oxygen heteroatomic terminations (prism faces (110)). If the surface as a whole was terminated like this, the particle would oxidize under reaction conditions [34–36] and hence exhibit no lifetime in the reaction. No catalytic activity is obtained on defect-free basal (0001) terminations. The left image illustrates that nanostructuring of carbon into onion-like carbon optimizes the abundance of stabilizing basal planes and hence leads to a maximum catalytic activity [37] per unit mass carbon.

6
The Semiconductor Approach

The advent of nanotechnology in catalysts has also led to extended activities in catalytic reactor design. The aim here is to replace large reactors with highly parallelized arrays of small reactors that are microfabricated. In this way the control of mass and energy flow can be managed to great extent and the technologies of semiconductor device fabrication may be applied to a rigorous physical preparation strategy for the active catalyst. Further, the gap between model systems susceptible to structural characterization and the scale-up to a large "chemical" system would vanish as the large system is a parallelized version of the small model. This approach was recently reviewed [38] and validated as promising development for future catalysis research. For complex multifunctional catalytic processes such as selective oxidation this seems a futuristic alternative due to the complexity of the catalytically active material that requires chemical means of preparation.

7
The Combicat Approach

The design principle of nanometer-sized objects with non-equilibrium surface structures seems to work well for metals and selected other compounds. In the large field of catalysis with metal oxides the concept does not work due to the lack of preparative methods to access active nanostructured particles with pre-defined properties. The common approach is to modify a given structure by chemical substitution. The active structure is always a "defect" structure with respect to the thermodynamically stable bulk oxide. It needs to be generated by very difficult-to-control thermal treatments called "calcination and activation", leading inevitably to phase mixtures and chemically complex materials referred to as "multi-phase-multi-element oxides" (MMO).

In recent years it has become possible to identify a structure of molybdenum sub-oxide as essential single phase [39, 40] in a variety of catalysts for a selection

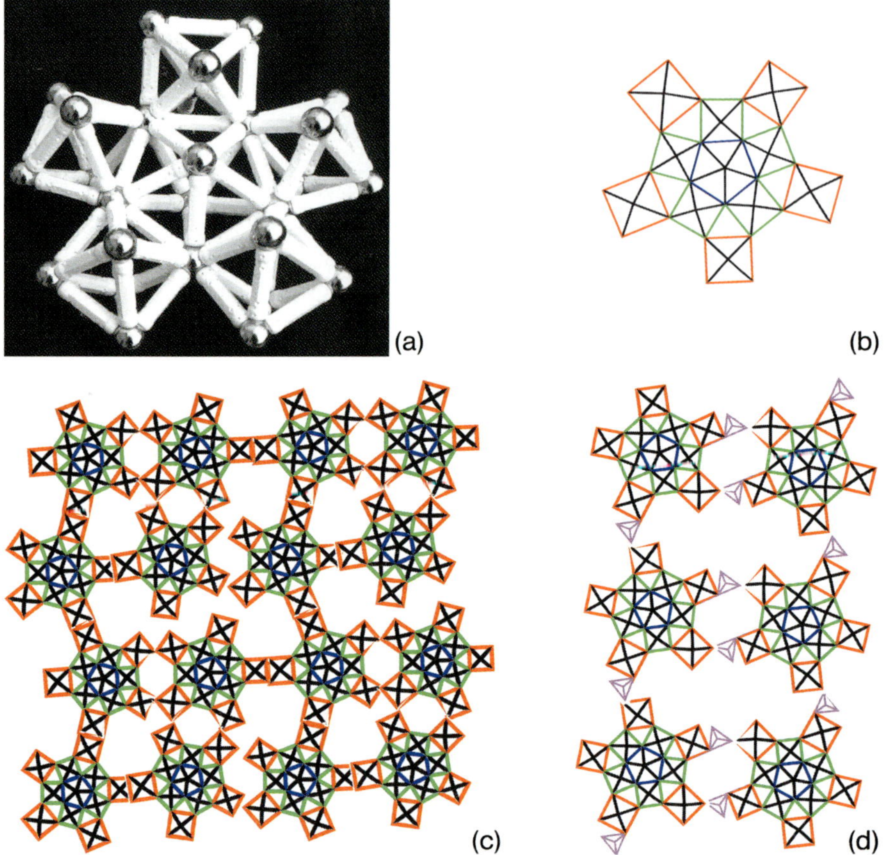

Fig. 5 The polyoxometallate functional unit in catalysts for selective oxidation of alkanes (a) can be realized with a variety of transition metal cations. The monomer (b) condenses during conventional preparation into a "glassy" state with close contact between the units (chains in c). By introducing suitable linkers a structure can be created (d) in which the units are isolated from each other and a hierarchical pore system will result. The nanostructure is of a layer type with structure-directing ions filling some of the channels to account for a stacking order.

of selective oxidation processes of small alkane molecules. The essential motif [41] is shown as mechanical model in Fig. 5a. It is a cluster of a pentagonal bipyramids surrounded by five distorted octahedra. The atoms shown in the model are oxygen, the metal (Mo, W, Nb, V) is at the center of each polyhedron. The object is about 1 nm in diameter and 0.41 nm thick. For catalytic function it seems essential that this active cluster is isolated [42, 43] from the matrix to prevent excessive flow of active oxygen atoms and of electrons. During conventional synthesis the cluster (Fig. 5b) is condensed into a highly disordered polymer [44] with some long-range order (chains in one direction, Fig. 5c). This inadequate structure needs to be defected [45] by calcination in dilute air creating a defective matrix Mo_5O_{14} and a second phase of ortho MoO_3. The latter material is detrimental for catalysis [45] as it accelerates the burning of all valuable initial products formed on the cluster.

If the correct nanostructure (Fig. 5d) could be achieved by using linker structures during the initial synthesis, such as in model cases with heteropolyoxomolybdates [46, 47], then the detrimental calcination may be omitted. The polymerization of the clusters into a stable nano-object with perfect isolation of the active sites is desired. Only if the final polymer is nanoscopic in size then sufficient dispersion (Fig. 1b) at stability against chemical reduction and sintering can be achieved.

In contrast to the empirical addition of heterocations [48] to the solution of monomeric anions, the Combicat approach utilizes detailed understanding to arrive at initial synthesis conditions allowing using the same element (Mo, Nb, V) for building the clusters and the linkers. The Combicat approach aims at the minimum of chemical complexity in contrast to, for example, the HPA structures. This is motivated by the severe stability problems encountered with practical reaction conditions: elevated temperatures, steam, and reducing conditions are aggressive to metal–oxygen chemical bonds involving two different metals with different degrees of polarization and differing redox properties. A homonuclear structure is the preferred option in order to minimize the decomposition processes during catalytic operation.

The nanostructuring of the linked clusters can be achieved by controlling the polymerization kinetics. This is required to optimize the transport of molecules and of energy to and from the active material. The necessary porosity and sufficient geometric surface area need to be incorporated without allowing the system to become too susceptible towards hydrothermal sintering during catalytic operation.

It is obvious that for executing control over the synthesis a set of independent variables is required. The identification and validation of these variables, such as concentration ratios of reagent, temperature, presence of auxiliary ions to stabilize complexation equilibria, mode of solidification (variable or constant concentration), extrinsic variables (reactor design and operation), post-precipitation treatments, drying, and thermal treatments is an immense effort being ideally suited for high-throughput experimentation (HTHE) strategies [49] adapted with special equipment to perform such controlled synthesis tasks. The usual processes of uncontrolled parallel precipitation in small quantities are unsuitable for such a task. Computer-automated larger scale reactors are required that allow the same amount of kinetic control of the synthesis reaction as it is exerted in the kinetics of catalytic performance tests. The results are diagrams of state for each unit operation allowing defining operation points for reproducible and scalable synthesis.

The general procedure is outlined in Fig. 6. The metal monomer is simultaneously linker and building block for a homo-polymer. This needs to be grown into nanoparticles and investigated for stability against crystallization into ortho MoO_3. If this cannot be achieved, the linkers may be chosen from a different ele-

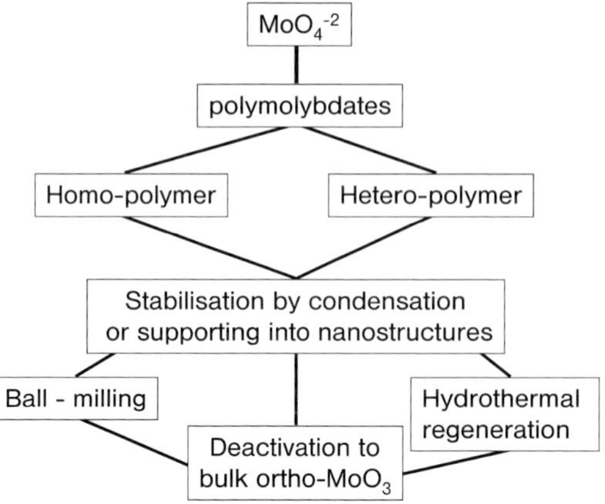

Fig. 6   Scheme of a preparation strategy to arrive at the desired nano-structure depicted in Fig. 5 d. Several alternatives are shown for the case that the homo-nuclear polymerization will not result in sufficiently active and stable materials. Even samples deactivated during synthesis may be re-activated by mechanical or chemical treatments providing ample chance to arrive at the desired structure.

ment much in the same way as glasses [50] are being produced. As the catalytic function is very difficult to asses from a structural analysis, it is mandatory to use HTHE catalytic screening in parallel with the material development in order not to develop the system towards maximum stability by depriving it from the essential [51, 52] metastable structural features (defects) that are necessary for selective oxidation catalysis [43, 53, 54]. The dual application of HTHE technology is thus vital for operating a rational nanostructuring strategy for such a complex functional material as a bulk selective oxidation catalyst.

8
Conclusions

Nanostructured objects are essential tools in shaping catalysis towards the needs of modern economy. It is the optimization between maximal exposure of metastable surface features and necessary stability of metastable functional material that can be guaranteed by nano-objects. Neither the size minimization nor the operations of collective electronic effects are relevant for catalysis, characteristics of "nano" that are so important in non-catalytic applications. Refined in-situ characterization methods for detecting the relevant local electronic features and rational synthesis protocols replacing the "black art" of catalyst manufacture are required to fully exploit the wealth of information from nanoscience for a new generation of catalyst development based on knowledge rather than empirical methods.

References

1 APARICIO, LM and DUMESIC, JA. *Ammonia Synthesis Kinetics: Surface Chemistry, Rate Expressions, and Kinetic analysis*, 1st ed. J.C. Baltzer AG **1994**.

2 BOUDART, M, DELBOUILLE, A, DUMESIC, JA, KHAMMOUMA, S, and TOPSOE, H. *J. Catal.* **1975**, *37*, 486.

3 BELL, AT. *Science* **2003**, *299*, 1688.

4 SOMORJAI, GA and MCKREA, K. *Appl. Catal. A* **2001**, *222*, 3.

5 EKARDT, W. *Application of the Jellium Model and its Refinements to the Study of the Electronic Properties of Metal Clusters.* Wiley, Chichester, UK **1999**.

6 RAO, CNR, KULKARNI, GU, THOMAS, PJ, and EDWARDS, PP. *Chemistry: A European Journal* **2002**, *8*, 29.

7 GATES, BC. *J. Mol. Catal. A* **2000**, *163*, 55.

8 FERRARI, AM, NEYMAN, KM, BELLING, T, MAYER, M, and ROSCH, N. *J. Phys. Chem. B* **1999**, *103*, 216.

9 IKEMOTO, Y, NAKANO, T, KUNO, M, and NOZUE, Y. *Physica B* **2000**, *281*, 691.

10 TAKASU, Y, UNWIN, R, TESCHE, B, TESCHE, AM, BRADSHAW, AM, and GRUNZE, M. *Surf. Sci.* **1978**, *77*, 219.

11 ERTL, G. *Elementary Steps in Ammonia Synthesis: The Surface Science Approach.* Plenum Press **1991**.

12 SOMORJAI, GA. *Progr. Surf. Sci.* **1995**, *50*, 3.

13 WANG, D, PENNER, S, SU, DS, RUPPRECHTER, G, HAYEK, K, and SCHLÖGL, R. *Mater. Chem. Phys.* **2003**, *81*, 341.

14 URBAN, J. *Cryst. Res. Technol.* **1998**, *33*, 1009.

15 GRYAZNOV, VG, HEYDENREICH, J, KAPRELOV, AM, NEPIJKO, SA, ROMANOV, AE, and URBAN, J. *Cryst. Res. Technol.* **1999**, *34*, 1091.

16 TELGHEDER, FW and URBAN, J. *Phys. Chem. Chem. Phys.* **1999**, *1*, 4479.

17 ERTL, G and FREUND, H.J. *Physics Today* **1999**, *52*, 32.

18 WINTTERLIN, J, VÖLKENING, S, JANSSENS, TVW, ZAMBELLI, T, and ERTL, G. *Science* **1997**, *278*, 1931.

19 HAMMER, B and NORSKOV, JK. *Surf. Sci.* **1995**, *343*, 211.

20 MAVRIKAKIS, M, BAUMER, M, FREUND, HJ, and NORSKOV, JK. *Catal. Lett.* **2002**, *81*, 153.

21 HOFFMANN, J, MEUSEL, I, HARTMANN, J, LIBUDA, J, and FREUND, HJ. *J. Catal.* **2001**, *204*, 378.

22 YUDANOV, IV, SAHNOUN, R, NEYMAN, KM, ROSCH, N, HOFFMANN, J, SCHAUERMANN, S, JOHANEK, V, UNTERHALT, H, RUPPRECHTER, G, LIBUDA, J, and FREUND, HJ. *J. Phys. Chem. B* **2003**, *107*, 255.

23 HARUTA, M. *Cattech* **2002**, *6*, 102.

24 HARUTA, M and DATE, M. *Appl. Catal. A* **2001**, *222*, 427.

25 MELZER, M, URBAN, J, SACK-KONGEHL, H, WEISS, K, FREUND, HJ, and SCHLÖGL, R. *Catal. Lett.* **2002**, *81*, 219.

26 CHAKRABARTI, A, HERMANN, K, DRUZINIC, R, WITKO, M, WAGNER, F, and PETERSEN, M. *Phys. Rev. B* **1999**, *59*, 10583.

27 CHEN, KD, BELL, AT, and IGLESIA, E. *J. Phys. Chem. B* **2000**, *104*, 1292.

28 HÄVECKER, M, KNOP-GERICKE, A, MAYER, RW, FAIT, M, BLUHM, H, and SCHLÖGL, R. *J. Electron. Spectrosc. Relat. Phenom.* **2002**, *125*, 79.

29 HERMANN, K, MICHALAK, A, and WITKO, M. *Catal. Today* **1996**, *32*, 321.

30 ABERMANN, R. *Mater. Res. Soc. Symp. Proc.* **1992**, 25.

31 SU, DS, RODDATIS, V, WILLINGER, M, WEINBERG, G, KITZELMANN, E, SCHLÖGL, R, and KNÖZINGER, H. *Catal. Lett.* **2001**, *74*, 169.

32 SU, DS. *Fresenius' J. Anal. Chem.* **2002**, *374*, 732.

33 MESTL, G, MAKSIMOVA, NI, KELLER, N, RODDATIS, VV, and SCHLÖGL, R. *Angew. Chem.: Int. Ed.* **2001**, *40*, 2066.

34 ATAMNY, F, BLÖCKER, J, DÜBOTZKY, A, KURT, H, LOOSE, G, MAHDI, W, TIMPE, O, and SCHLÖGL, R. *J. Mol. Phys.* **1992**, *76*, 851.

35 ATAMNY, F, BLÖCKER, J, HENSCHKE, B, SCHLÖGL, R, SCHEDEL-NIEDRIG, T, KEIL, M, and BRADSHAW, AM. *J. Phys. Chem.* **1992**, *96*, 4522.

36 ATAMNY, F, SPILLECKE, O, and SCHLÖGL, R. *Phys. Chem. Chem. Phys.* **1999**, *1*, 4113.

37 KELLER, N, MAKSIMOVA, N, RODDATIS, VV, SCHUR, M, MESTL, G, BUTENKO, YV, KUZNETSOV, VL, and SCHLÖGL, R. *Angew. Chem.: Int. Ed.* **2002**, *41*, 1885.

38 ARAKAWA, H, ARESTA, M, ARMOR, JN, BARTEAU, MA, BECKMAN, EJ, BELL, AT, BERCAW, E, CREUTZ, C, DINJUS, E, DIXON, DA, DOMEN, K, DUBOIS, DL, ECKERT, J, FUJITA, E, GIBSON, DH, GODDARD, WA, GOODMAN, DW, KELLER, J, KUBAS, GJ, KUNG, HH, LYONS, JE, MANZER, LE, MARKS, TJ, MOROKUMA, K, and NICHOLAS, KM. *Chem. Rev.* **2001**, *101*, 953.

39 DIETERLE, A, MESTL, G, JAGER, J, UCHIDA, Y, HIBST, H, and SCHLÖGL, R. *J. Mol. Catal. A* **2001**, *174*, 169.

40 LINKE, D, WOLF, D, BAERNS, M, TIMPE, O, SCHLÖGL, R, ZEYSS, S, and DINGER-DISSEN, U. *J. Catal.* **2002**, *205*, 16.

41 KNOBL, S, ZENKOVETZ, GA, KYUKOVA, GN, OVSITSER, O, NIEMEYER, D, SCHLÖGL, R, and MESTL, G. *J. Catal.* **2002**, *215(2)*, 177.

42 GRASSELLI, RK. *Topics in Catalysis* **2002**, *21*, 79.

43 DEROUANE-ABD HAMID, SB, CENTI, G, PAL, P, and DEROUANE, EG. *Topics in Catalysis* **2001**, *15*, 161.

44 TYTKO, KH and GLEMSER, O. *Adv. Inorg. Chem. Radiochem.* **1976**, *19*, 239.

45 OVSITSER, O, UCHIDA, Y, MESTL, G, WEINBERG, G, BLUME, A, JAGER, J, DIETERLE, M, HIBST, H, and SCHLÖGL, R. *J. Mol. Catal. A* **2002**, *185*, 291.

46 LEE, JK, MELSHEIMER, J, BERNDT, S, MESTL, G, SCHLÖGL, R, and KOHLER, K. *Appl. Catal. A* **2001**, *214*, 125.

47 GAYRAUD, PY, STEWART, IH, DEROUANE-ABD HAMID, SB, ESSAYEM, N, DEROUANE, EG, and VEDRINE, JC. *Catal. Today* **2000**, *63*, 223.

48 STERN, DL and GRASSELLI, RK. *J. Catal.* **1997**, *167*, 550.

49 SCHLÖGL, R. *Angew. Chem.: Int. Edn* **1998**, *37*, 2333.

50 WERNER, H, TIMPE, O, HEREIN, D, UCHIDA, Y, PFÄNDER, N, WILD, U, and SCHLÖGL, R. *Catal. Lett.* **1997**, *44*, 153.

51 SCHLÖGL, R, KNOP-GERICKE, A, HAVECKER, M, WILD, U, FRICKEL, D, RESSLER, T, JENTOFT, RE, WIENOLD, J, MESTL, G, BLUME, A, TIMPE, O, and UCHIDA, I. *Topics in Catalysis* **2001**, *15*, 219.

52 RESSLER, T, TIMPE, O, NEISIUS, T, FIND, J, MESTL, G, DIETERLE, M, and SCHLÖGL, R. *J. Catal.* **2000**, *191*, 75.

53 BOUCHY, C, PHAM-HUU, C, HEINRICH, B, DEROUANE, EG, DEROUANE-ABD HAMID, SB, and LEDOUX, MJ. *Appl. Catal. A* **2001**, *215*, 175.

54 RESSLER, T, WIENOLD, J, JENTOFT, RE, and GIRGSDIES, F. *Eur. J. Inorg. Chem.* **2003**, 301.

Biomimetic Nanoscale Structures on Titanium

Ralf-Peter Franke

1
Introduction

Medical diagnostic and therapeutic tools and systems can contact the body from the outside (e.g., the scanner head of an ultrasound imaging system) or from the inside (e.g., endoscopic devices). These tools and systems comprise medical devices ranging from instruments, blood bags with connective tubing, temporary implants (e.g., cardiac pacemakers), to permanent implants (e.g., joint prostheses, artificial heart valves, stents).

The importance of medical devices and procedures for the treatment of human diseases has constantly grown over the years. This is especially true for the aging populations in Europe and North America where more than 25% of the people are expected more than 65 years old by 2020. Also the incidence of coronary heart disease in the central European population exceeds 70% in those more than 65 years old. Of similar importance is the 40% incidence of osteoarthritis in the knees in people over 70 years old.

Some years ago WHO stated that coronary stenting and hip and knee reconstruction are going to be the most important medical treatments in the next 15 years. With the focus now set on regenerative therapies the medical treatments will change. On the one hand the chances to apply the growing knowledge of cellular genetics and physiology, of histodynamics and organogenesis, successfully in medical therapy get better by the day. On the other hand there is a strong movement to overcome "open access surgery" by minimally invasive surgery, which already is the gold standard in gall bladder surgery by now. Minimally invasive surgery strongly reduces the surgical access trauma and has a big potential to have best results in a patient in a minimum of time thus complying with the European "MatMed" initiative to perform medical therapies "cheaper, quicker, and better".

The Nano–Micro Interface: Bridging the Micro and Nano Worlds
Edited by Hans-Jörg Fecht and Matthias Werner
Copyright © 2004 WILEY-VCH Verlag GmbH & Co. KGaA, Weinheim
ISBN: 3-527-30978-0

2
Biocompatibility

In regenerative and substitutive therapies foreign body materials get into contact with tissues and fluids of the body. There is a general consensus that these body foreign materials need to be biocompatible and biofunctional. Biocompatibility has two sides: histocompatibility and hemocompatibility.

Histocompatibility means that medical device materials, especially those of implants, do not damage or destroy the surrounding tissues, but do integrate into them. Our understanding is that cells from the surrounding tissues must have a chance to contact the implant surface and adhere to it. These cells then should be able to bond the body tissues tightly to the implant surface so that mechanical load can be transferred between tissues and implants. There are ideas about how cells can arrive at the implant surface, how they adhere, and what conditions favor bonding of foreign surfaces to body tissues. Here, the concept of *biomimicry* plays a growing role. In the context of biomaterials, forcign body surfaces are madc cell- and tissue-friendly by creating micro- and nanoscale structures on the implant surfaces that mimic biological features such as adhesive tissue structures and receptor ligand structures.

The result of acceptable integration of the titanium mesh covered hip stem of a modular hip prosthesis is shown in Fig. 1. Cells from surrounding tissues evidently contacted the implant surface and bridge the gap between bone and implant by mineralized bone tissues. The implant thus was successfully integrated into the surrounding tissues. Other analytical methods show that there can be chemical bonding between the newly developed bone and the titanium mesh covering the hip stem.

It is necessary to remember that on the whole histocompatible surfaces can also interact strongly with bacteria so that biofilms develop on these surfaces. Great

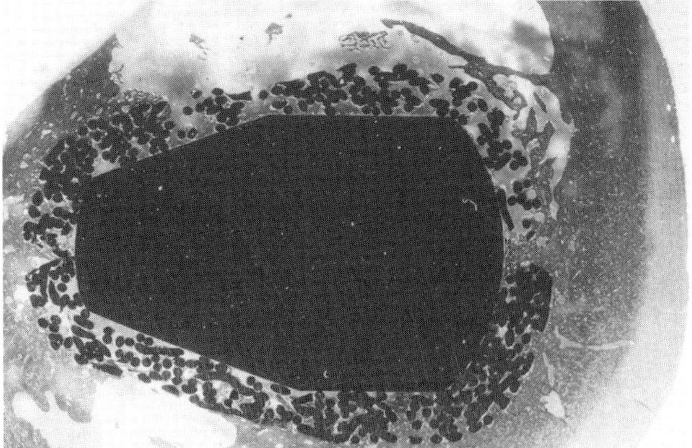

Fig. 1 Scanning electron microscopy image of the stem of a titanium hip implant integrated in the surrounding bone.

care has to be taken to avoid adherence of bacteria and the development of biofilms. Infection from contaminated implants necessitates re-operation.

Hemocompatibility creates some opposite demands compared with histocompatibility. Where histocompatibility asks for strong interaction between tissues and implant and for stable and tight bonding, hemocompatibility requires that there is no strong interaction between body fluids, especially blood, and the foreign surfaces. As with histocompatibility, of course, these materials also need to be non-toxic/non-damaging. So it is obvious that hemocompatible materials are needed for medical devices designed for the contact with blood. Blood clotting and thrombus formation must be avoided in order not to obstruct the blood flow at the region of medical interventions and not to cause thrombus induced damage in tissues farther away, such as the brain and heart.

Hemocompatibility is essential for a great number of devices such as blood bags and connective tubing/fittings for electrodes and connective wires of heart pacemakers, ultra filtration membranes in hemodialysis devices and heart lung machines, vascular prostheses, heart valve prostheses and so on.

In some cases hemo- and histocompatibility are necessary as in the case of vascular prostheses and special applications such as the implementation of missing ventricles in newborns.

Fig. 2 shows the malformed heart of a baby that died several months after the implantation of an artificial left ventricle. Histocompatibility is essential for the tight integration of the polymer graft into the myocardial tissue. Hemocompatibility is crucial for the contact of the graft with the blood pumped in the heart.

Problems with these conventional solid implants, among others, arise from the degradation of the implants. All implants are degraded in the human body more or less quickly. Often the degradation results in gaps between implant and surrounding tissue thus endangering the necessary tight integration between implant and tissues. The principles of remodeling in the interface between implant and tissues after implant degradation are not completely understood. There are emerging concepts to deal with the situation that also exploit biomimicry.

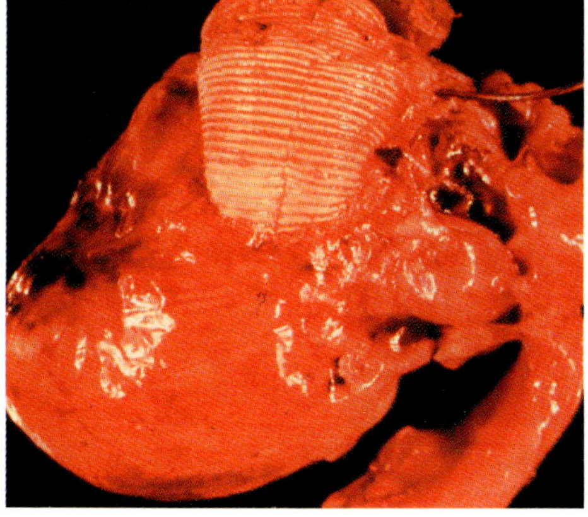

Fig. 2 Heart of a baby who died after implantation of a polymer substitute of the missing left ventricle.

Different concepts are applied to enhance the integration of implants into surrounding tissues. Many concepts apply signaling sequences derived from biology and medicine. These implants are called biohybrids. Others apply purely artificial materials of different classes (polymers, ceramics, metals), either one material alone or in combination as composites.

Assuming that the right cells move towards the biomaterial surfaces, adhere there, and start to develop the tissues necessary to couple the implant to the surrounding tissue, a new problem occurs. Will the newly developing tissues be supplied with oxygen and nutrients, as are the body's regular tissues. Clearly in regular tissues every 30–100 μm a small blood vessel will be found supplying surrounding tissues.

Fig. 3 shows the proximal part of the tibial bone, the tibia head, as most of readers probably will not have seen it before. The blood vessels, even the smallest ones, were filled with a low viscosity oligomer of a resin, which then polymerized and solidified. Afterwards the mineral parts were dissolved by mild acid treatment and the soft tissue components were enzymatically digested. The procedure is called "corrosion casting". As can be seen almost the whole tibia bone seems to be made of blood vessels. That explains why so many implants failed in the past. Newly developed interface tissues necessary to integrate the implants lacked the blood vessels needed to keep tissues alive. And bone is a truly living tissue, as can readily be seen by corrosion casting.

Another item is very important here. Only in some cases has the development of new blood vessels, neoangiogenesis, around implants been described. It is feasible that these vessels even resembled original blood vessels, but then probably were not functional. That means they were not perfused, they did not supply oxygen and nutrients. There may be many reasons for non-perfused blood vessels. A major reason is the local absence or the damage of endothelial cells, the inner lining cells of blood vessels as shown in Fig. 4.

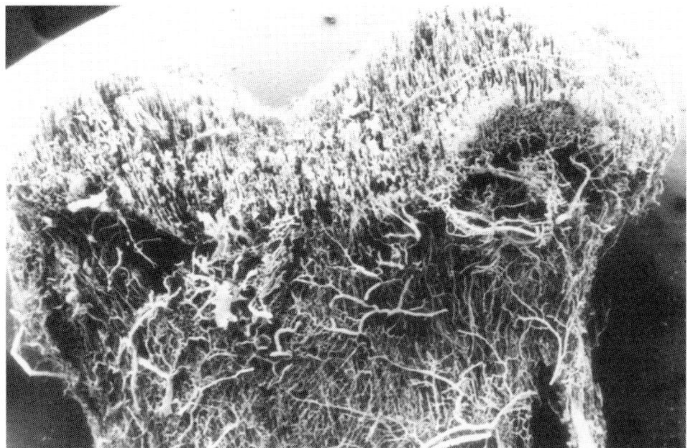

Fig. 3 Rabbit tibia head after corrosion casting of blood vessels (light microscopy).

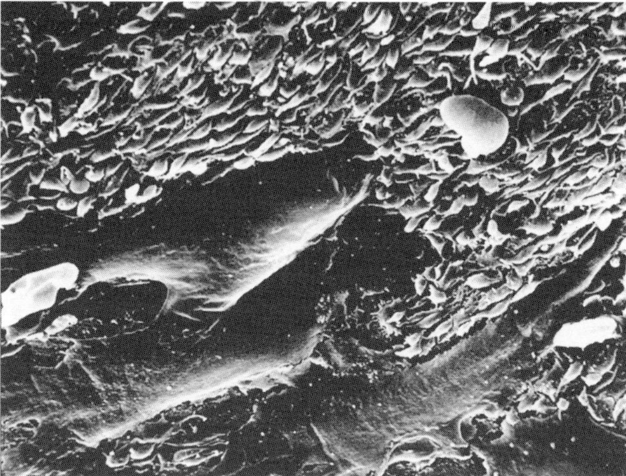

Fig. 4 Intimal surface of blood vessel *in vivo* after removal of endothelial cells by catheterization and adherence of platelets (SEM).

Absent or not functional endothelial cells cannot establish non-thrombogenicity, the crucial condition for blood flow in blood vessels.

What can be done to engineer, firstly

- functional, well-perfused interface tissues around conventional implants, and secondly
- bigger volume substitutes of tissues/organs

when the necessary knowledge to switch on the organigenetic parts of the genome is still not available?

The factors that determine tissue integration of implants are:

- materials
  - chemical composition of bulk
  - mechanical characteristics

- surface of material
  - geometrical structure
  - chemical structure
  - surface charge

- porosity
- corrosion
- degradability
- contamination (biofilms)
- micromovement
- mechanical load

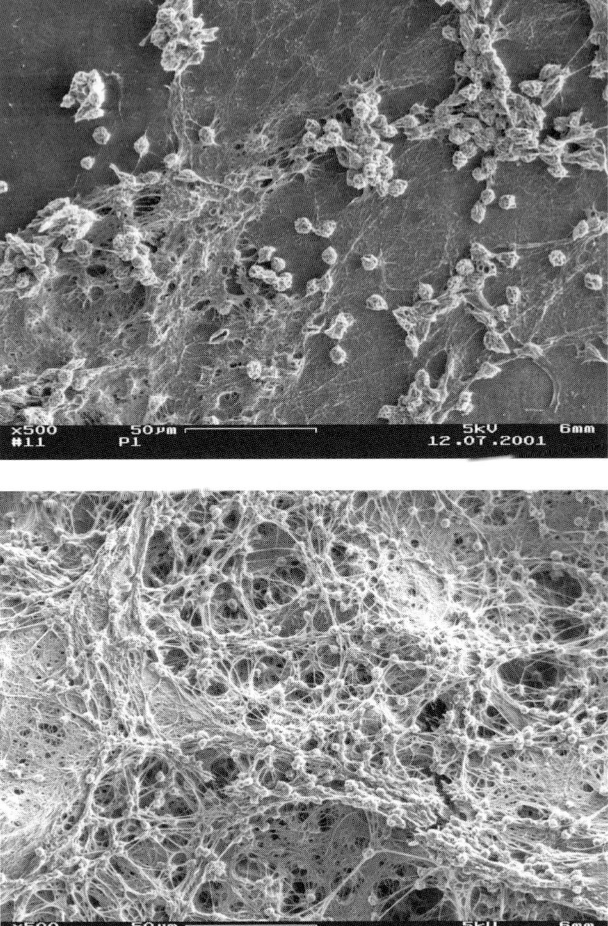

Fig. 5 Influence of implant materials with comparable chemistry and different surface roughness on tissue integration.

It is not completely understood how these determinants influence the fate of the implant in the body. Boyan et al. recently reviewed the influences of implant topography, chemical composition, surface structure, and electrical surface charge [1]. It can be assumed that a combination of determinants will exert major effects. Now, we have to be aware that during implantation the implant contacts various tissues and body fluids and will arrive at the implantation site covered with substances from these fluids.

To address this question we started to put polymer samples representing the same chemistry, but with different surface roughness, in cell culture media with serum supplements. It turned out that the rougher surfaces exhibited faster and more con-

Fig. 6 Titanium sample with machined surface topography, used for the evaluation of glycoprotein binding (fibronectin/vitronectin) to sample surface *in vitro*.

densed matter from the culture media. Samples of the same kind were also implanted in mice subcutaneously in the throat and neck regions. The rougher surfaces again drew more accretions from body fluids/blood and were covered with more cells than the smoother ones at the time of explantation, as shown in Fig. 5.

So here was a clear hint that the same chemistry, but the greater roughness really interacted with blood/body fluid components and also with cells from a host organism. The same was found thereafter with samples representing the same ceramic chemistry, but different surface roughness.

The approach was changed then, when samples from different materials were produced with the same surface roughness. The samples were machined as shown in Fig. 6 (titanium sample) and then incubated in cell culture media for different periods of time from 2 s to hours.

After such a short time already big differences in condensed matter from the fluid at the solid surfaces were evident. These differences were even stronger after one hour. We were especially interested in glycoproteins such as fibronectin and vitronectin, which can act such as glues and connect ("nectins") cells with a substrate. After immunohistochemical processing of the samples after their incuba-

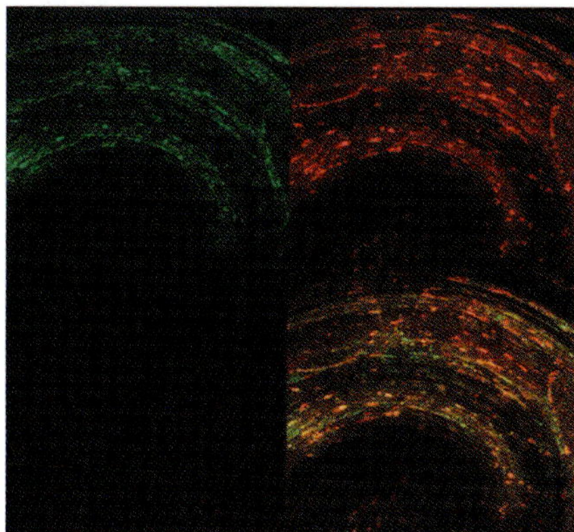

Fig. 7 Fibronectin (red) and vitronectin (green) binding to CoCr surface *in vitro*.

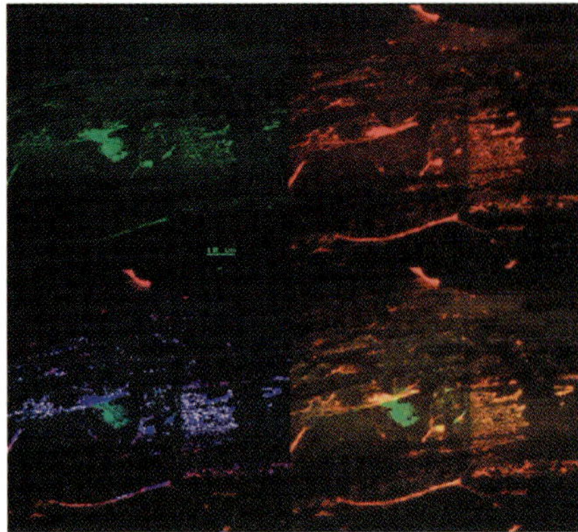

Fig. 8 Fibronectin (red) and vitronectin (green) binding to PEEK surface *in vitro*.

tion in the culture media they were viewed in confocal laser scanning microscopy (CLSM). The observed differences were striking. Fig. 7 shows fibronectin and vitronectin binding to a CoCr surface showing a stronger coincidence with the more coarse lines of the machining profile.

Different binding patterns resulted as shown in Fig. 8. In these experiments the materials containing calcium phosphates generally showed stronger and more homogenous effects than other materials.

Based on available results we implanted samples made of different materials in mice and rats, either in soft tissue environments (throat or neck regions) or opposed to bone (os temporale). In all implantation sites we found nectins in the interface tissues between implant and surrounding tissues. An expected and interesting result was that different materials opposed to bone resulted in different staining patterns of matrix fibronectin in the interface matrix of bone and implants as viewed by CLSM and shown in Fig. 9.

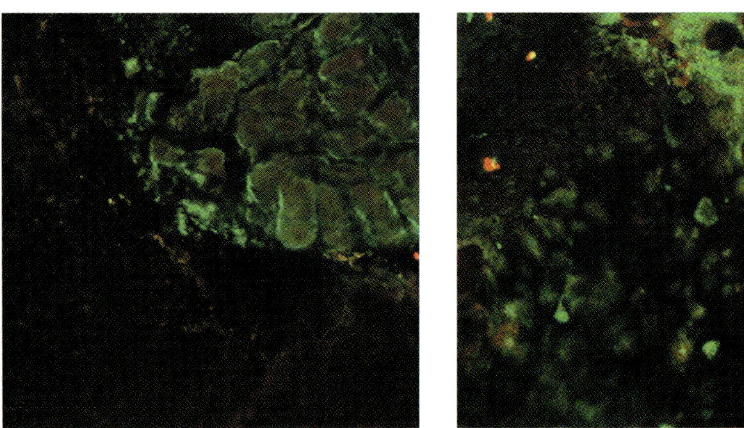

(a)   (b)

Fig. 9 Confocal laser scanning microscopy assessment of matrix fibronectin (green fluorescence) content of different materials tested *in vivo*.

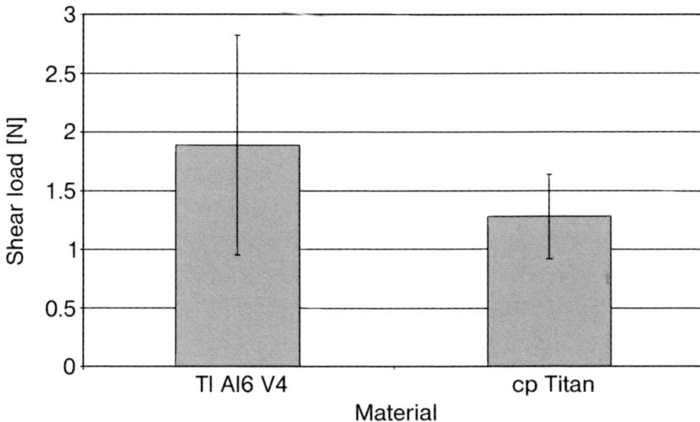

Fig. 10 Maximum shear load of bone bonded cp Ti/TiAl$_6$V$_4$ samples after explantation out of wild type NMRI Mice.

Then we wanted to know whether these differences in interface matrix composition had a meaning for the integration of implants in tissues especially in bone tissues. The head bones with opposed material samples were explanted after the sacrifice of the mice and mounted on a mechanical testing device to assess the shear strength of the samples bonded to the bone. As shown in Fig. 10 and Tab. 1, TiAl$_6$V$_4$ was a material with a slightly higher microroughness, which resulted in a slightly higher shear strength in the head bone of wild type NMRI mice.

It is important to note that surface roughness is not only modified by binding of constitutes of body fluids/blood, but also by contact with bone as shown in Fig. 11.

In the case of contact between head bone and TiAl$_6$V$_4$ the macroroughness (measured with profilometry) became less whereas microroughness (measured with AFM on enzymatically surface digested post explant bone) increased.

With all these hints on the importance of microstructures ("microroughness") on the integration of implants in tissues and of course aware of the sizes and grouping of cell-membrane-placed receptor structures we looked for ways to place attractive chemical molecules in micrometer-sized groups with a nanometer scale

Tab. 1 Assessment of surface roughness by means of atomic force microscopy (AFM) and of profilometry (chromatographic aberration).

Material	Method	New R_a [μm]	After explantation R_a [μm]
cp Ti	Profilometry	0.829	0.788
	AFM	0.0356	0.0401
TiAl$_6$V$_4$	Profilometry	0.934	0.806
	AFM	0.0713	0.0427

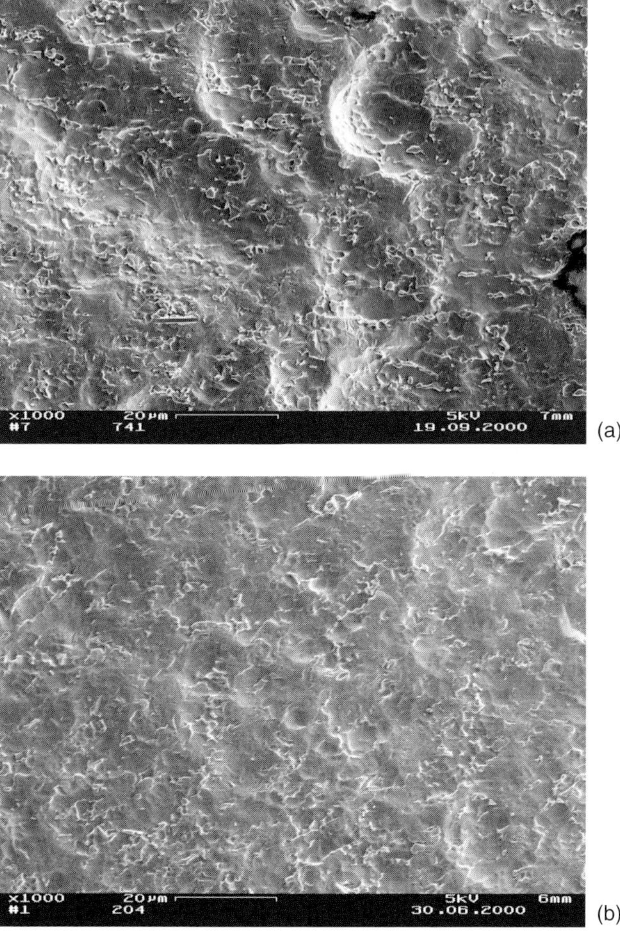

Fig. 11 Surface structure of TiAl$_6$V$_4$ samples before (a) and after (b) implantation at the os temporale of wild type NMRI mice.

structure of molecules within the groups. All this should happen in a tissue-friendly environment. As published recently [2, 3] there are ways to establish microstructured groups/clusters of molecules with different nanoscale structures depending on the slow (Fig. 12) or the fast evaporation (Fig. 14) of aqueous nanosphere suspensions where the nanoscale structure can be assessed by AFM as shown in Fig. 13.

It is fascinating to select appropriate attractive chemicals and try to make these cell specific and possibly specific for certain differentiation routes of cells, so that a certain histodynamical development of substitutes tissues can be determined.

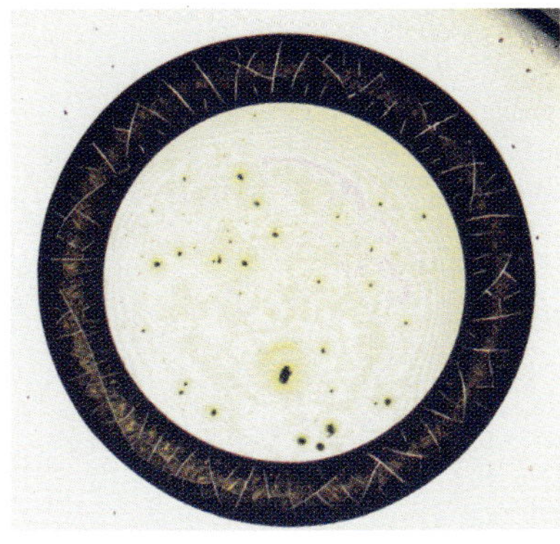

Fig. 12 Light microscopy of slow evaporation of a 60 nm aqueous nanosphere suspension on mirror polished titanium.

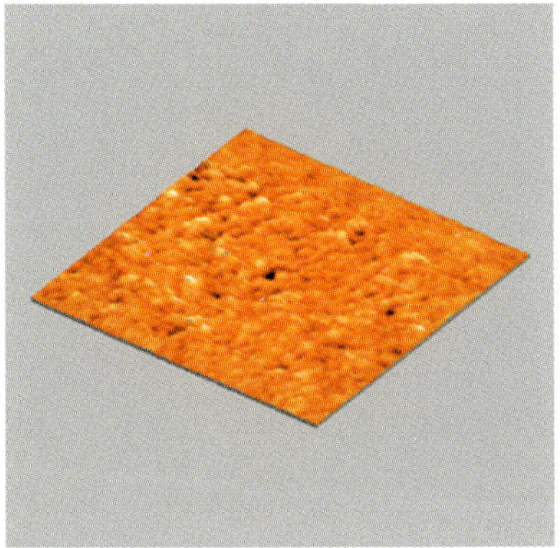

Fig. 13 AFM scan of nanosphere suspension on mirror polished titanium in the central field (within the ring structure) of 8×8 μm side length.

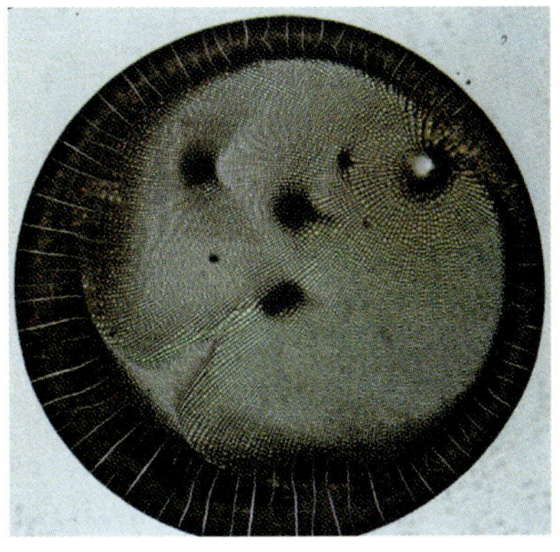

Fig. 14 Light microscopy of fast evaporation of a 60 nm aqueous nanosphere suspension on mirror polished titanium.

References

1 K. Kieswetter, Z. Schwartz, D. D. Dean, B. D. Boyan, *Crit. Rev. Oral Biol. Med.* **1996**, *7*, 329–345.
2 A. Sommer, R. P. Franke, *Nanoletters* **2003**, *3*, 321–324.
3 A. Sommer, R. P. Franke, *Nanoletters* **2003**, *3*, 573–575.

Microwave-Driven Hydrothermal Synthesis of Oxide Nanopowders for Applications in Optoelectronics

Witold Lojkowski, Agnieszka Opalinska, Tomasz Strachowski, Adam Presz, Stanislaw Gierlotka, Ewa Grzanka, Bogdan Palosz, Wieslaw Strek, Dariusz Hreniak, Larisa Grigorjeva, Donats Millers, Federica Bondioli, Cristina Leonelli, and Edward Reszke

1
Introduction

Nanocrystalline powders for optoelectronic applications have attracted much interest. It is expected that when the size of the particles becomes smaller than a characteristic interaction length for the physical processes controlling luminescence, new properties will be observed, and thus new physical phenomena with potential for practical applications discovered. In semiconductor particles, specific effects such as a blue shift of luminescence can be observed when the crystal size is comparable to the exciton radius [1–3]. In oxide nanopowders, an established fact is the effect of grain size on the relaxation times of the luminescence centers caused by their interactions with surfaces [4]. For such non-conducting materials the size effects are connected with the diffusion range of the excited states compared with the particle radius. Furthermore, in small particles an increased lattice relaxation around luminescence centers may be possible, leading to changes of relaxation mechanism [4] or possibility to accommodate different valence states from bulk crystals [5]. The effect of the grain size on the relaxation time could be exploited in optoelectronic devices or X-ray scintillators [4, 6].

For applications it is necessary to develop non-expensive methods for the production of nanocrystalline powders with controlled grain sizes and chemical compositions and their subsequent compaction. There is a number of techniques for producing nanometric oxide materials such as sol-gel [7], hydrothermal [8], co-precipitation [9], combustion [10], or gas-phase condensation methods [11]. One of the novel techniques for preparing nanopowders with a narrow distribution of grain-size is a microwave (MW) hydrothermal method [12]. In this chapter we show in more detail this technique and the luminescence and scintillating properties of the resulting powders.

Application of microwaves (MW) in chemical synthesis has recently attracted considerable attention [13, 14]. The advantage of using MW as an energy source is primarily the possibility of carrying out the processes at a much higher rate than during conventional heating. In addition, reactions have been reported that are impossible to perform without microwaves, where the energy is transferred directly to the particles suspended in a solution [15].

The Nano–Micro Interface: Bridging the Micro and Nano Worlds
Edited by Hans-Jörg Fecht and Matthias Werner

In this chapter we give examples of production of luminescent nanocrystalline powders with controlled grain size, and the effect of the grain size on their luminescence properties.

2
Experimental Methods

2.1
The Reactor for the Synthesis of Nanopowders

The reactions were carried out in a microwave reactor (Ertec, Poland) where the microwave energy is transmitted to approximately 100 ml of an aqueous solution of the reagents contained in a tight Teflon container. The container can withstand an interior pressure up to 10 MPa because it is inside a steel pressure vessel. The power dissipated in the fluid can be coupled to a pressure or temperature sensor and reaches 400 W. Therefore a power density of on average up to 4 W/mL is achieved uniformly in the reaction fluid. The system allows the temperature of the fluid to rise to 250 °C, which is a typical reaction time, in 5 min or less. Fig. 1 shows a typical thermogram of a reaction. Such fast and uniform heating of the fluid would not be possible by conventional heating. We like to stress that the system makes it possible to carry out hydrothermal synthesis in high-purity conditions and with controlled well process parameters, among the most important which is precise control of the reaction time.

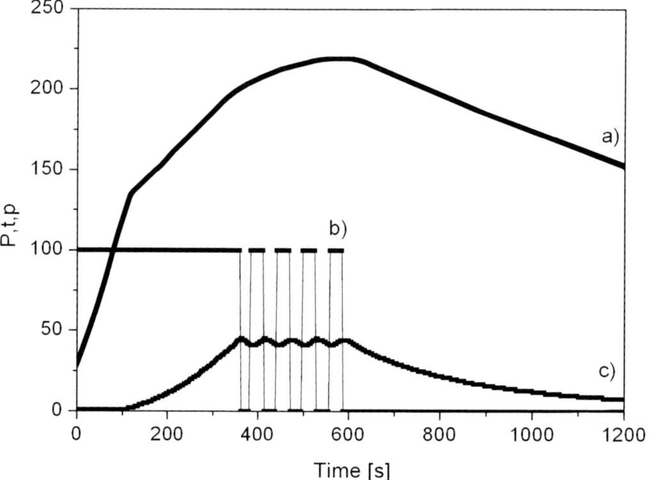

Fig. 1 A typical thermogram of the process of synthesis in the microwave high-pressure reactor: (a) temperature, (b) power, which in this experiment is coupled to the pressure, (c) pressure.

2.2
Hydrothermal Synthesis of ZnO, ZrO$_2$, and Zr$_{1-x}$Pr$_x$O$_2$

The microwave hydrothermal synthesis method [16] and the modification using a high-pressure microwave reactor [12, 17]. The typical ramp time to a pressure of 4 MPa was 5 min and the cooling down time was 10 min. When the reaction was completed the solid and solution phases were separated by filtering and the solids were washed free of salts with distilled water and isopropanol. One run produced approximately 0.5 g of powder.

Zinc oxide was prepared from a zinc chloride (0.1 M) solution by the addition of urea (weight ratio to zinc chloride: 1:2, 1:3, 1:4, and 1:5) [18].

Zirconia powders containing praseodymium in the 0–18 mol-% range were obtained by adding praseodymium(III) nitrate (Pr(NO$_3$)$_3$H$_2$O, Carlo Erba) to a 0.5 M ZrOCl$_2$ aqueous solution. The solutions were neutralized with NaOH 1 M to pH 10:40 ml [12]. Tab. 1 gives the synthesis parameters of the samples used in subsequent investigations of luminescence decay times.

Tab. 1 Specific surface area calculated from BET data for 0.05% mol Pr^{3+} ZrO$_2$ powders (S$_{BET}$, specific surface area; d, average size of crystallites).

Synthesis conditions	S$_{BET}$ [m^2/g]	d [nm]
t = 30 min	100.3	10.5
t = 40 min	82.7	13.0
t = 60 min	78.6	13.5
t = 60 min and T = 500 °C	63.1	17.0
t = 60 min and T = 600 °C	46.0	23.0
t = 60 min and T = 700 °C	20.2	53.0
t = 60 min and T = 800 °C	17.4	61.5

Variables of synthesis: t, time under pressure in autoclave (5.5 atm) and T, annealing temperature after the hydrothermal process.

2.3
Characterization of the Powders

Dry powders were characterized by X-ray diffraction (XRD), using a Siemens D5000 diffractometer. The mean size and morphologies were assessed by scanning electron microscopy (SEM, model LEO 1530) and transmission electron microscopy (TEM, model JEM 2010, JEOL). For TEM studies the powders were dispersed in distilled water and a drop of suspension was placed onto a copper grid fitted with a transparent polymer and then dried. The grain size was assessed using three methods:

- the nitrogen adsorption method (BET, Model Gemini 2360, Micromeritics Instruments Corp.) where the average grain diameter d_{BET} was calculated by means of the equation $d_{BET}=6/(DS_{BET})$ where S_{BET} is the specific surface of the powder and D is the density,
- from the broadening of XRD peaks using the Debye-Scherrer formula [18], and
- using the equation derived by Pielaszek [19], which also evaluates the grain size distribution of nanopowders.

The pore size distribution in the powders was measured using the nitrogen desorbtion method.

2.4
Sol-Gel Synthesis of YAG doped with 1% Nd

YAG ($Y_3Al_5O_{12}$) nanopowders co-doped with 1% mol Nd^{3+} and 5% mol Yb^{3+} ions have been synthesized using the Pechini method [7].

2.5
Investigations of Luminescence Properties

Photoluminescence spectra were measured at room temperature with a OceanOptics Spectrometer SD2000 with a resolution of 0.3 nm. Emission decays were measured using Jobin-Yvon TRW 1000 spectrophotometer and a photomultiplier (Hamamatsu R928) by means of a Tektronics TDS 380 oscilloscope. The 308 nm line of excimer laser (LPX100) was used as the excitation source.

For the study of the scintillating properties, each sample was prepared by slightly pressing the nanocrystalline powders into a small stainless steel cell. The cell with the sample was placed in a vacuum chamber. The luminescence decay kinetics was recorded for different wavelengths over the whole spectrum. Luminescence was excited by a pulsed electron beam (270 keV, 10 ns) and analyzed using the grating monochromator (MDR-2, LOMO), detected by photomultiplier tube and output signal was displayed on Tektronix 5052 digital oscilloscope. The time-resolution of the equipment was ~ 15 ns.

2.6
Sintering

Sintering experiments were carried out using a high-pressure system with a toroid type high-pressure anvils. Fig. 2 shows a diagram of the experiment. The powders to be sintered were annealed in order to remove the gas absorbed at the surfaces, compacted at room temperature into pellets and subsequently inserted in a graphite tube of 5 mm innernal diameter. The graphite tube serves as a heater. The set up is inserted into a gasket and introduced between the anvils of the high-pressure press. The system is calibrated so that the temperature is set as a

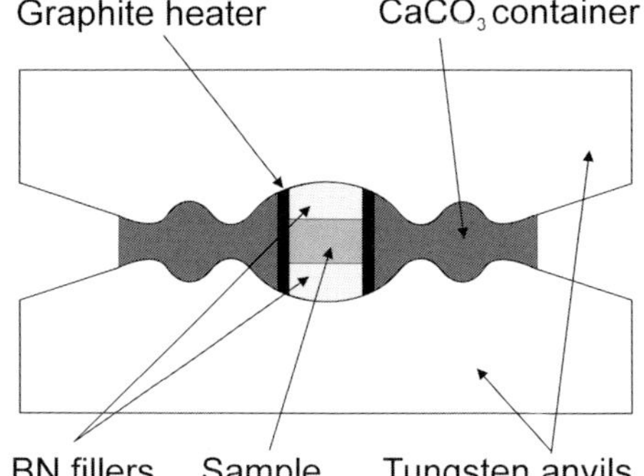

Fig. 2 Diagram of the arrangement of the sample, furnace, and gasket in the high-pressure vessel for high-pressure sintering.

function of current flowing between the upper and lower anvil. The powders are sintered under pressures up to 8 GPa and at temperatures up to 1500 °C. The procedures are described in detail elsewhere [20].

3
Results and Discussion

3.1
The Properties of the Powders

3.1.1 ZrO$_2$ and Zr$_{1-x}$Pr$_x$O$_2$ Powders

ZrO$_2$ powders with average grain size 8 and 10 nm were obtained for the reaction pressure 5 and 8 MPa, respectively. Fig. 3 shows the effect of the reaction pressure on the degree of the crystallinity of the powders. The crystallinity is defined as the weight fraction of the crystalline oxide in the reaction product. The degree of crystallinity increases with annealing temperature after the synthesis, which is presumably connected with evaporation of hydroxides. Since hydroxyl groups lead to a non-radiative decay of luminescence, increasing the reaction pressure leads to powders with higher quality for optoelectronic applications. Fig. 4a shows a TEM image of the powder with 1% Pr produced under 4.8 MPa and Fig. 4b shows the powder produced under 8 MPa pressure. Both powders show a similar shape, with some facets and rounded corners, but the powder produced under 8 MPa has a somewhat larger dimension. Although the degree of crystallinity is a continuous function of the pressure of synthesis, the grain size is not. Fig. 5 shows an increase of the grain size from approximately 7 to 11 nm (as measured from XRD peaks broadening) in the pressure range 4–5 MPa, and a stabilization of the grain size above that pressure. It follows that most likely the increase of the degree of crystallinity takes place by transformation of the structure of some grains with increasing reaction temperature and pressure, without substantial grain growth. This crystallization process is accompanied by formation of the monoclinic phase, which is not stable in the particles with grain size below 6 nm produced at

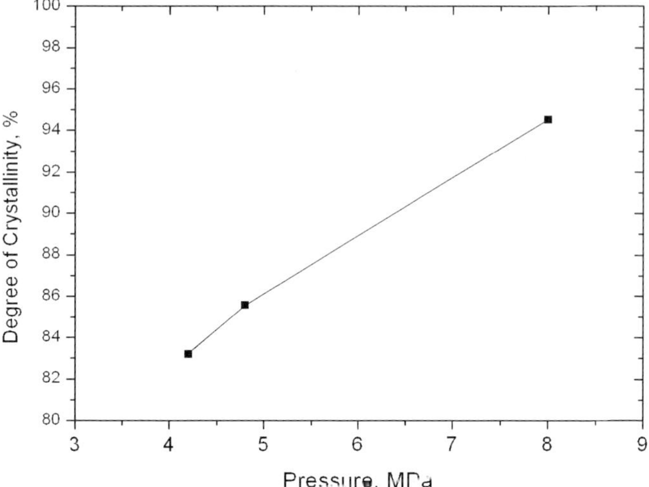

Fig. 3 The effect of synthesis pressure on the degree of crystallinity for the ZrO$_2$ sample doped with 1 mol-% Pr.

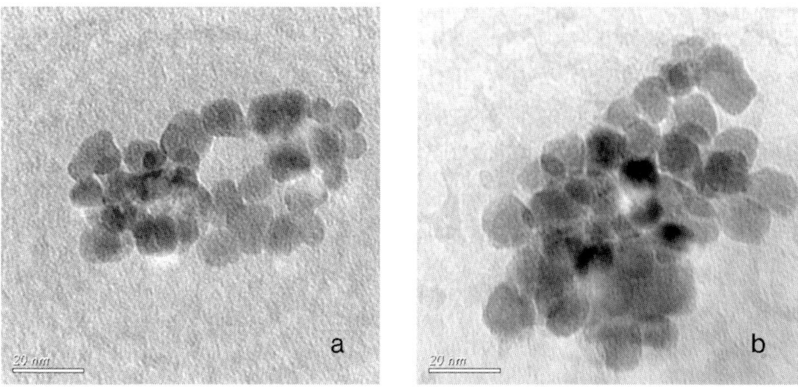

Fig. 4 TEM images of ZrO$_2$: 1 mol-% Pr powders produced at (a) 4.8 and (b) 8 MPa.

4.2 MPa but can be observed in particles with grain size of about 10 nm (Fig. 6). The synthesis at pressures above 4.5 MPa leads to a two phase product, with a monoclinic and tetragonal phase. The ratio of both phases depends on the grain size, which in turn depends on the Pr content and synthesis conditions. Fig. 7 shows the grain size distribution for both phases for a sample produced under pressure of 5.5 MPa. It is seen that both phases have a similar grain size. This indicates that there is a mixture of powders with different phase structure rather than two phases in single grains. The average grain size of the tetragonal phase is 10 nm. Fig. 8 shows the pore size distribution in powders produced at 5.5 MPa for

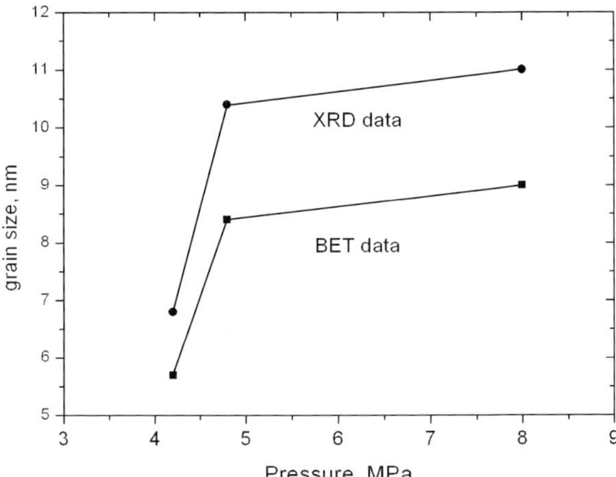

Fig. 5   The grain size as a function of pressure of synthesis for ZrO_2 doped with 1 mol-% Pr.

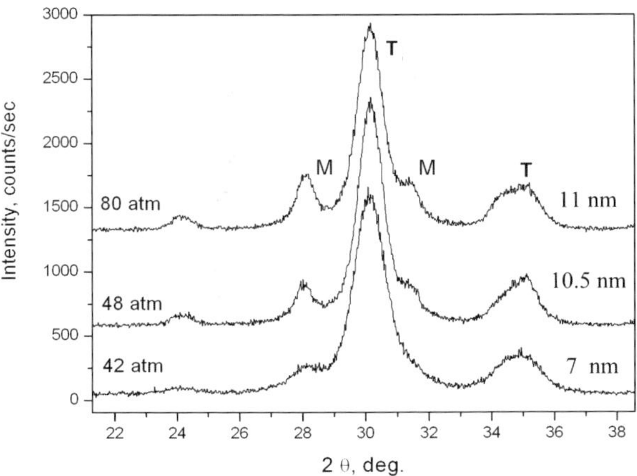

Fig. 6   XRD spectra as a function of the synthesis pressure for ZrO_2 doped with 1 mol-% Pr: M indicates the location of the peaks corresponding to the monoclinic structure and T the tetragonal one. On the right side the grain size determined from XRD data is indicated.

15 min and annealed at 500 °C for 30 min as measured by the nitrogen desorption method. The pore size distribution is indicative of the grain size distribution. Both methods lead to consistent results: for a grain size of about 10 nm the half height width of the grain size distribution peak is about 8 nm.

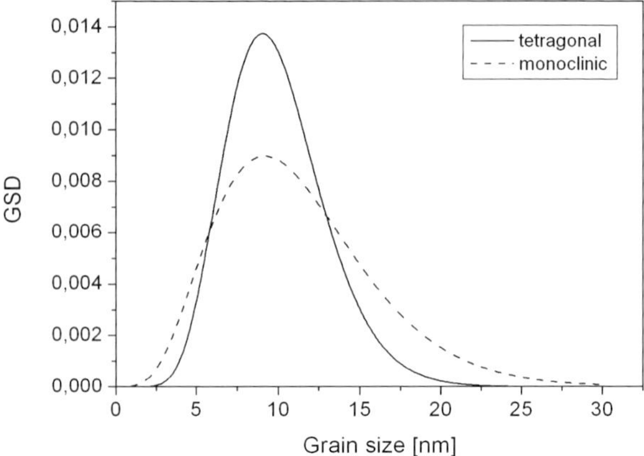

Fig. 7 Grain size distribution in the ZrO_2 sample produced at 5.5 MPa.

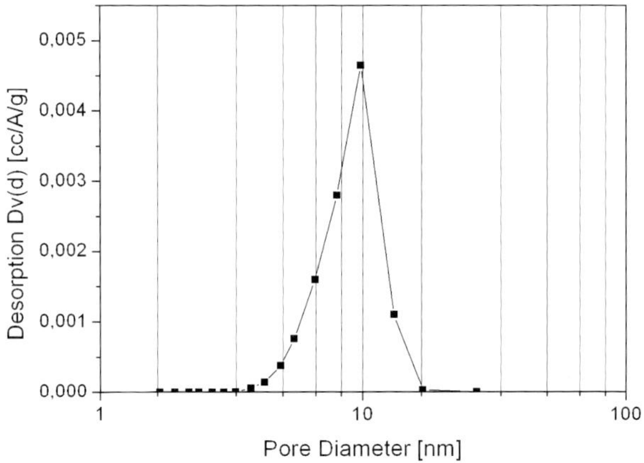

Fig. 8 The pore size distribution in the ZrO_2 powder doped with 1 mol-% Pr and annealed at 500 °C in air for 30 min.

The relative content of both phases depends on the Pr content and synthesis conditions. Fig. 9 shows the XRD of samples produced at 5.5 MPa for 15 min for three Pr contents: 0.1, 1, and 2 mol-%. With increasing Pr content the amount of monoclinic phase decreases and of the tetragonal phase increases, as already observed [21]. However, the effect of Pr content can be an indirect one, that means an increased Pr content slows down the grain growth and in smaller grains the tetragonal phase of ZrO_2 is stable.

Fig. 10 shows the effect of annealing temperature and Pr content on the grain size. For Pr contents above 1 mol-% the grain growth rate is substantially decreased comparing to samples with a small Pr content. Fig. 11 shows the microstructure of a powder with 0.5% Pr produced at 5.5 MPa and after the thermogra-

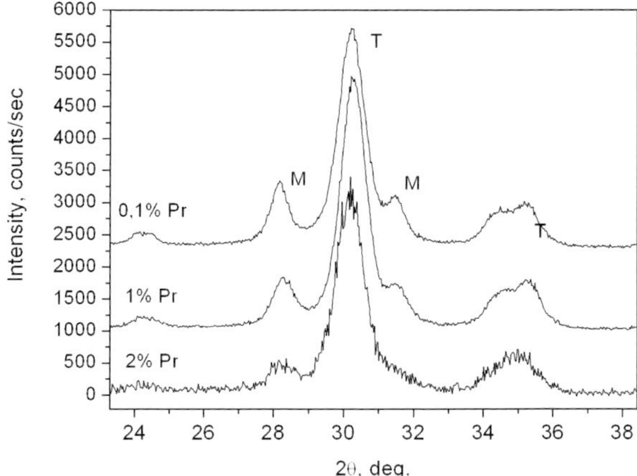

Fig. 9 XRD of samples produced at 5.5 MPa for 15 min for three Pr contents: 0.1, 1, and 2 mol-%.

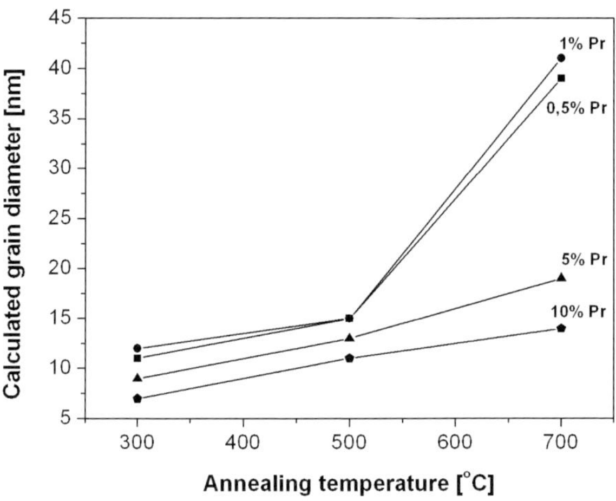

Fig. 10 Effect of annealing on the grain size of $Zr_{1-x}Pr_xO_2$ as a function of Pr content. The grain size was estimated from BET measurements.

vimetry run up to 1000 °C. It is seen that the SEM images correspond qualitatively with the results of X-ray and BET investigations as far as the narrow size distribution. This shows that the grain size of ZrO_2 powders and their phase composition can be controlled by varying the preparing conditions, annealing temperature and Pr content.

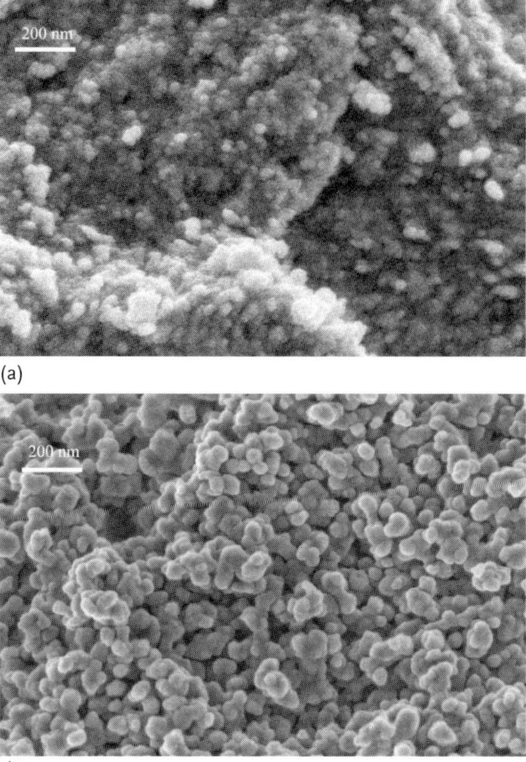

(a)

(b)

Fig. 11 SEM images of the powders with 0.5 mol-% of Pr: (a) after synthesis at 5.5 MPa and (b) subsequent annealing at 1000 °C for 30 min.

3.1.2 ZnO Powders

Fig. 12 shows the effect of reaction conditions on the grain size of ZnO nanopowders. A nucleation period of 4 min is seen. During the next 2 min the grain size increases from approximately 20 to 120 nm. Hence a precise control of the reaction time permits to control the grain size.

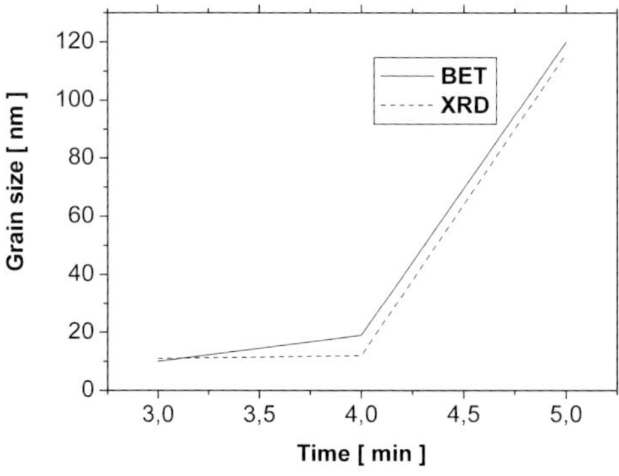

Fig. 12 Effect of reaction time on the average grain size in ZnO. The grain size was estimated from BET and X-ray diffraction measurements.

Fig. 13 SEM images of agglomerates of zinc oxide powders after 3 min of synthesis.

The particles build large aggregates in form of self-assembled flowers (Fig. 13).

3.2
Investigations of Luminescence

3.2.1 Luminescence of $Zr_{1-x}Pr_xO_2$

The emission spectra and emission decay curves were recorded under UV excitation ($\lambda_{exc} = 308$ nm). Fig. 14 shows the emission spectra as a function of the synthesis conditions. The weak blue ($^3P_0 \rightarrow {}^3H_4$) and strong red ($^1D_2 \rightarrow {}^3H_4$) lumi-

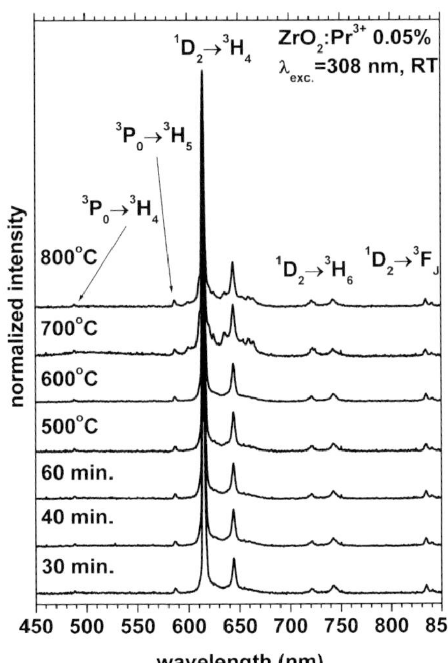

Fig. 14 Room temperature emission spectra from samples of ZrO_2 0.05 mol-% Pr prepared under different conditions, given in Tab. 1.

nescence from Pr^{3+} ions were observed for all samples. The red luminescence decreases with increasing concentration of Pr^{3+} ions. This is caused by concentration quenching due to the $[^1D_2,^3H_4] \rightarrow [^1G_4,^3F_{3,4}]$ cross-relaxation process [5]. Fig. 15 shows the effect of grain size of the particles on the relaxation time for the $(^1D_2 \rightarrow {}^3H_4)$ in the samples with 0.05 mol-% Pr. It is seen that with decreasing grain size the luminescence decay time decreases. This can be explained by a decrease in the density of the defects in bigger grains and removal of hydroxyl group and, in consequence, a lower probability for non-radiative transitions due to OH vibrations. For samples heated above 700 °C, additional weak emission bands appear around 600–670 nm (Fig. 14) corresponding to Pr^{3+} sites in monoclinic structure of ZrO_2. It is important that the samples heated in air at 800 °C still show very intensive long lifetime emission of Pr^{3+}.

The results of investigations of luminescence induced by the 270 kV electron beam correspond well with the above observations. The luminescence decay kinetics displays two stages. One corresponding to ~ 10–20 ns and second one to about 1 μs. For low concentrations of Pr the long time relaxation is observed only for grain size above 50 nm. This suggests that the luminescence center is responsible for slow luminescence decay in ZrO_2:Pr is Pr^{3+}. The incorporation of Pr^{3+} in the ZrO_2 lattice needs a charge compensation and this may occur more easily on the nanocrystal surface. This unexpected stabilization of Pr^{3+} in ZrO_2 structure has been observed only for nanocrystalline material produced by this method and it is

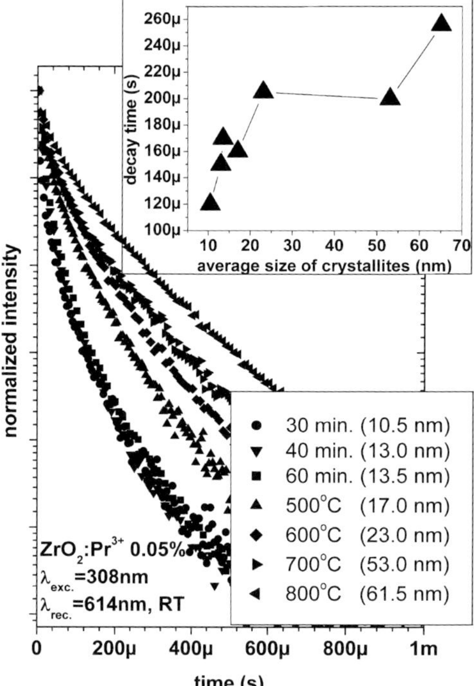

Fig. 15 Decay time curves of ZrO_2 0.05 mol-% Pr samples as a function of grain size or preparation conditions, as given in Tab. 1.

absent in the $Pr:ZrO_2$ solid solutions with large crystal size [22]. Furthermore, the segregation of Pr^{3+} on the surface of the crystals may be one of the reasons of their smaller growth rate comparing to pure ZrO_2.

3.2.2 Cathodoluminescence of ZnO

The spectra of luminescence excited by a pulsed electron beam shows two main bands peaking at ~ 620–650 nm and ~ 375–390 nm for all ZnO nanocrystals studied. The band at ~ 620–650 nm is known as due to defects in the oxygen sub-lattice, possibly vacancies. The band at ~ 375–390 nm is from radiative decay of excitons. It was observed in our experiments that the luminescence intensity ratio between these bands strongly depends on the nanocrystals production technology.

The band in the range 375–390 nm is complex (Fig. 16) with at least two sub-bands contributing to it. For all the powders investigated, with grain size 10, 20, 30, and 40 nm, a main band is observed at ~ 375 nm. With a decrease in grain size from 50 to 10 nm, a sub-band appears at ~ 390 nm. Both bands are presumably of excitonic origin. The ratio of the luminescence intensities for these two sub-bands depends on the nanocrystals size. The relatively high intensity of luminescence from the sub-band at ~ 390 nm is observed for the smallest size nano-

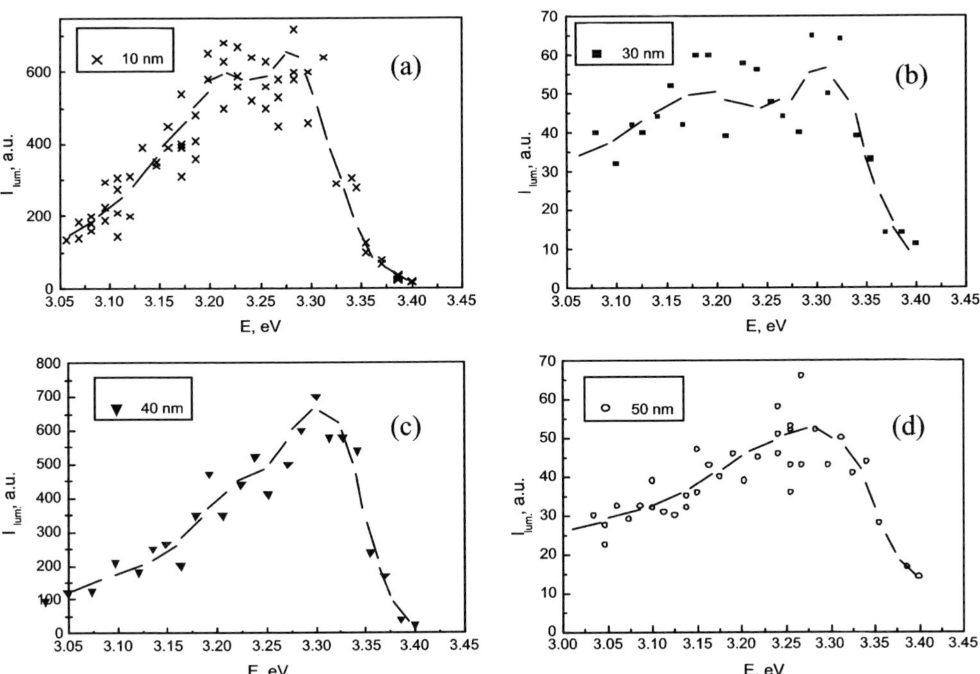

Fig. 16 Cathodoluminescence spectra of ZnO as a function of grain size: (a) 10 nm, (b) 30 nm, (c) 40 nm, (d) 50 nm.

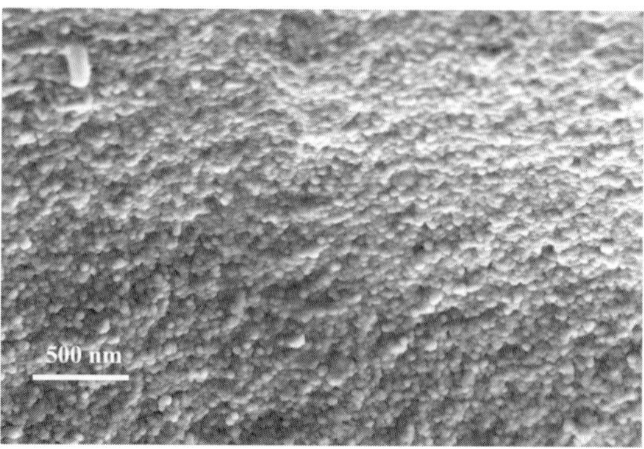

Fig. 17 SEM image of the fracture surface of the sintered semi-transparent sample made of YAG nanopowders co-doped with 1 mol-% Nd and 5 mol-% Yb.

Luminescence of (Nd,Yb):YAG under 514.5 nm

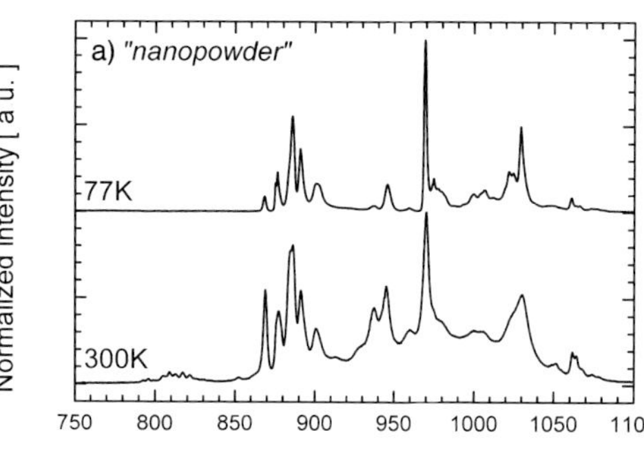

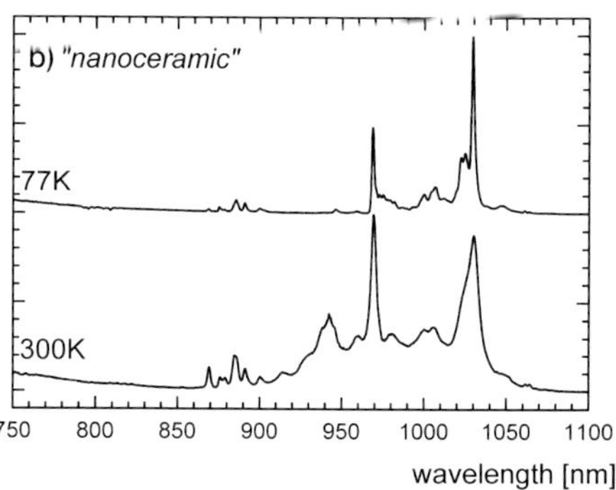

Normalized intensity [a u.]

Fig. 18 Room and liquid nitrogen temperature emission spectra of YAG samples co-doped with 1 mol-% Nd and 5 mol-% Yb: (a) before sintering, (b) after sintering.

crystals. Their intensity is the same for grain size 10 nm. We suppose that the observed bands are of excitonic nature, and the $\sim$390 nm sub-bands is connected with excitons bound to surface defects. The decay kinetics of the exciton luminescence is very fast. Previously [23] it was shown that the life time is 0.4 ns and therefore decay is faster than the time resolution of the equipment used in our experiments, which is about 10–15 ns.

Our experiments have shown that the content of defects responsible for luminescence at $\sim$620–650 nm depends on the nanocrystals production method. As far as the luminescence at $\sim$375–390 nm is concerned, it displays a contribution from bound excitons, which increases with decreasing grain size.

3.3
Luminescence of Sintered and Not Sintered YAG Nanocrystals

A semi-transparent sample was made from YAG nanopowders co-doped with 1 mol-% Nd and 5 mol-% Yb by hot-pressing at 7.7 GPa and at temperatures below 1000 °C. Fig. 17 shows the fracture surface of the sample after sintering. It is seen that the grain size in the sintered sample is about 50 nm. Fig. 18 shows the luminescence spectra for both a non-sintered powder pellet and after sintering. The luminescence spectrum of the sintered material is similar to that observed for YAG single crystals. As far as the powders are concerned, a band in the wavelength range 800–900 nm is observed, which is caused by a temperature rise in the isolated particles due to a poor thermal conductivity of such a pellet. Additionally, much efficient energy transfer from Nd^{3+} to Yb^{3+} ions in nanoceramic compared to nanopowder has been observed. A detailed analysis of such behavior needs further complementary research to study the luminescence decay measurements and time-resolved spectroscopy.

4
Conclusions

The microwave driven hydrothermal synthesis makes it possible to control the reaction times precisely and thus also the grain size during synthesis of ZnO. An increase of pressure leads to powders with less hydroxide groups comparing to low temperature/pressure synthesis routes. The combination of the two techniques permits to best exploit their advantages: high temperature of the process and fast heating and cooling, in high-purity conditions.

Example of ZrO_2 doped with Pr shows that the luminescence centers in nanocrystalline powders may have a different structure from bulk materials. Pr^{3+} ions, which would not be stable in a bulk material, are stable by being associated with the surface of the powder particles, and influence both their growth rate and luminescence properties. Similarly, excitoning states bound to the defects near surfaces of ZnO nanocrystals may influence the luminescence spectra for grain sizes of about 10 nm.

Interaction of the excited states with surfaces leads to very short luminescence decay rates of about 10 ns, which in addition can be controlled by varying the grain size. This opens perspectives for new scintillating materials with short and controlled relaxation times.

The nanopowders can be sintered using high-pressure techniques and the grain size can be preserved in the nanometer range. The sintered YAG : Nd,Yb ceramics displays similar luminescence spectra to single crystals.

Acknowledgements

The work was carried out within the Network of Centers of Excellence Nanostuctured Materials" and the European Commission Excellence Centers "CAMART" (Contract No ICA1-CT-2000-7007) and High Pressure (Contract No ICA1-CT-2000-7005) and COST action D10. Studies were partially supported from the Polish Committee for Scientific Research (KBN) under Grant No. 4 T08A 046 22 and Latvian Council of Sciences, grant 01.0813 and UNESCO for fellowship grant according to letter No. 3250003426. One of us (D. Hreniak) is holder of the scholarship of the Foundation for Polish Science (FNP). The authors are grateful to Prof. Horst Hahn for the measurements of nitrogen desorption.

References

1 M. L. Steigerwald, A. P. Alivisatos, J. M. Gibson, T. D. Harris, R. Kortan, A. J. Muller, A. M. Thayer, T. M. Duncan, D. C. Douglass, and L. E. Brus, *J. Am. Chem. Soc.* **1988**, *110*, 3064.

2 A. Fojtik, H. Weller, U. Koch, and A. Henglein, *Ber. Bunsenges. Phys. Chem.* **1984**, *88*, 969.

3 M. Zacharias, L. X. Yi, J. Heitmann, R. Scholz, M. Reiche, and U. Gösele, *Solid State Phenomena* **2003**, *9*, 95.

4 U. Herr, H. Kaps, and A. Konrad, *Solid State Phenomena* **2003**, *9*, 85.

5 A. Opalinska, D. Hreniak, W. Lojkowski, W. Strek, A. Presz, and E. Grzanka, *Solid State Phenomena* **2003**, *9*, 141.

6 D. Millers, L. Grigorjeva, A. Opalinska, and W. Lojkowski, *Solid State Phenomena* **2003**, *9*, 135.

7 D. Hreniak and W. Strek, *J. Alloys Comp.* **2002**, *346*, 183.

8 M. Zawadzki, J. Wrzyszcz, W. Strek, and D. Hreniak, *J. All. Comp.* **2001**, *323/324*, 279.

9 W. Yuren, L. Kunquan, W. Dazhi, W. Zhonghua, and F. Zhengzhi, *J. Phys.: Condens. Matter* **1994**, *6*, 633.

10 E. Zych, D. Hreniak, and W. Strek, *J. Phys. Chem. B* **2002**, *106*, 3805.

11 D. K. Williams, H. Yuan, and B. M. Tissue, *J. Luminescence* **1999**, *83/84*, 297.

12 F. Bondioli, A. M. Ferrari, S. Braccini, C. Leonelli, G. C. Pellacani, A. Opalinska, T. Chudoba, E. Grzanka, B. Palosz, and W. Lojkowski, *Solid State Phenomena* **2003**, *9*, 193.

13 M. Willert-Porada (Ed.) *Application of Microwaves in Materials Science, Chemical Processing and Solid State Chemistry*, Shaker, Aachen, **1998**.

14 S. Komarneni, M. C. D'Arrigo, C. Leonelli, G. C. Pellacani and H. Katsuki, *J. Am. Ceram. Soc.* **1998**, *81*, 3041.

15 R. Kerner, O. Palchik, and A. Gedanken, *Chem. Mater.* **2001**, *13*, 1413.

16 F. Bondioli, A. M. Ferrari, C. Leonelli, C. Siligardi, and G. C. Pellacani, *J. Am. Ceram. Soc.* **2001**, *84*, 2728–2730.

17 T. Strachowski, E. Grzanka, B. Palosz, A. Presz, L. Slusarski, and W. Lojkowski, *Solid State Phenomena* **2003**, *9*, 189.

18 H. P. Klug and L. E. Alexander, *X-Ray Diffraction Procedures for Polycrystalline and Amorphous Materials*, Wiley-Interscience, **1974**.

19 R. Pielaszek, PhD Thesis *Diffraction Studies of Nanocrystals Subjected to High Pressure*, Faculty of Physics, Warsaw University, October **2002**.

20 E. A. Ekimov, A. G. Gavriliuk, B. Palosz, S. Gierlotka, P. Dluzewski, E. Atianin, Yu. Kluev, A. M. Naletov, P. Biczyk, A. Grzegorczyk, and A. Presz, *Appl. Phys. Lett.* **2000**, *7*, 954.

21 A. M. Ferrari, A. B. Conradi, F. Bondioli, U. A. Tamburini and A. F. Gualtieri, *Materials Science Forum* **2000**, *321–324*, 932.

22 T. H. Etsel and S. N. Flengas, *Chem. Rev.* **1970**, *70*, 339.

23 G. Xiong, J. Wilkinson, J. Lyles, K. B. User, and R. T. Williams, *Radiation Effects and Defects in Solids* **2003**, *158*, 83.

New Approach to Improve the Piezoelectric Quality
of ZnO Resonator Devices by Chemomechanical Polishing

Jyrki Molarius, Martin Kulawski, Tuomas Pensala, and Markku Ylilammi

1
Introduction

The main application for FBARs is in telecommunications, especially in mobile phones. There is an ever-ongoing development of making smaller and smarter mobile phones than before. One phone should be able to work around the world on different frequencies, which of course increases the number of filters needed in a given phone. The number of RF filters in a mobile phone ranges from three to seven, which results in world market of about 2 billion pieces per year [1]. Because of the big size of the telecommunications market also the research activity around the world on the subject of FBAR has recently been high [1–6]. There are two ways to make phones smaller; shrinking the size of individual devices and increasing integration. FBAR in the current form achieves the first goal by its capability to produce very small, thin, and light filters. Comparison with the current market leaders of surface acoustic wave filters shows that FBAR filters have steeper filter skirts, smaller temperature coefficient of frequency (TCF), smaller chip size and better power handling capability [1, 6]. All these differences are significant, and furthermore the comparison is between mature SAW technology and the first generation of commercial FBAR devices: SMR type (solidly mounted resonator, see below) or bridge type. FBAR also gives good promise for future integration with RF circuits as materials, processing, and thermal budget can easily be designed to be compatible with CMOS processing, for example. The FBAR structure lends itself also to other applications (gas- and pressure-sensors have been suggested). The performance of these sensors using ZnO as piezolayer and SMR type structure with mirrors has been calculated and found to be excellent [7].

There are two basic FBAR structures; namely bridge (also called membrane) resonators or solidly mounted (also called mirror) resonators. The piezoelectric layer is vibrating between two electrodes and in bridge type device the substrate and the resonator are decoupled by an air gap. The air gap, in principle, provides almost ideal isolation, but in practice the lower electrode as well as the supporting structure cause losses. The bridge structure has been utilized to make filters using the resonators in a ladder configuration [5]. Bridges or membranes are usually made by surface (or bulk) micromachining, which is often considered limiting the yield and productivity. Nevertheless there is a commercial producer for bridge-type FBAR passband filters for frequencies around 2 GHz [6].

Mirrors for isolating the resonator from the substrate was first proposed by Newell [8] in 1965 and then 30 years later developed to modern SMR structure by Lakin [9]. Here the resonator is isolated from the substrate by an acoustical mirror, build from alternating layers of high and low acoustical impedance materials, whose thicknesses are a quarter of the acoustic wavelength at the operation frequency. Depending of the choice of materials typically two to four layer pairs are needed. If molybdenum is used as the high impedance material and SiO_2 as the low impedance material three layer pairs are needed, but substituting Mo with W, two pairs are sufficient for adequate resonator substrate isolation. Mirrors can also be made of completely insulating materials such as AlN and SiO_2, which would give the benefit of easier processing as the mirror would not need any patterning. Metals in the mirror layers need patterning otherwise the parasitic capacitances associated with the conductive metal layers would cause problems in the device operation. Heavy metals have a high acoustic impedance and a high acoustic reflectivity, but a mirror made of AlN/SiO_2 requires four reflective pairs.

For making devices using bulk acoustic waves the quality of the piezolayer is paramountly important. In this chapter we are concentrating on ZnO, but same is true to other piezoelectric materials such as AlN or PZT $\{Pb(ZrTi)O_3\}$. Zinc oxide is chosen because of its high acoustic coupling coefficient ($k_{mat}=0.282$, $k^2=7.95\%$) [5]. Lead zirconate titanate PZT promises the highest coupling coefficient at 0.28–0.5, after annealing, but very low Q-values (only 18 at 2.3 GHz) [3]. Aluminum nitride AlN on the other hand has a smaller temperature coefficient of frequency (TCF) (~ 25 ppm) than zinc oxide (~ 50 ppm). The smaller TCF of AlN in some FBAR filter applications can compensate the benefit of the higher acoustic coupling coefficient of ZnO. The longitudinal sound velocity in AlN is 10 400 m/s, which is over 60% higher than in ZnO, 6400 m/s. As the thickness of the piezolayer largely determines the frequency of the device, it is advantageous, at low frequencies, to have low sound velocity as the films will be thinner and therefore faster to deposit resulting in increased productivity. At high frequencies, the effective coupling becomes low with a piezomaterial with a low sound velocity because of the small thickness of the layer. This would indicate that ZnO is better at low frequencies and AlN at high frequencies.

The most important parameter determining the quality of piezoelectric layer (AlN or ZnO) is a strong preferred orientation of the film. For FBAR devices operating in the longitudinal wave mode this is (0001). As the crystal is thus growing c-axis perpendicular to the substrate the hexagonal basal plane is interacting with the seed layer. ZnO films can and have been deposited by several different ways, such as CVD (chemical vapor deposition), laser ablation, and PVD methods (physical vapor deposition). In the PVD methods magnetron sputtering, either by rf-sputtering from ZnO-target or by dc or pulsed dc in reactive mode from zinc target in oxygen containing atmosphere, have been extensively used as the temperatures remain low during deposition and there is good compatibility with standard semiconductor device processing. Beside the sputtering parameters themselves, other issues affecting the piezoelectric film quality are the seed layer and the sputtering environment (affecting contamination). If the seed layer is used as the bottom electrode of the device, it has to be highly conducting to keep the electrical losses to a minimum. Therefore it is advantageous to separate these functions to a highly conducting bottom electrode and to a separate seed layer, which can be optimized to promote piezolayer growth and acoustical properties of the FBAR stack. Device quality ZnO can be grown on several metals, at VTT this has been realized on gold and molybdenum [4, 5, 10, 11]. It has been shown that good quality AlN, which has the same wurtzite-type crystal structure as ZnO, can be grown on different seed materials [12]. Löbl [13] and S.-H. Lee [14] for AlN and J.B. Lee [15] for ZnO identified the surface roughness of the seed layer as the decisive factor on piezoelectric film quality; the smoother the seed layer, the better the piezoelectric film quality. The film quality was measured by X-ray rocking curve of ZnO film and the FWHM (full width half maximum) of the (0002) peak was shown to correlate with the effective acoustic coupling coefficient, k_{eff} [3]. K_{eff} on the other hand determines the bandwidth and insertion loss of a filter [16]. Quality factors, Q, are determined at both series and parallel resonance. They are calculated according to the IEEE standard [17]. The reason for the correlation between surface roughness and film quality is quite simple. Since the deposition takes place at low temperature and with low particle energy, adatoms on the surface have low mobility and they do not reach the energetically favored positions for the growth in the preferred orientation. On a smooth substrate the movement of the adatoms is easier and this results in improved piezolayer quality. Control of the surface roughness of a given metal can only be achieved to a certain degree with adjustment of the sputtering parameters, as other film properties, especially stresses in the film, will also be strongly affected. Therefore a better approach is needed.

CMP is usually used to planarize device structures, as seen in the diagram in Fig. 1. The wafer is held on a carrier and pressed with a defined force against a micro-porous polyurethane polishing cloth, which is glued on the rigid polishing

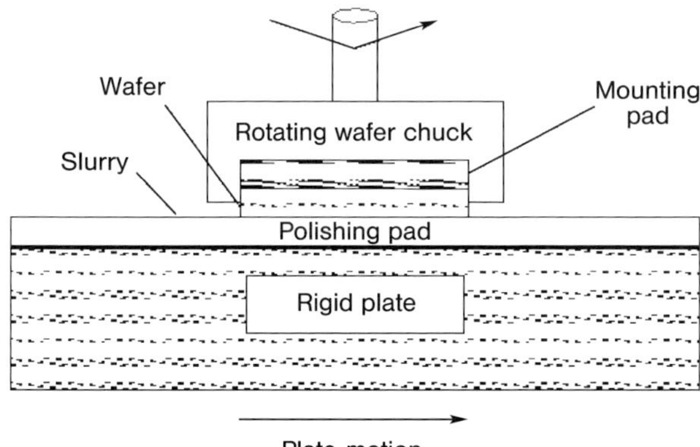

Fig. 1 Diagram of chemomechanical polishing (CMP).

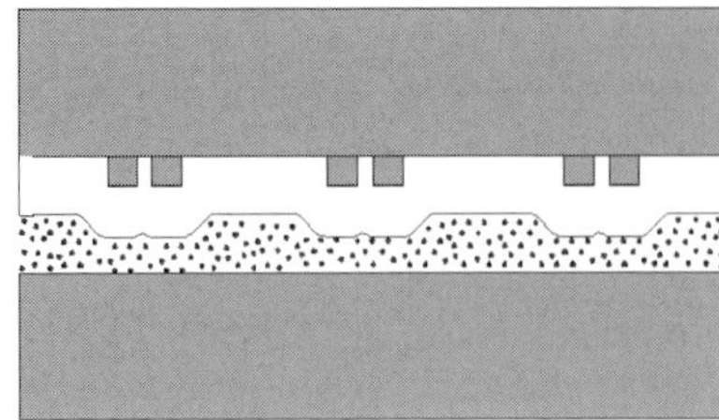

Fig. 2 Principle of CMP removal.

platen of the tool. While the platen and the chuck are rotating a suspension with adjusted pH-value of deionised water (DIW) and abrasive particles (slurry) is dispensed on the platen. The abrasive particles are made from silica or ceria with diameters in the 50 nm range. During polishing the particles are accelerated by the micro-pattern of the polishing pad and impinge on the surface of the wafer, thus weakening the strength of the atomic network. The pH-adjusted liquid can penetrate into the weakened network and dissolve atomic clusters from it. Owing to the height variation of the patterned surface a different local pressure is applied to elevated and lower areas of the wafer leading to increased removal on the higher regions. This leads to a planarization of the pattern on a large scale. In Fig. 2 the principle of the removal is presented in detail. In this chapter the emphasis is on surface smoothing by CMP, which can be achieved with a modified CMP planarization process.

Any kind of CMP, however, will lead to a heavy contamination of the polished substrates with particles left on the surface. Since these particles have strong adhesion a special cleaning is an important issue after polishing. Beside soft etching methods and standard cleaning with megasonic agitation we are currently qualifying a special post-CMP cleaner, which scrubs the surface with a soft PVA-brush (polyvinyl alcohol) and removes particles also by mechanical means. Beside the particles a slurry often contains metallic contamination. Thus by including chemistry to the post-CMP cleaning process care has to be taken for lowering the metal contamination down to the stringent levels for CMOS compatible production. It is the aim of this chapter to develop the CMP smoothing and apply it to FBAR to achieve high quality ZnO for resonators and filters.

2
Experimental

We have used 100 mm (100)-oriented silicon wafers as substrates. Wafers with 1–10 Ω cm resistivity were used for structural characterization of the deposited zinc oxide films. High resistivity wafers (>500 Ω cm) were chosen as substrates for processing resonators and filters to eliminate the parasitic effects associated with the semiconducting silicon [4]. Films were sputtered in a cluster tool (Von Ardenne CS 730 S) from 200 mm diameter round targets. In this system metal films are deposited in a multitarget chamber with dc-magnetrons, but to minimize cross-contamination, ZnO is sputtered in a dedicated single target chamber. Reactive dc-magnetron sputtering from zinc target (purity of 99.995%) in argon/oxygen atmosphere is utilized in all experiments. Both gases have purity of 99.9999%. The flow rates of the gases were rationed with massflow controllers and the oxygen content was always set to 41 vol.-%. The loadlocked cluster type sputtering system was pumped with oil free turbodrag and diaphragm pumps to below 5×10^{-5} Pa before sample processing.

Resonators were SMR-type fabricated on quarter wavelength acoustical mirrors, consisting of alternating layers of high (W) and low (SiO_2) acoustic impedance materials. We used a simplified resonator process, where only the top electrode is patterned leaving all other layers (mirror, bottom electrode and ZnO) to cover the whole wafer. Contact to the bottom electrode is done capacitively. The film thicknesses for the resonator stack were calculated by an in-house developed one-dimensional modeling program. Tungsten has the highest known acoustic impedance and the impedance ratio $Z_{high}:Z_{low}$ for $W-SiO_2$ pair is $7.7:1$. This ensures good acoustic isolation with only two layer pairs. In case of $Mo-SiO_2$ the ratio is $4.8:1$ and consequently one needs an extra pair. Metals were sputter deposited by dc-magnetron and PECVD (plasma enhanced chemical vapor deposition) was used for SiO_2. Our resonator and filter fabrication is explained in more detail elsewhere [4].

Smoothing was done on a Strasbaugh 6 DS-SP planarizer. Standard processing consumables such as pads and slurries were varied in the experiments to achieve the best surface smoothing. Polishing times were varied in the range 30–90 s. Cleaning was done in a batch cleaning in a quartz sink using megasonic agitation for 10 min with pure DIW at 55 °C and subsequent spin-drying for particle removal. Additionally in some cases a SC-1 bath was used for enhanced particle removal.

The film morphology was studied using a digital scanning electron microscope (SEM, Leo 1560). Surface roughness and topology were measured by atomic force microscopy (AFM, Digital Instruments Dimension 3100). FilmTek 4000 spectrophotometry tool was used for optical characterization of ZnO. For electrical characterization of the completed devices Agilent Technologies 8753 network analyzer was utilized.

3
Results and Discussion

Schematic cross sections of the film stacks for both bridge and solidly mounted resonator are shown in Fig. 3. In surface micromachining the air gap is formed by dissolving a sacrificial layer from underneath the bridge, such as copper below silicon nitride bridge [5]. SMR schematic in Fig. 3 has the designed layer thicknesses, whereas the lowest high-Z layer in the actual scanning electron microscopy (SEM) cross section micrograph of a FBAR is too thick. Fortunately it does not affect the mirror performance substantially. This can be deciphered from Fig. 4, which shows the relative calculated displacement amplitude in a resonator

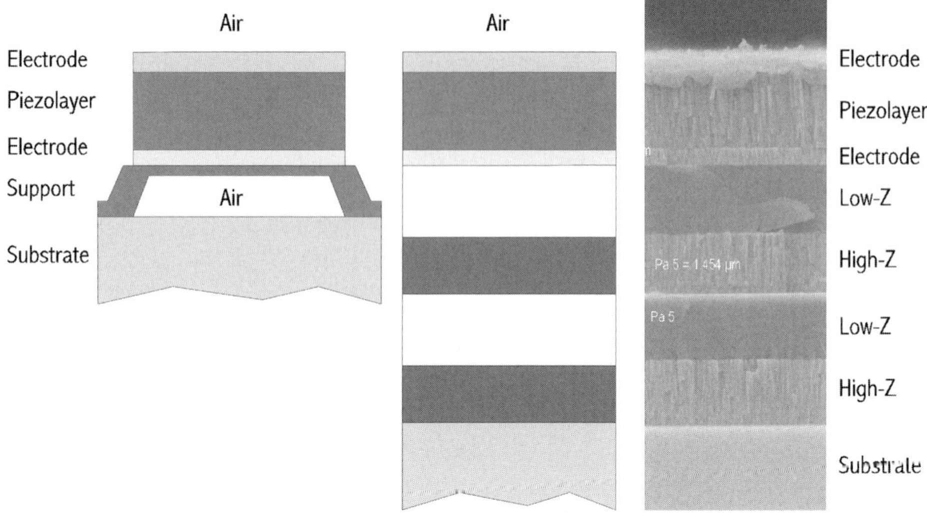

Fig. 3 Bridge and mirror resonator film stacks with cross section SEM of the mirror FBAR.

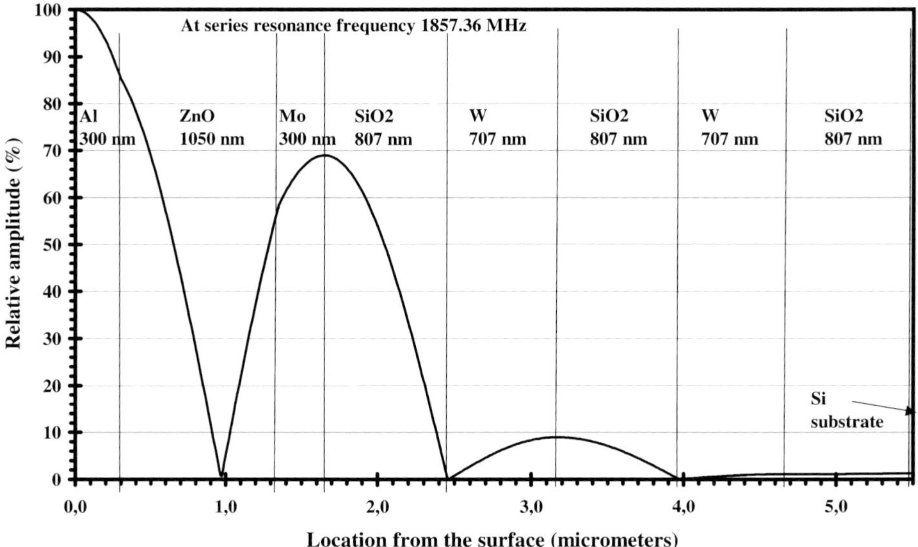

Fig. 4 Relative displacement in a resonator stack at 1857 MHz. Free surface is on the left.

stack with silicon dioxide/tungsten mirror. The relative displacement at the substrate is <2% of the maximum at the series resonance frequency of 1857 MHz; therefore it is not necessary to increase the number of mirror layers in this stack. The highly columnar structure of the metals in the mirror and bottom electrode

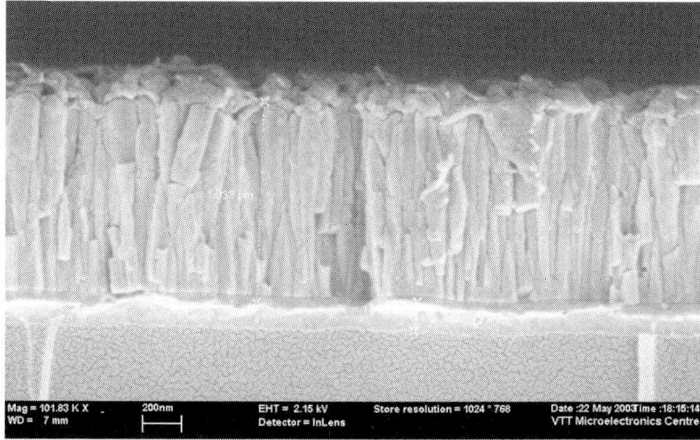

Fig. 5 Cross section SEM micrograph with non-CMP ZnO
piezoelectric film on resonator.

is clearly depicted in Fig. 3, as well as the amorphous nature of silicon dioxide. Top electrode is sputtered aluminum and it has been deformed during sample cleaving. The ZnO piezolayer is also highly columnar as desired for strongly preferred orientation. But as can be seen in Fig. 5 the ZnO film on a wafer, which is not CMP smoothed is porous and the ZnO surface appears to be very rough.

ZnO is transparent, therefore we have measured the samples optically with Filmtek 4000 from ultraviolet to infrared wavelengths (450–1650 nm) [18]. A typical index of refraction (n) and extinction coefficient (k) as function of the wavelength are shown in Fig. 6. Index of refraction at 633 nm varies on different films from 1.910 to 1.921, which is quite close to the bulk value of 1.997 [19]. In optical respect our ZnO films seem to be of fairly good quality even without CMP smoothing.

It is not feasible to smooth the bottom electrode metal layer by CMP, due to contamination issues. Gold, one of our electrode materials, is the most feared yield killer in microelectronics. Therefore we decided to smooth the top mirror layer SiO_2 as this is compatible with other work done on our CMP tool. A CMP smoothing process for the silicon dioxide was developed and the topological properties of the films were measured with AFM. As one can see in AFM nanographs in Fig. 7, the surface appearance of PECVD deposited SiO_2 film changes completely in CMP. During the removal of 70–80 nm of oxide by CMP the rms roughness is dramatically reduced from 4–5 nm to <0.3 nm. After the first trials with actual FBAR samples a smooth surface was achieved, but lots of particles were detected on the surface even after post-CMP cleaning. This was resolved by adding a SC-1 cleaning step at 55 °C in the post-CMP cleaning procedure. In the actual FBAR samples we found holes in the oxide. In the overall view the large number of holes can be seen in the optical micrograph (Fig. 8 a), a close up reveals that the holes go through the oxide (Fig. 8 b). This was further confirmed with stylus measurements. It seems that holes originate from defects on the thick tungsten high Z film. These defects in turn cause flaws on the growing SiO_2 and the SC-1 cleaning solution (ammonium and hydrogen peroxide) attacks these resulting in holes.

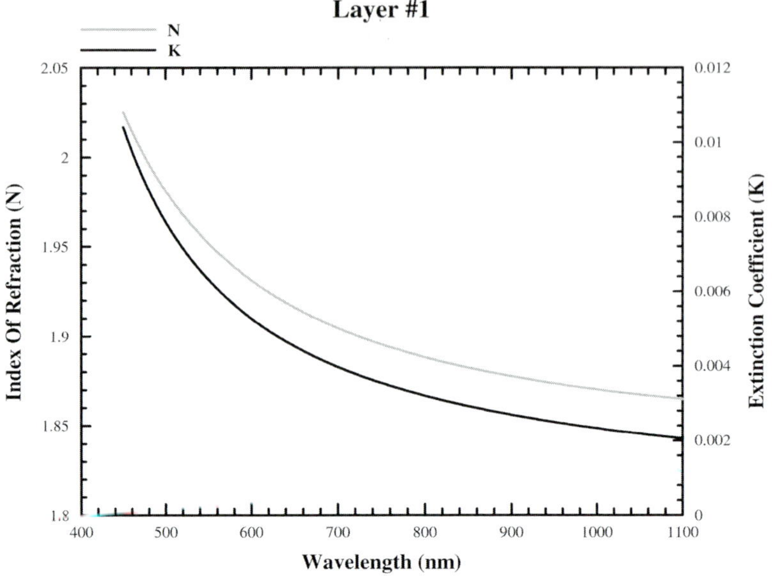

Fig. 6 Filmtek spectrophotometry results of the index of refraction and extinction coefficient as function of wavelength.

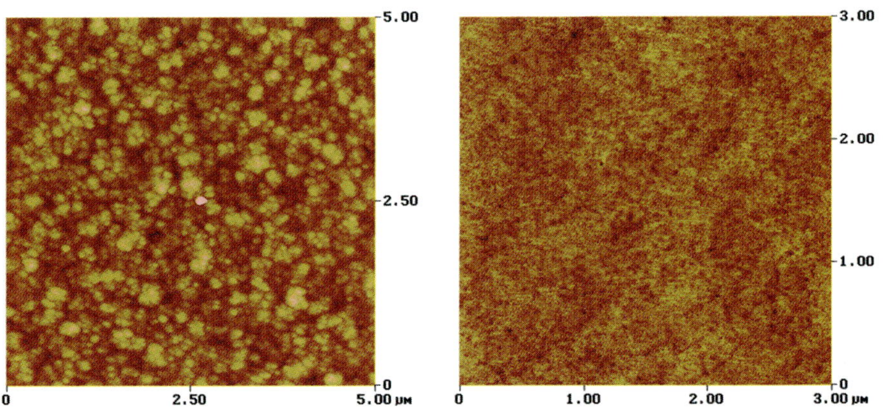

Fig. 7 AFM nanographs of SiO$_2$: (a) before and (b) after CMP smoothing.

Another issue with CMP smoothing is how our resonator stacks survive through the CMP. This is illustrated in Fig. 9, where a stylus trace (in fact a cross section, compare Fig. 3) of the resonator stack is shown before and after the CMP smoothing. The corners are rounded during CMP, but a large center area of the mirror stack remains flat and uniform. It does seem to be possible to reduce the rounded area further by CMP process development. When including CMP into the FBAR process flow, one has to take into account not only the removal of the

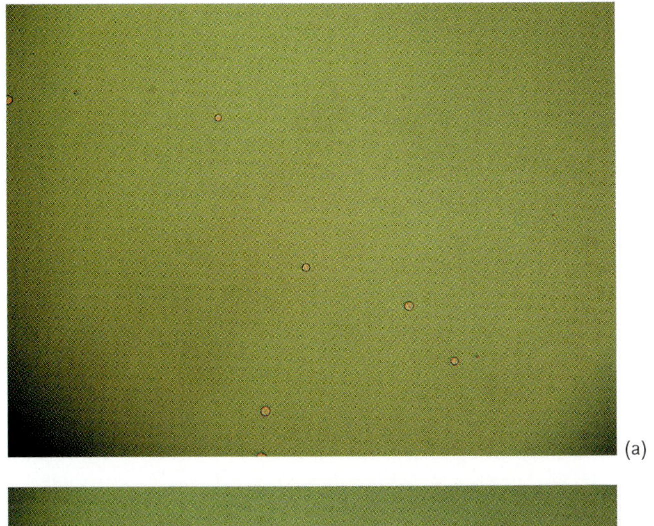

(a)

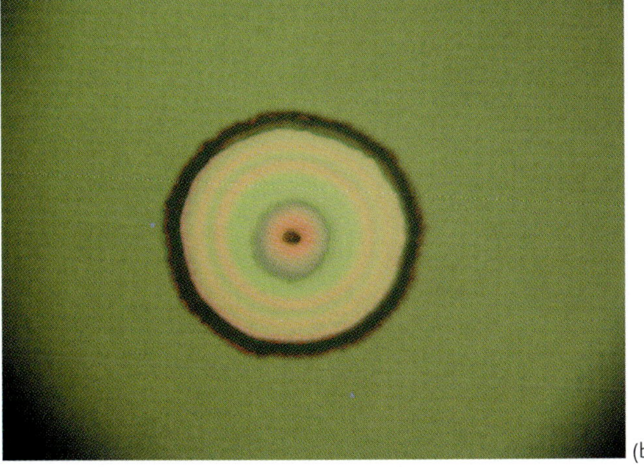

(b)

Fig. 8 Optical micrograph of SiO$_2$ surface after post CMP cleaning:
(a) overall view, (b) close up of a hole.

oxide by CMP, but also the rounding of the corners by allocating some extra area on the resonator mirrors. Although we loose some chip "real estate" future integration of CMP smoothing into the filter processing seems feasible.

In Fig. 10 AFM nanographs of zinc oxide surface in the old process without CMP smoothing of the silicon oxide and with CMP are shown. Surface roughness has been reduced from 23 to 4.4 nm. The ZnO film (non-CMP) in cross section SEM micrograph in Fig. 5 is from the same sample as in Fig. 10a. The cross section microstructure and surface roughness are also closely related in case of the CMP smoothed ZnO film, where smooth surface also results in dense and featureless SEM cross section (not shown here).

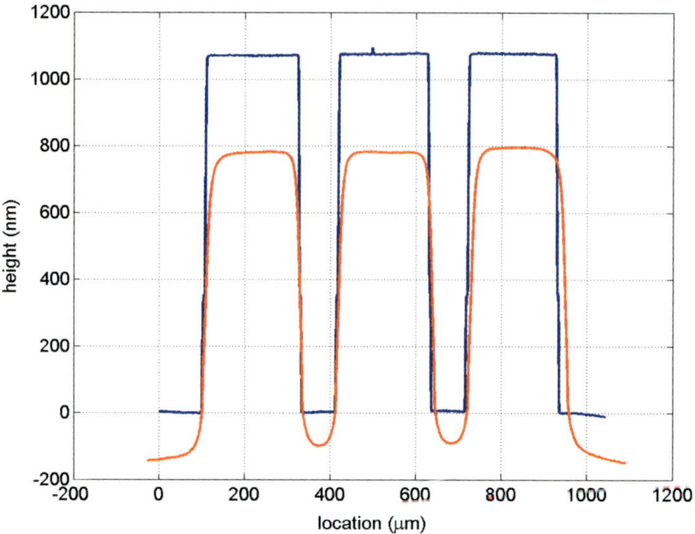

Fig. 9 Dektak stylus trace of the FBAR stack before (blue) and after (red) CMP smoothing.

Smoothing seems to result in better ZnO smoothness as expected and has been reported before [16]. In the smoothest ZnO the rms roughness we have achieved with CMP is 4.4 nm, which compares favorably with the best achieved smoothness of non-CMP ZnO, where rms roughness is 13 nm. In a recent work [20] very smooth, rms roughness 1.06 nm ZnO film is reported on an actual SMR FBAR with gold electrodes. ZnO roughness was measured by AFM on the as-deposited samples and it is plotted as function of the deposition run in Fig. 11. Data show that the smoothest ZnO is achieved on those wafers where the top SiO_2 has been smoothed by CMP and the worst roughness was on non-CMP wafers. Another trend is that the ZnO on the non-CMP resonators is getting better with time (more runs). Owing to the instabilities in the ZnO sputter deposition and small number of samples, evidence however is not unambiguous.

The quality of the piezolayer (and the whole resonator) is measured by the effective coupling coefficient and the Q-values at series and parallel resonances. In Fig. 12 the coupling coefficient as function of surface roughness is presented. Two lowest points can be explained by the dc-bias on the substrate during sputtering, which is obviously detrimental to the piezolayer quality. Unfortunately only one CMP-smoothed; wafer survived through the processing (the smoothest one). At the parallel resonance (Qp) Q-values range from 70 to 307, and at the series resonance (Qs) from 69 to 446 without showing any correlation to the ZnO surface roughness. the CMP-smoothed resonator had Qp of 108 and Qs of 339. Due to our simplified resonator process this structure is not optimal for Q-value determination, but nevertheless our Q-values are comparable to the one (201) reported recently [20].

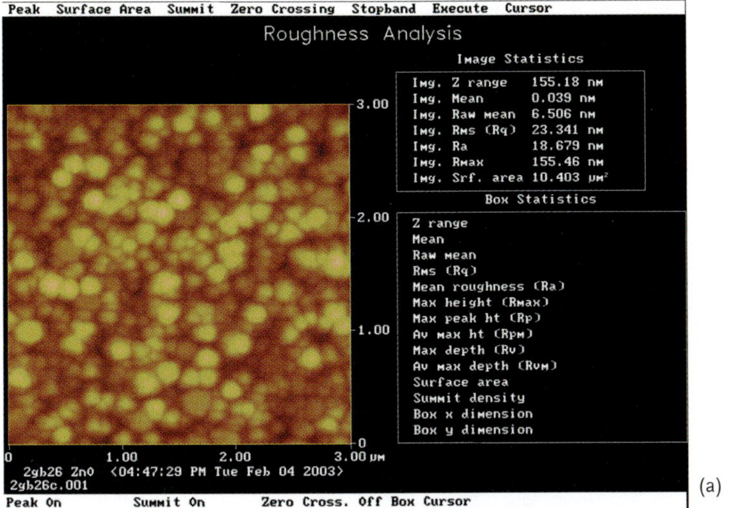

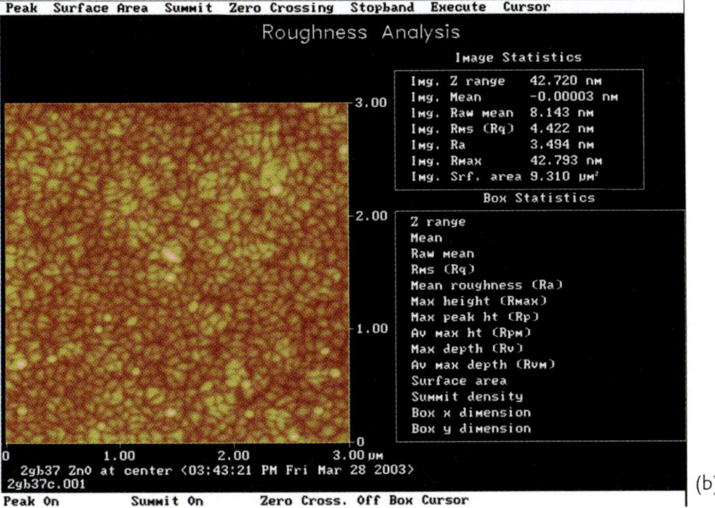

Fig. 10 AFM nanographs of ZnO surface after sputter deposition:
(a) non-CMP, (b) CMP smoothing applied to SiO_2.

One has to keep in mind that the wavelength of the acoustical waves in ZnO in these resonators is 3 µm, but the scale of the surface roughness is in (tens of) nanometers. Therefore strong physical coupling between the acoustical waves and the surface roughness was not expected. One can speculate that a smoother surface would result in smoother and denser ZnO, because of the effect on film growth during sputtering. This has not been completely verified, as the evidence is inconclusive at this stage.

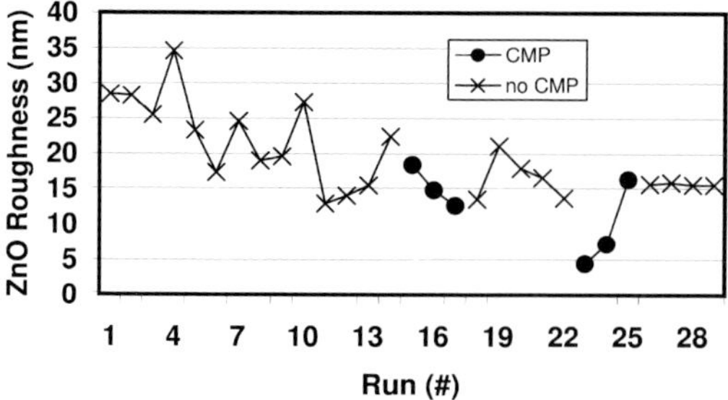

Fig. 11 ZnO surface roughness as measured by AFM after sputtering runs.

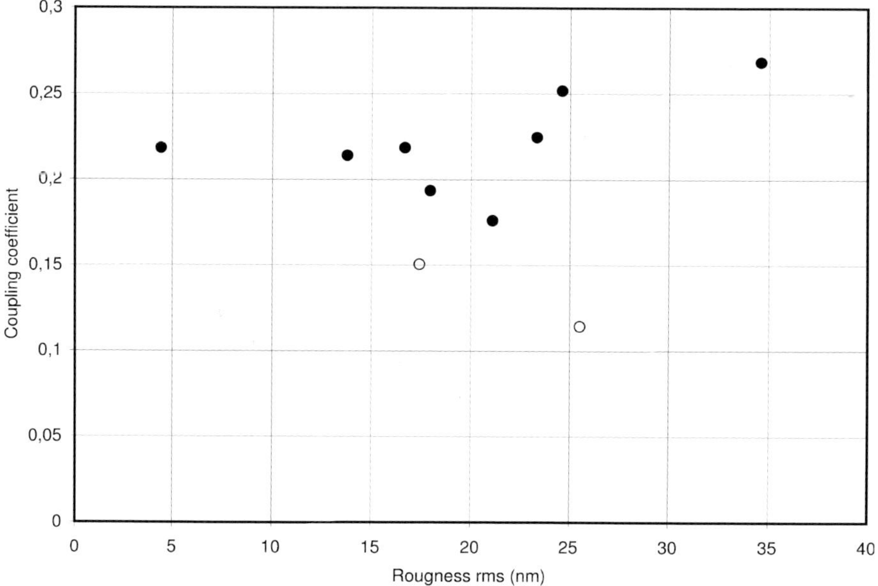

Fig. 12 Acoustic coupling coefficient correlation to ZnO surface roughness. Open circles denote DC-bias on the substrate during sputtering.

4
Conclusions

CMP smoothing was introduced to FBAR (thin film bulk acoustic wave resonator) technology and it was used in fabrication of SMR-type resonators. The ZnO film morphology and microstructure were studied with AFM and SEM, respectively. Optical properties were sampled with spectrophotometry. It was shown that the smoothing of the top mirror SiO_2 improves the surface roughness of the zinc oxide. The roughness achieved with CMP smoothing was 4.4 nm (rms), which compares favorably with 13 nm without CMP. The quality of the ZnO piezolayer was evaluated by measuring the acoustic coupling coefficient and Q-values at parallel and series resonance. ZnO was piezoelectrically good and the CMP smoothing shows good prospects for future integration in FBAR process flow.

Acknowledgments

Special thanks are due to Ms. Meeri Partanen for sample processing and Kimmo Solehmainen for Filmtek measurements.

References

1 AIGNER R., ELLÄ J., TIMME H.-J., ELBRECHT L., NESSLER W., MARKSTEINER S., in: *Proc. IEEE International Electron Devices Meeting (IEDM)*, San Francisco, 9–11 December **2002**.

2 LAKIN K., *IEEE Ultrasonics Symp.* **2002**.

3 LÖBL H. P., KLEE M., METZMACHER C., BRAND W., MILSOM R., DEKKER R., LOK P., *Mater. Chem. Phys.* **2003**, *79*, 143–146 and *Mater. Sci. Eng. B* **2002**, 1–4.

4 KAITILA J., YLILAMMI M., MOLARIUS J., ELLÄ J., MAKKONEN T., *IEEE Ultrasonics Symp.* **2001**, 223.

5 YLILAMMI M., ELLÄ J., PARTANEN M., KAITILA J., *IEEE Trans. Ultrasonics, Ferroelectrics, Frequency Control* **2002**, *49*, 535.

6 LARSON III J. D., RUBY R. C., BRADLEY P. D., WEN J., KOK S.-L., CHIEN A., *IEEE Ultrasonics Symp.* **2000**, 869.

7 MANSFELD G. D., KOTELYANSKY I. M., *IEEE Ultrasonics Symp.* **2002**.

8 NEWELL W. E., *Proc. IEEE* **1965**, *53*, 575–581.

9 LAKIN K., KLINE G., and McGARRON K., *IEEE Trans. Microwave Theory and Techniques* **1995**, *43*, 2933.

10 MOLARIUS J., KAITILA J., PENSALA T., YLILAMMI M., *J. Mater. Sci.: Mater. Electron.* **2003**, *14*, 431–434.

11 MOLARIUS J., YLILAMMI M., US Patent 6,521,100, **2003**.

12 IRIARTE G. F., BJURSTRÖM J., WESTLINDER J., ENGELMARK F., KATARDJIEV I. V., *IEEE Ultrasonics Symp.* **2002**.

13 LÖBL H. P., KLEE M., MILSOM R., DEKKER R., METZMACHER C., BRAND W., LOK P., *J. Eur. Ceram. Soc.* **2001**, *21*, 2633–2640.

14 LEE S.-H., LEE J. K., YOON K. H., *J. Vac. Sci. Technol. A* **2003**, *21*, 1.

15 LEE J. B., KWAK S. H., KIM H. J., *Thin Solid Films* **2003**, *423*, 262–266.

16 LAKIN K., KLINE G., and McGARRON K., *IEEE Ultrasonics Symp.* **1992**, 471–476.

17 *IEEE Standard on Piezoelectricity*, ANSI/ IEEE Standard 176, **1987**.

18 US Patent 5 999 267, **1999**.

19 JELLISON Jr. G. E., BOATNER L. A., *Phys. Rev. B* **1998**, *58*, 3586–3589.

20 LEE J. B., KIM H. J., SOO GIL KIM, HWANG C. S., HONG S.-H., SHIN Y. H., LEE N. H., *Thin Solid Films* **2003**, *435*, 179–185.

Self-Assembled Semiconductor Nanowires

Theodore I. Kamins

1
Introduction

For more than three decades the functionality of integrated circuits has increased exponentially, as noted by Gordon Moore, who summarized his observations as "Moore's Law." This increased functionality has been achieved primarily by reducing the size of the electronic devices and the interconnections between them, until today the minimum features are much smaller than the wavelength of light. However, as feature sizes decrease, short-channel effects and inter-device interactions become increasingly troublesome. The cost of the fabrication facilities also increases exponentially ("Moore's second law"), limiting the number of companies that can economically participate in the continued evolution of conventional integrated-circuit technology. The cost of a conventional fabrication facility was approximately US\$ 3 billion in 2003 and continues to increase. New approaches are needed to allow the increasing functionality predicted by Moore's law to continue within sustainable economic limits.

To expand the functionality beyond that of conventional electronics and possibly to replace some of the electronic components, non-conventional microsystems have been developed. These microsystems add the capability of mechanical functions to the basic substrate. However, the non-planar features of these systems and their different processing requirements make integration with conventional electronics difficult. More recently, as scaling of conventional electronics reaches its technical and economic limits, several "nanoelectronic" approaches to memory and computation have been proposed. Nanoelectronics offers the possibility of new concepts that extend the capabilities of electronics and bypass the increasingly severe constraints imposed on conventional integrated-circuit technology.

One of the major costs associated with conventional electronics is the fine-scale lithography needed to form ever smaller critical dimensions; a modern stepper can cost more than US\$ 20 million. The high cost of this lithography motivates the exploration of "self-assembly" and "directed-assembly" techniques that use thermodynamic forces and kinetic control to form small features without fine lithography. Both equi-axed and one-dimensional features can be formed by self-assembly.

The Nano–Micro Interface: Bridging the Micro and Nano Worlds
Edited by Hans-Jörg Fecht and Matthias Werner
Copyright © 2004 WILEY-VCH Verlag GmbH & Co. KGaA, Weinheim
ISBN: 3-527-30978-0

One-dimensional "nanowires" may provide the bridge connecting conventional electronics, microsystems, and nanoelectronics. They may also serve as the electrodes of nanoelectronic elements such as molecular switches within the nanoelectronic portion of the system. The nanowires can contain modulated elements (transistors). The small diameter of the nanowire makes modulation by surface charges especially effective. Thus, nanowires can be key elements in sensors, especially when made selective by adding functional coatings to the surfaces of the nanowires. Similarly, the conductance of a nanowire can be modulated by an externally applied electric field, allowing it to serve as the channel of an MOS transistor. Possible optical applications have also been demonstrated. Complex nanowires of III–V semiconductors can contain modulated light-emitting regions [1, 2]. Nanowires of other compound semiconductors have been shown to act as lasers [3, 4].

Because all of the proposed approaches to nanoelectronics require major changes in technology, they are most likely to become practical if they can be implemented gradually: first augmenting conventional electronics, and then assuming a greater role in the overall circuitry. If nanoelectronics and conventional electronics are to coexist for an interim period, with nanoelectronic components placed on top of the conventional silicon electronics, interconnections between them must be formed under conditions compatible with the conventional electronics already present.

2
Growth

"Metal-catalyzed" nanowire growth [5, 6] is especially attractive for forming self-assembled, one-dimensional nanostructures. In addition to avoiding expensive lithography, the surfaces of these structures are formed by growth, rather than by etching, which can cause crystal damage. In metal-catalyzed nanowire growth, small regions ("nanoparticles") of an active metal are first prepared (Fig. 1). They are then exposed to a gaseous source of (usually) silicon under conditions where the normal growth rate of silicon is very low. The metal nanoparticle locally accelerates the reaction so that silicon atoms are deposited on the nanoparticle. The silicon atoms diffuse through or around the nanoparticle and precipitate on the underlying substrate, pushing the nanoparticle up and leaving a silicon column of similar diameter (a nanowire) behind. The nanowires can be grown from catalysts placed randomly or selectively on the substrate, or they can possibly be grown from metals already present in the underlying conventional electronics or microsystems, efficiently providing the desired interconnections. The nanowires can be virtually defect-free, crystalline silicon, with the metal catalyst remaining at the tip of the growing wire.

When the nanowires are to be grown on a partially fabricated substrate already containing electronic and mechanical elements, the temperatures at which the nanowires can be formed and further processed is constrained. To be compatible

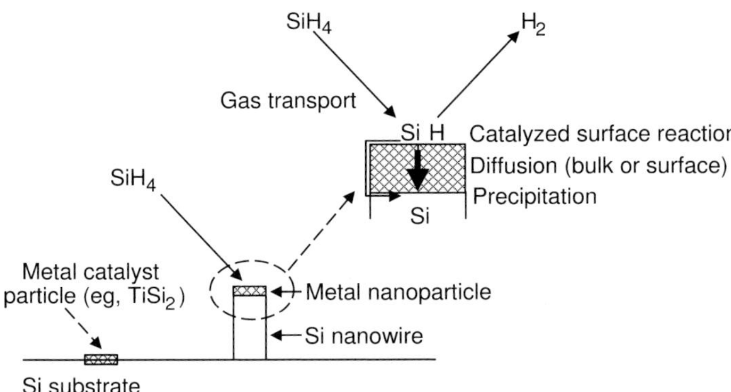

Fig. 1 Concept of metal-catalyzed nanowire growth on metal nanoparticles.

with a partially processed substrate containing diffusions and selected silicides and barrier layers, the temperature should be less than 700 °C. To be compatible with a fully metallized substrate, the temperature should remain below 400 °C.

The choice of the metal catalyst is critical to nanowire growth. The catalyst should accelerate decomposition of the silicon-containing gas and promote nanowire growth. It should ideally be compatible with silicon IC technology, have a low solid solubility in silicon, and not introduce deep energy levels into the silicon bandgap. The absence of mid-gap energy levels is especially important when the nanowires are to be integrated with conventional silicon electronics because the minority-carrier generation-recombination rate U in a semiconductor depends on the energy of the impurity states, as shown in Eq. (1) [7]

$$U = \frac{pn - n_i^2}{\left[p + n + 2n_i \cosh\left(\dfrac{E_t - E_i}{kT} \right) \right] \tau_0} \tag{1}$$

where p and n are the hole and electron densities, respectively, n_i is the intrinsic carrier density, E_t and E_i are the impurity energy level and the mid-gap energy level, respectively, and $\tau_0 = 1/(N_t v_t \sigma_n)$ with N_t being the impurity density, v_{th} being the thermal velocity of the carriers, and σ_n being the capture cross section. Thus, the degradation of the minority carrier properties depends approximately exponentially on the energy separation of the impurity energy level from mid-gap. When the impurity level is at mid-gap, both electrons and holes interact efficiently with the impurity; when the impurity level is displaced from mid-gap, the interaction with either electrons or holes decreases rapidly, so the impurity level is less effective in electron-hole generation or recombination, which requires interaction with both types of carriers. If the nanowires are used independently from a silicon circuit and only majority-carrier devices are considered, a catalyst such as gold, which introduces mid-gap levels into the silicon bandgap, may be useful. However, a catalyst with energy levels well separated from mid-gap will be more readily accepted.

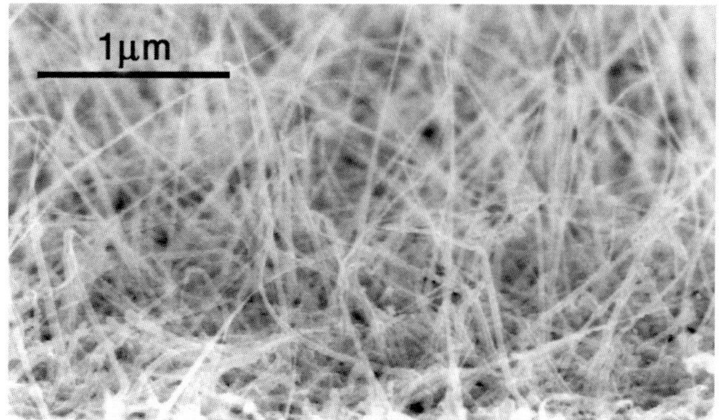

Fig. 2 Si nanowires grown at 620 °C using Ti catalyst nanoparticles.

Titanium is attractive as a catalyzing material for silicon nanowires [8] because it is compatible with conventional integrated-circuit fabrication, does not introduce allowed states near the middle of the Si bandgap, and has a low solid solubility in Si. We have grown silicon nanowires using titanium as the catalyst (Fig. 2) at temperatures considerably below 700 °C (typically at ~600 °C), compatible with an unmetallized silicon CMOS integrated circuit. As with most growth techniques, however, obtaining an adequately clean surface to allow selective and directed growth will be a significant challenge.

Selective growth on only the desired metal catalyst will require manipulation of the gas phase so that uncatalyzed, two-dimensional growth on other regions of the substrate is avoided. For example, catalyzed growth of silicon to form the nanowires can be much faster than normal growth of polysilicon on oxide, but a non-negligible amount of silicon may deposit on the oxide or on exposed regions of silicon during nanowire growth when SiH_4 is used. The nucleation on oxide can be suppressed by including a Cl-containing species in the gas phase. A non-negligible ratio of normal to catalyzed growth will also allow growth on the sides of the nanowires as they become longer, leading to a tapered nanowire. (The tapered shape may be desired in some applications, but is undesirable for many of the most widely discussed uses.) For growth at somewhat lower temperatures, disilane Si_2H_6 may be a useful silicon source.

To allow integration with a fully metallized IC, the growth temperature of the nanowires needs to be reduced below 400 °C. This requirement for a lower growth temperature suggests germanium as a potential candidate for the nanowires. However, nucleating one-dimensional growth of germanium nanowires is more difficult than nucleating Si nanowires. Gold is often used as a catalyst for germanium nanowire growth. Although gold is not readily compatible with silicon technology, it allows demonstration of germanium nanowire growth [9, 10] and may be useful when compatibility with Si ICs is not required.

The bulk liquid eutectic temperature of the Au-Ge system is approximately 360 °C, so that the nanoparticle can be in the liquid state during nanowire growth

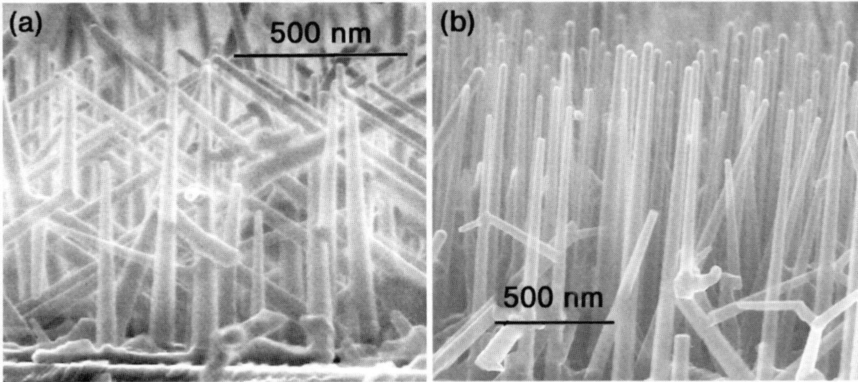

Fig. 3 Ge nanowires grown on: (a) Si(001), (b) Si(111)-oriented Si wafers.

at temperatures in this range, allowing vapor-liquid-solid (VLS) growth. The small size of the nucleating nanoparticles is likely to reduce the eutectic temperature somewhat. We have grown Ge nanowires over the temperature range 310–380 °C, which would be compatible with standard CMOS technology. Fig. 3a shows Ge nanowires grown on a (001)-oriented Si substrate at approximately 320 °C, which is optimum under the particular conditions employed. The nanowire growth direction appears to be [111], and many of the nanowires grow epitaxially oriented with respect to the [111] directions of the substrate. On (001) substrates, the nanowires grow in preferred directions corresponding to the [111] directions of the substrate. On (111)-oriented wafers, many of the nanowires grow vertically, as shown in Fig. 3b.

3
Positioning Nanowires

To use nanowires as interconnecting elements between different parts of a total system comprising microelectronics, microsystems, and nanoelectronics, the nanowires must be grown at locations predetermined by the underlying electronics and microsystems, as shown conceptually in Fig. 4a. Materials already present in the underlying structure may possibly be used to catalyze nanowire growth so that the nanowires can be self-aligned to, and intimately connected with, the underlying components. More likely, catalyzing nanoparticles will be selectively added to the desired regions of the structure by placing them in vias or by controlling the shapes of the underlying surfaces to position the nanoparticles at surface irregularities, as shown in Fig. 4b.

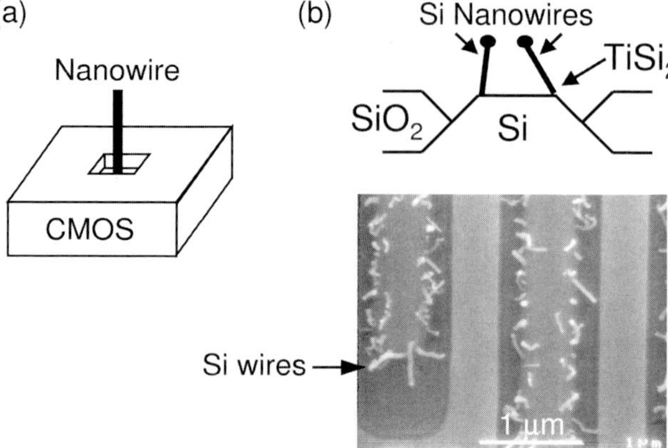

Fig. 4 (a) Diagram of nanowire self-aligned to underlying conventional electronics. (b) Si nanowires grown on patterned substrate, showing self-alignment on a one-dimensional feature.

3.1
Stability

To complete the fabrication of other electronic and mechanical components, further processing is necessary, and the nanowires must remain stable during this processing. We have examined the limits of the nanowire stability in various ambients at elevated processing temperatures [11]. Because of their large surface-to-volume ratio, the major instability is expected to arise from surface diffusion of the silicon atoms. During air exposure after deposition, a thin native oxide forms on the nanowires. This native oxide retards surface diffusion and increases the

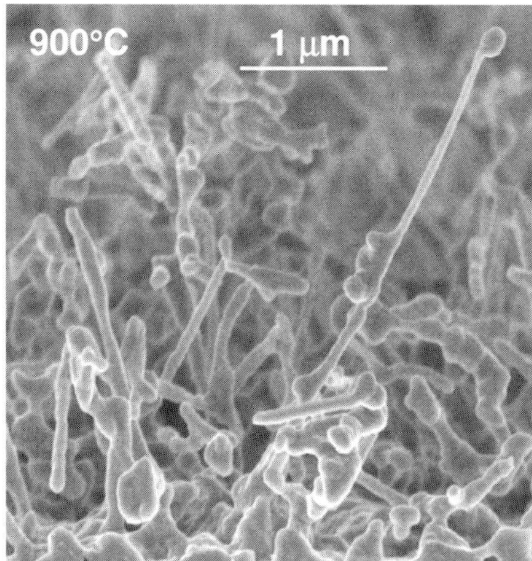

Fig. 5 Deformation of Ti-catalyzed Si nanowires without a passivating surface oxide after annealing at 900 °C in H$_2$.

stability of the nanowires during subsequent heat treatments in an inert ambient. Under these conditions, the wires are stable to temperatures above 950 °C. In a reducing ambient with no native oxide present, the exposed silicon atoms can move on the surface at temperatures above 850 °C. At higher temperatures, tapered wires evolve into wires with segments of constant diameter, as shown in Fig. 5, demonstrating the nanoscale version of a classic instability [12, 13]. Stability at 850 °C is adequate for fabrication of the additional components.

4
Transistors and Sensors

In addition to providing interconnections between different functional blocks, self-assembled nanowires can also be used directly as microsystems: especially as gas sensors [14]. When a nanowire is used as a sensor, gas molecules adsorbed on the outer surface of a passivated or functionalized nanowire modulate the conductivity of the nanowire. The small diameter of the nanowire makes modulation by surface charge especially effective. A fixed gate can similarly be used to electrically control the nanowire conductance. Applications as gas sensors and as transistors require consideration of the interface and near-surface properties of the nanowire device. In the following discussion, we can consider a gas sensor as a transistor with the gating function provided by the gas being sensed. The potential application of Si nanowires to nanoelectronics is very attractive because it can build on the extensive Si technology that has been developed over the past 50 years.

4.1
Interface Control and Insulator Material

The most critical task in fabricating useful transistors within nanowires will be controlling the interfaces so that the gate voltage can be effective in modulating the conduction of the nanowire. Approximately forty years ago, the IC industry learned to control the Si/SiO_2 interface. Achieving this control allowed fabrication of the stable and reproducible MOS transistors that currently dominate electronics. Control of the interface has been further improved in the intervening years, as indicated by the multi-billion MOS transistors fabricated each week. Forming useful devices in silicon nanowires must build on this vast knowledge of Si surfaces and interfaces to obtain the high-quality interface that is critical to stable and reproducible devices.

In particular, the native oxide formed on exposing a clean Si surface to air at room temperature is of poor quality and contains many trapping sites that will shield the gate field and prevent the applied gate voltage from inducing or depleting carriers in the conducting channel. Because the surface is silicon, high quality SiO_2 can be grown, and nearly ideal interfaces can be formed (considering, of course, that different crystal planes bound the nanowire). High-permittivity dielectrics presently being developed for bulk Si transistors can be applied to nanowires. Again, obtaining control of the interface is critical, and nanowire devices can benefit from the ongoing work directed at mainstream silicon technology.

4.2
Device Stability

To be useful, the characteristics of the nanowire must be stable for thousands of hours (ideally 10 years). Conventional techniques of high-quality oxidation and moderate-temperature passivation with hydrogen should improve the characteristics to the desired values.

4.3
Doping

Doping a Si nanowire during its growth will be different from doping a Si layer during normal chemical vapor deposition (CVD). The catalyzed decomposition of the Si containing gas is unlikely to be accelerated at the same rate as the decomposition of the doping gas, which may or may not be accelerated by the metal catalyst. (In conventional CVD, the gas-phase B/Si composition differs from that in the growing Si layer.) Even when dopant atoms are successfully added to the nanowire, they may segregate to the surface, rather than remaining as bulk dopant atoms within the Si crystal. Dopant segregation is expected for the *n*-type dopants, especially. The *p*-type dopant boron is less likely to segregate to the surface.

4.4
Device Physics

Nanowire transistors can operate in one or more of several modes. If the wire is moderately doped and has a small diameter, the channel can be depleted by a gate voltage, changing a normally ON transistor to the OFF state. Depending on the interface charges, the threshold voltage can be either positive or negative. To be able to deplete the entire diameter of the nanowire, the dopant concentration must be relatively low. The series resistance in the unmodulated region of the nanowire used to interconnect devices will be correspondingly high. Consequently, this mode of operation will require a careful (and perhaps unacceptable) trade-off between the conductance of the interconnecting region and the ability to deplete the modulated region.

Other modes of operation induce mobile carriers in a thin channel adjacent to the gate insulator. The channel can be of the same type as the wire (accumulation mode) or of the opposite type (inversion mode, as in a conventional, bulk MOS transistor). Because of the limited cross section of the nanowire, accumulation-mode operation is practical, as in a fully depleted silicon-on-insulator (SOI) MOSFET; however, the OFF current needs to be carefully controlled by controlling interface charges at *all* surfaces of the nanowire.

When the conductance of a region of the nanowire is to be modulated, the trade-off between the resistance of the interconnecting regions and the ability to modulate the conductance of the nanowire must be considered. If the doping in the modulated region of the nanowire and the interconnecting element is the same, then the nanowire is normally conducting, and the entire thickness of the nanowire must be depleted of carriers by the gate to turn off the conduction. The maximum depletion thickness limits the dopant concentration for a given nanowire diameter. For a nanowire with dopant concentration N, Poisson's equation can be written (within the depletion approximation) in cylindrical coordinates to correspond to the geometry of the nanowire

$$\frac{1}{r}\frac{d^2}{dr^2}(r\phi) = -\frac{qN}{\varepsilon} \tag{2}$$

where ϕ is the potential, N is the ionized dopant concentration, and ε is the permittivity of silicon.

Within the depletion approximation, Eq. (2) can be solved for the maximum depletion-region thickness before surface inversion. This thickness is the maximum distance into the nanowire that the mobile carriers can be depleted by an applied gate voltage or by surface charge to turn off the transistor. Additional gate voltage or surface charge creates mobile carriers of the opposite conductivity type near the surface of the silicon, but does not deplete additional mobile carriers from the neutral regions of the wire. For the cylindrical geometry the maximum depth $r_{d\,max}$ of the depletion region is

$$r_{d\,max} = \sqrt{\frac{12\varepsilon\phi_B}{qN}} = \sqrt{3}x_{d\,max} \tag{3}$$

which is a factor of $3^{1/2}$ greater than the one-dimensional, planar maximum depletion region thickness $x_{d\,max}$ [7].

If the gate completely surrounds the nanowire (the most effective geometry for gate control of the nanowire), $r_{d\,max}$ must be greater than half the nanowire diameter to be able to deplete the entire channel and turn off the transistor. The resulting limit on the dopant concentration is shown in Fig. 6 as a function of the

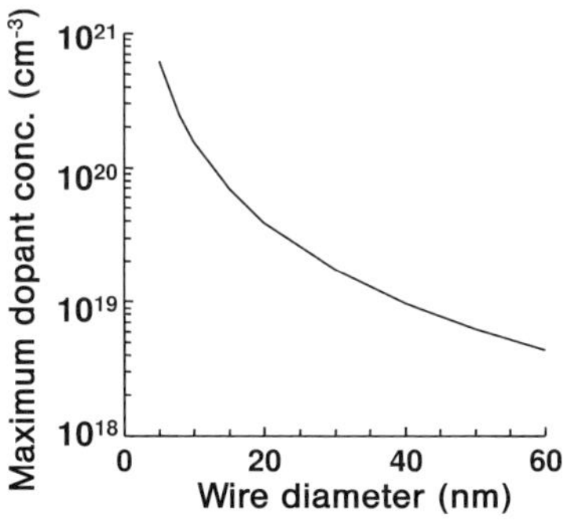

Fig. 6 Calculated maximum dopant concentration allowing complete depletion as a function of nanowire diameter.

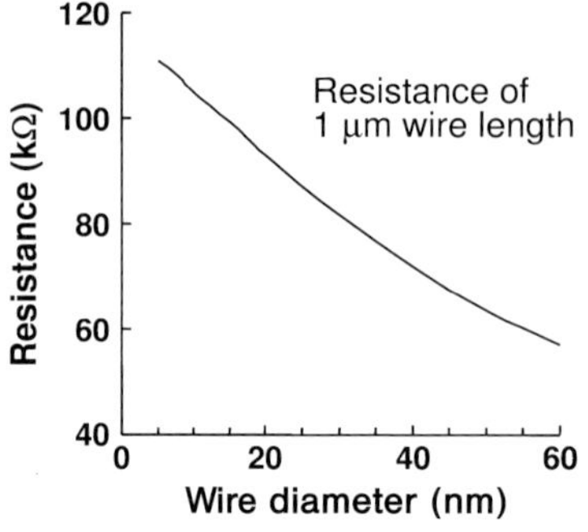

Fig. 7 Resistance of a 1 µm length of a nanowire doped according to Fig. 6 as a function of nanowire diameter.

nanowire diameter. For a 20 nm-diameter nanowire the maximum depletion region thickness must be greater than 10 nm, and the simple estimate of Eq. (3) shows that the dopant concentration is limited to approximately 4×10^{19} cm^{-3}. The limited dopant concentration has several consequences. First, the number of dopant atoms in a short segment of the nanowire (the transistor channel) is small; if the channel length is equal to the wire diameter, the number of carriers in the channel is about 250, with a resulting statistical fluctuation in the number of carriers approaching 10%. Second, the resistance of the nanowire is very high, as shown in Fig. 7. Even for a 1 µm length of the nanowire, the resistance is of the order of 10^5 Ω. With a reasonable voltage, the current is about 10 µA. Even though the capacitance of the next gate is small, the charging time through the nanowire is of the order of several ps, as shown in Fig. 8 (solid line). As the nanowire becomes thinner, the maximum allowable dopant concentration increases, so the resistance does not increase very rapidly with decreasing nanowire diameter. The capacitance that must be charged decreases, and the charging time improves significantly. However, dealing with high resistances and low currents complicates circuit design. In addition, the gate length may be limited to a minimum value by short-channel effects, rather than continuing to scale with the nanowire diameter. The charging time then increases as the nanowire diameter further decreases, as shown by the dashed line in Fig. 8 for a gate length limited to 20 nm.

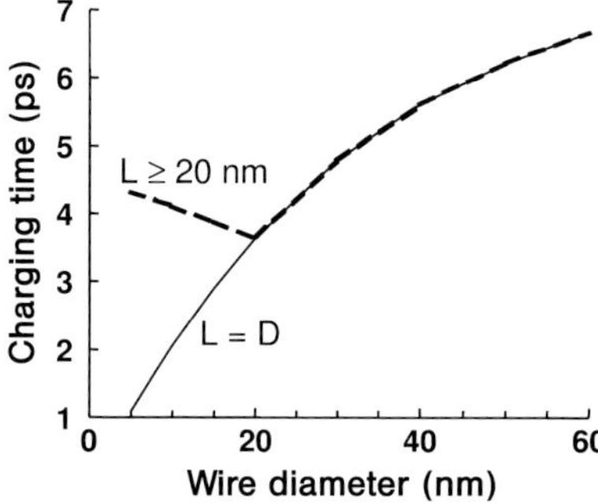

Fig. 8 Time to charge a succeeding transistor through a wire of resistance shown in Fig. 7 as a function of nanowire diameter.

Alternatively, the doping in the interconnecting regions and in the modulated regions can be controlled independently (at the expense of more complex processing). This approach removes the trade-off between the conductivity of the interconnecting regions of the nanowire and conductance modulation in the transistor channel. Again using the example of a 20 nm diameter nanowire, the resistance can be decreased to the order of 1–10 kΩ for a 1 µm length of the nanowire. The higher current decreases the charging time of the next gate through the nanowire to a sub-ps value that will probably not be the dominant limitation on circuit speed. The analysis assumes, however, that an adequate transistor can be designed with an arbitrarily short channel length.

4.5
Mobility

Nanowires with diameters greater than 10 nm are expected to conduct classically. If the dopant atoms are within the silicon nanowire, ionized impurity scattering of carriers will limit the mobility, as in bulk silicon, so the mobility should be similar in a nanowire and in bulk silicon. In bulk silicon transistors, the mobility is further limited by surface scattering at the Si/dielectric interface, especially if the interface is rough or scatters diffusely. For example, nitrogen at a Si/SiO_2 interface causes additional scattering and degrades the surface mobility in a conventional MOS transistor. Similarly, scattering at a rough interface will degrade the mobility in a nanowire device with inadequately formed dielectrics and poorly controlled interfaces.

If the dopant atoms are segregated to the exterior of the nanowire but still contribute mobile carriers to the nanowire, ionized impurity scattering may be reduced. However, the diameter of the nanowire is not markedly different from the thickness of SOI layers being investigated for MOS transistors; thus, a similar effect should be seen in thin SOI devices if it is to be expected in nanowire devices. To the best of our knowledge, significantly enhanced mobility has not been reported for MOS transistors in thin SOI.

4.6
Contacts

Methods of making low-resistance ohmic contacts to silicon are well known and can be employed in making contacts to nanowires. Either the work function of the metal is chosen to produce an accumulation layer in the Si near the contact, or the Si is heavily doped so that the depletion region associated with the surface Schottky barrier is thin enough to allow easy tunneling. Equally important, interface layers between the metal and the Si must be controlled so that they do not limit the contact resistance. Good ohmic contact thus requires control of the native oxide on the silicon surface and its effective removal before deposition of the contact metal.

One non-conventional aspect of the contacts is the possibility of direct contact from the substrate to the nanowire through the grown interface between the substrate and the nanowire. Preliminary studies have shown that ohmic or rectifying contact can be made at this junction by choosing the dopant type in the substrate and nanowire appropriately [15].

5
Conclusion

In this chapter, we have seen how self-assembled nanowires can help integrate and interconnect conventional electronics, microsystems, and nanoelectronics, as well as forming portions of nanoelectronic systems themselves. Challenges remain in forming the nanowires under conditions compatible with the complementary components of the total system. The vast experience of conventional Si technology underlying the integrated-circuit industry can be used to expedite development and integration of nanowire devices.

Acknowledgment

This work was partially supported by DARPA agreement MDA972-01-3-0005.

The author thanks Dr. R. Stanley Williams for useful discussion and encouragement during this work.

References

1 M. T. Bjorkz et al., *Nano Letters* **2002**, *2*, 87.

2 M. S. Gudiksen et al., *Nature* **2002**, *415*, 617.

3 M. H. Huang et al., *Science* **2001**, *292*, 1897.

4 X. Duan et al., *Nature* **2003**, *421*, 241.

5 J. Westwater et al., *J. Vac. Sci. Technol. B* **1997**, *15*, 554.

6 A. M. Morales and C. M. Lieber, *Science* **1998**, *279*, 208.

7 R. S. Muller and T. I. Kamins with M. Chan, *Device Electronics for Integrated Circuits,* 3rd ed., Wiley **2003**.

8 T. I. Kamins, R. Stanley Williams, D. P. Basile, T. Hesjedal, and J. S. Harris, *J. Appl. Phys.* **2001**, *89*, 1008.

9 D. Wang and H. Dai, *Angew. Chem. Int. Ed.* **2002**, *41*, 4783.

10 T. I. Kamins, X. Li, and R. Stanley Williams, *Nano Letters* **2004**, *4*, 503.

11 T. I. Kamins, X. Li, and R. Stanley Williams, *Appl. Phys. Lett.* **2003**, *82*, 263.

12 F. A. Nichols and W. W. Mullins, *Trans. Met. Soc. AIME* **1965**, *233*, 1840.

13 Lord Rayleigh, *Proc. London Math. Soc.* **1878**, *10*, 4.

14 Y. Cui, Q. Wei, H. Park, and C. M. Lieber, *Science* **2001**, *293*, 1289.

15 Q. Tang et al., unpublished.

3D Nanofabrication of Rutile TiO$_2$ Single Crystals with Swift Heavy-Ions

Koichi Awazu, Ken-ichi Nomura, Makoto Fujimaki, and Yoshimichi Ohki

1
Introduction

Titanium dioxide (TiO$_2$) exists in three crystalline polymorphs: rutile (tetragonal), anatase (tetragonal), and brookite (orthorhombic). Rutile electrodes are of interest in the field of semiconductor–electrolyte interfaces and a great amount of work has been done on photocatalysts [1, 2] and solar cells [3]. TiO$_2$ in both rutile and amorphous phases has another potential used in the field of integrated optics and photonic devices for telecommunications. Because rutile TiO$_2$ single crystals are uniaxial, they have a large birefringence and an excellent chemical resistance. TiO$_2$ in the amorphous phase shows low optical loss and high refractive index in the vicinity of wavelengths of 1.5 μm. In this respect, amorphous TiO$_2$ is doped in amorphous SiO$_2$ for planar optical waveguides and arrayed waveguide gratings (AWG) for optical communications [4]. Amorphous TiO$_2$, which has an optical bandgap at 3 eV, is also available as a coating material for the windows of Hg and Xe arc lamps [5].

A new model has emerged in the last decade, in which the band-structure concepts of solid-state physics are applied to radio, microwave, and optical waves [6]. This has led to the invention of photonic crystal structures for controlling electromagnetic waves in two or three dimensions applied for outstanding optical devices, such as superprisms [7, 8], sharp-bend optical wave-guides [9], and high-Q cavities [10]. Among the numerous materials, semiconductors such as silicon, GaAs, and InP have been chosen as research materials because of the abundant know-how about their microfabrication for integrated circuits. One of the most used methods for making microstructures is reactive ion etching (RIE), however, roughness and ripple patterns on the side wall of microstructures, RIE-lag, and etch stop are problems that have been frequently observed [11]. As a result, optical transmittance in AWG is 1000 times lower than in conventional optical fibers, though the compounds of both devices are germanium- or titanium-doped silica glass. It is assumed that the significant optical loss is related to the sidewall roughness of the core in AWG. In other words, nanofabrication with an accuracy within several nanometers is required for optical components for telecommunications. Another difficulty is that materials subjected to etching must react with etchant gas, for example, XeF$_2$ is the gas phase in the RIE process or the solid

The Nano–Micro Interface: Bridging the Micro and Nano Worlds
Edited by Hans-Jörg Fecht and Matthias Werner
Copyright © 2004 WILEY-VCH Verlag GmbH & Co. KGaA, Weinheim
ISBN: 3-527-30978-0

compound can be deposited as a residue. For example, it has been desired to develop a planar waveguide in the form of an erbium-doped optical amplifier for telecommunications, but it is difficult to obtain the ideal microstructure because erbium fluorides generated in the RIE process are solids [12].

Besides the main stream of photonic crystal research based on semiconductors, photonic crystals based on TiO$_2$ have also been reported [13]. Microfabrication techniques for the material have not been established yet, but this material offers many advantages. Reflection loss due to connection between photonic crystals and silica based optical waveguides would be reduced because refractive index values of (2.4($\perp c$ axis) and 2.7(//)) in a rutile TiO$_2$ single crystal are close to those in silica, rather than in semiconductors such as silicon, GaAs, and InP [14]. Furthermore, optical transmission loss of TiO$_2$ is ten times lower than in silicon in the vicinity of 1.5 µm as mentioned previously [15]. New structures of TiO$_2$ are inspired by the 3D geometry of man-made crystals, such as opal structures of TiO$_2$ by sol-gel methods and microstructures of TiO$_2$ by interference photolithography methods [16]. In the sol-gel methods, however, shrinkage and cracking of the gel on drying and heating prevent the production of a nanostructure with a precise contours. In order to achieve nanofabrication by photolithography, precision is determined by the wavelength and coherence of the light source.

Since TiO$_2$ remains a high priority material for photonic crystals, it is desired to fabricate a nanostructure with nano-order flatness and precision. As mentioned, we could not find the solution from the RIE method. Since 1996, the author and coworkers have demonstrated nanofabrication with swift heavy ions. In many kinds of solid materials, the passage of swift heavy ions creates a heavily damaged cylindrical or conical region, which is known as a latent track. It has been known that a latent track turns into a visible track on etching in amorphous SiO$_2$, for example [17] and polyimide [18], because of the enhanced etch rate in the track. Side walls in the etched track seemed very flat, of the order of a few nanometers observed by scanning electron microscopy (SEM). In this chapter, we investigated the preparation of a microstructure with nanoflatness on both bottom and side walls in rutile single crystals by swift heavy ions. The irradiated part was well dissolved in 20% hydrofluoric acid and the pristine part was not dissolved. Furthermore, as an unexpected phenomenon, a 3D structure was created in rutile TiO$_2$ single crystal.

2
Experimental

The samples used in the present experiment are (100) rutile TiO$_2$ single crystals (purity >99.99%, density: 4.25 g/cm^3) synthesized by the Verneuil flame-fusion method. They were cut into plates 500 µm thick and polished. The root-mean-squared roughness of the sample was less than 0.8 nm, estimated by atomic force microscopy (AFM). In order to examine the dependence of crystal orientation, rutile single crystal plates with (111) and (001) faces also used. Ion irradiation using the 12 MV tandem accelerator at the Tandem Accelerator Center, University of

Tab. 1 Changed beam energy and charge after penetrating each foil.

Ion species	Energy [MeV]	H_{up} [µm]	D_s [µm]
I	78.8	0.5	4.6
Br	120	0.9	8.1
	50	0.4	4.0
Cu	110	0.9	8.1
	84.5	0.7	6.3
	15	0.1	0.9
Ti	100	1.0	9.6

Tsukuba (UTTAC) was performed at room temperature in a vacuum with a residual pressure below 1×10^{-3} Pa.

In order to avoid overlap of the latent tracks, a very low accumulated dose, typically 1×10^{10} cm^{-2} was achieved by diffusing the ion beam with a free-standing foil in forward scattering geometry. Simultaneously changed beam energies as well as ion charges are summarized in Tab. 1. For example, an initial beam of 120 MeV I^{11+} was turned into 78.8 MeV I$^{22.7+}$ through Al foil of 3.3 µm. As another method, a direct beam without foils was employed. The accuracy of the forward scattering method and the direct beam were within 10 and 50%, respectively.

Structural change induced by ion irradiation was investigated by X-ray diffraction (XRD) measurements with a Rigaku FR-MDG apparatus. A scanning electron microscope (SEM, Hitachi S-2500CX) and a high-resolution electron microscope (HREM, Hitachi H-9000NAR) were also employed to observe sample surfaces. Chemical etching was performed with 20% hydrofluoric acid (20% HF) at room temperature.

3
Results

To determine the etching rate of rutile (100) single crystal subjected to ion irradiation, we must know the etching rate of pristine rutile (100) single crystal. A rutile (100) single crystal surface was half masked with the conventional resist for lithography, then etching with 20% HF was performed. It was found that etching with 20% HF was not observed, even after one week, with the surface profiler for which the detection limit was a few nanometers.

Fig. 1a shows an AFM image of rutile (100) single crystal surface subjected to 120 MeV Br ion irradiation at an accumulated dose of 8.0×10^{13} cm^{-2} through a 13 µm thick gold stencil mask with array of 32 µm$\times$32 µm open squares. Volume expansion of the irradiated parts was observed as upheavals. The step height of upheaval was eventually saturated against accumulated dose. Saturated step heights (H_{up}) are summarized in Tab. 2.

Tab. 2 Values of the D_s and H_{up} in each ion implantation.

Ion species and acceleration energy	D_s [μm]	H_{up} [μm]
78.8 MeV I	4.6	0.5
120 MeV Br	8.1	0.9
50 MeV Br	4.0	0.4
110 MeV Cu	8.1	0.9
84.5 MeV Cu	6.3	0.7
15 MeV Cu	0.9	0.1
100 MeV Ti	9.6	1.0

(a)

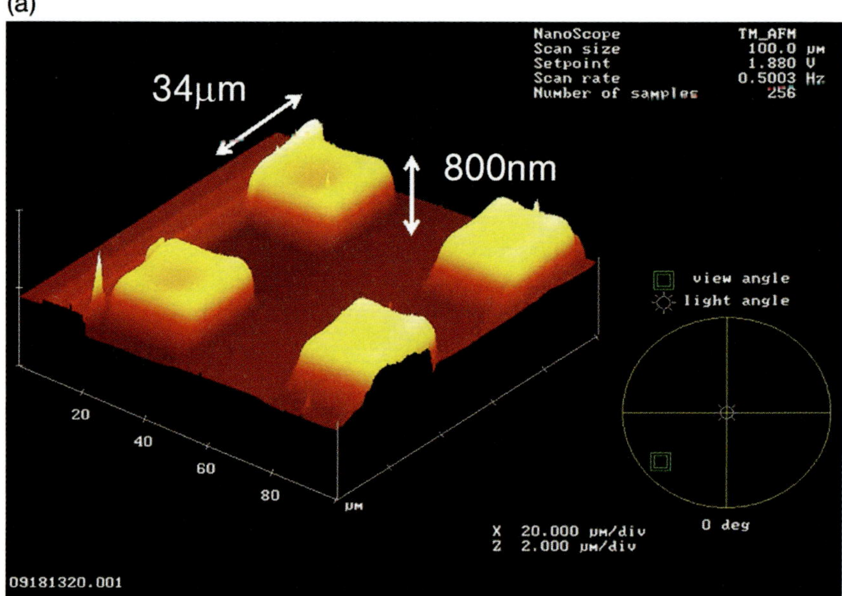

(b)

Fig. 1 (a) An AFM image of a TiO$_2$ surface subjected to 120 MeV Br ion irradiation at an accumulated dose of 8×10^{13} cm^{-2} through 13 μm thick gold mask. (b) SEM image of the TiO$_2$ surface etched by 20% hydrofluoric acid for 40 min. The sample was irradiated with 120 MeV Br ion irradiation to a dose of 8.0×10^{13} cm^{-2} through a 13 μm thick gold mask.

The irradiated rutile (100) single crystal was then immersed in 20% HF for 40 min and SEM observation was shown in Fig. 1b. It is noticed that edges of the hole were rough because of the roughness of a present stencil mask. Also noticed that the bottom of the structure seemed to be very flat. In order to distinguish whether the obtained roughness precisely appeared the roughness of patterns on a stencil mask, a cleaved silicon single crystal was used as a mask with a straight edge.

Fig. 2a shows an SEM image of a rutile (100) single crystal surface subjected to 110 MeV Cu irradiation at an accumulated dose of 8.0×10^{13} cm^{-2}, subsequently soaked in the 20% HF for 40 min. At a glance, an 8.1 µm step was seen between the irradiated and non-irradiated regions. It was also noticed that the surface of the irradiated region as well as the sidewall were very smooth. An AFM observation of the etched part was depicted in Fig. 2b. Roughness was estimated within 2.5 nm.

Fig. 3a shows the etched depth (D) for rutile (100) single crystals subjected to 84.5 MeV Cu irradiation at an accumulated dose of 5.0×10^{13}, 1.0×10^{13}, and 7.0×10^{12} cm^{-2} followed by etching with 20% HF against etching time. Values of D started with negative values, which corresponded to the upheaval height for irradiation. For an accumulated dose of 5.0×10^{13} cm^{-2}, D increased linearly until $D = 6$ µm and saturated with accumulated dose. Hereafter, the saturated D value was referred to as D_e. With decreasing accumulated dose from 5.0×10^{13} cm^{-2} to 7.0×10^{12} cm^{-2} via 1.0×10^{13} cm^{-2}, values of etching rate as well as D_e were also decreased.

The dependence of etching rate on surface orientation was examined with (100), (111), and (001) rutile single crystals subjected to 84.5 MeV Cu irradiation at an accumulated dose of 5.0×10^{13} cm^{-2} and depicted in Fig. 3b. The value of D was increased linearly and saturated against etching time. The value of D_e for (100) single crystal was similar to that for (111) single crystal and higher than for (001) single crystal.

Fig. 4 shows the evolution of D_e against accumulated dose of 84.5 MeV Cu ion. One can apparently define three phases:

- the accumulated dose up to 5×10^{12} cm^{-2} during which evolution of D_e was less than the detection limit of the present method,
- from 5×10^{12} cm^{-2} to 1×10^{13} cm^{-2} during which D_e commenced to increase, and
- over 1×10^{13} cm^{-2} during which D_e was saturated in the vicinity of 6 µm.

D_s, which was defined as saturation depth of D_e (Fig. 4a), for each ion in the various energies examined in this experiment is summarized in Tab. 2. Notice that the values of H_{up} were always around 10% of the value of D_s.

Fig. 5 shows the evolution of XRD spectra of rutile (100) single crystals subjected to the 84.5 MeV Cu ion irradiation. The spectrum (1) represents the XRD spectrum of the pristine sample. Two peaks at 39.2° and 84.3° in the spectrum (1) for pristine one have been assigned to the (200) and (400) planes of the rutile phase. XRD spectra of rutile single crystal subjected to the 84.5 MeV Cu ion irradiation at an accumulated dose of (2) 4.0×10^{13} cm^{-2}, (3) 7.0×10^{13} cm^{-2}, and (4)

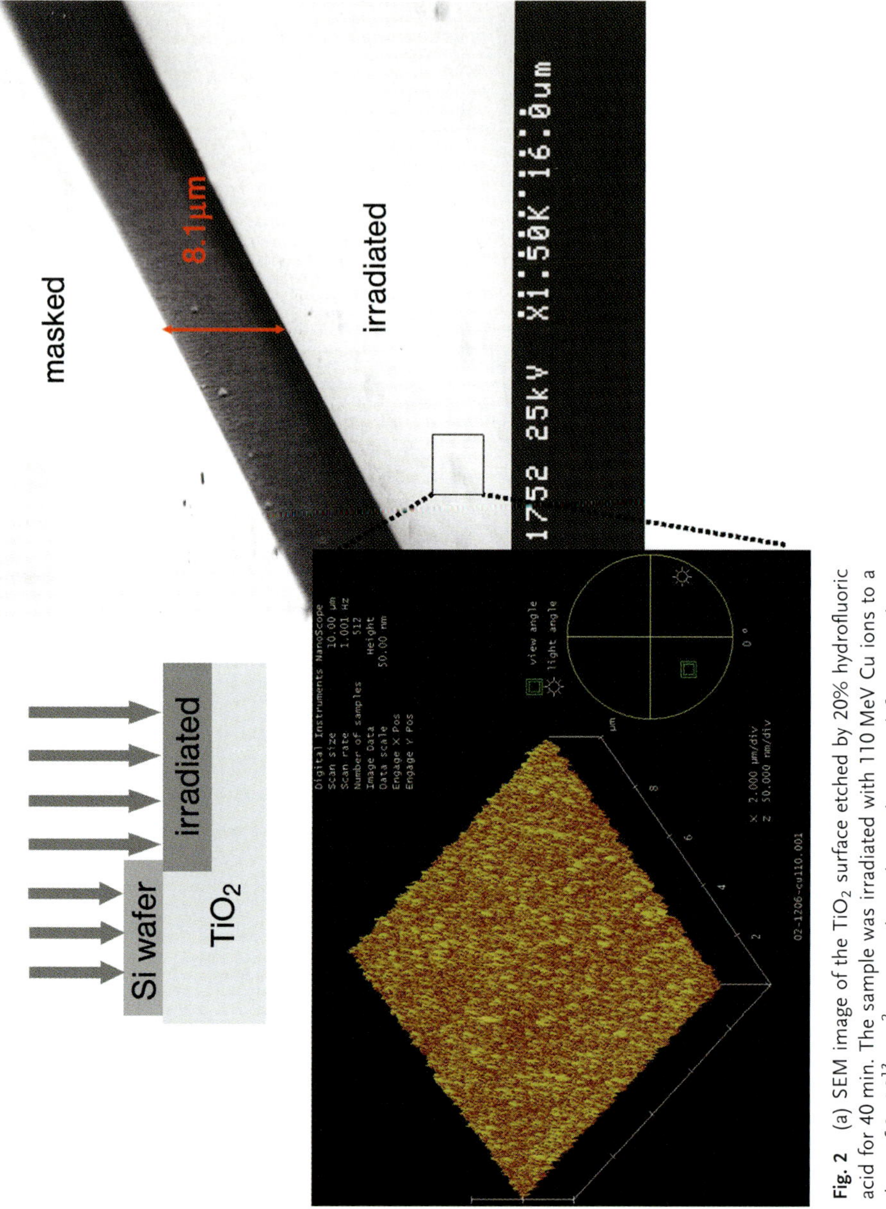

Fig. 2 (a) SEM image of the TiO₂ surface etched by 20% hydrofluoric acid for 40 min. The sample was irradiated with 110 MeV Cu ions to a dose of 8×10^{13} cm^{-2} prior to the etching. The upper left area in the image was masked with a 1 mm thick cleaved silicon plate during the irradiation. (b) An AFM image of the etched surface in (a).

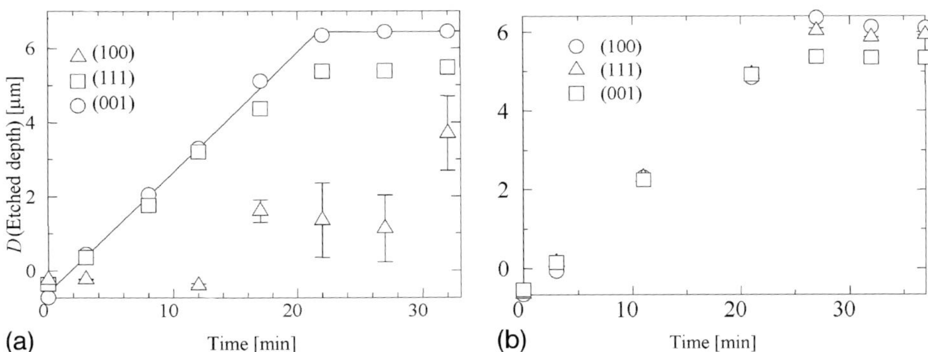

Fig. 3 (a) Correlation between the etching time and the D_e in the samples irradiated with 84.5 MeV Cu ions to doses of 5.0×10^{13} cm^{-2} (open circles), 1.0×10^{13} cm^{-2} (open squares), and 7.0×10^{12} cm^{-2} (open triangles). The solid line is the result of a least-square-fitting for the data obtained in the samples irradiated at an accumulated dose of 5.0×10^{13} cm^{-2}.
(b) Dependence on surface orientation of the values of D against etching time. All samples were irradiated with 84.5 MeV Cu ions at an accumulated dose of 5.0×10^{13} cm^{-2} prior to etching.

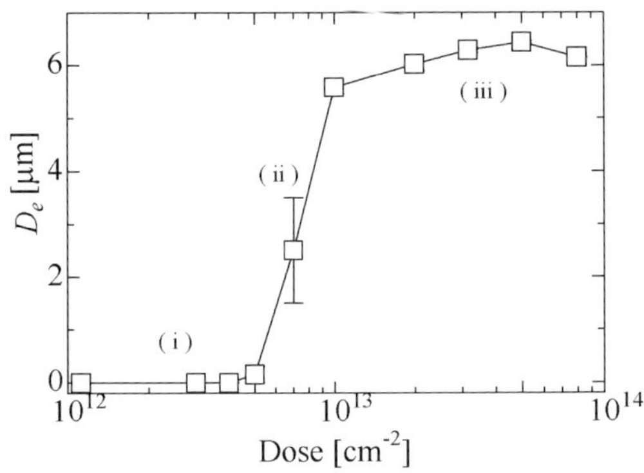

Fig. 4 Correlation between the D_e and the dose of 84.5 MeV Cu.

5.0×10^{13} cm^{-2} were also shown. Intensity of two peaks for (200) and (400) planes was decreased with increasing of accumulated dose. Noticed that new peaks were generated at 38.3° and 82.5° in the spectra (2). With increasing accumulated dose, the intensity of new peaks was decreased with shifting their peak positions toward the smaller angular side (spectrum (3)), and finally these peaks disappeared completely (spectrum (4)).

Fig. 6 shows HREM bright images at different scales of the rutile (100) single crystal subjected to the irradiation of 84.5 MeV Cu ions at an accumulated dose of 1.0×10^{10} cm^{-2}, which was an appropriate dosage to observe individual latent

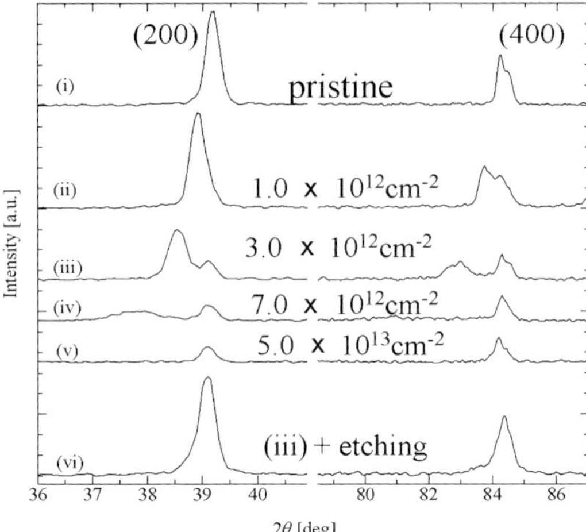

Fig. 5 XRD spectra of the as-received sample (1), after the
84.5 MeV Cu Ion irradiation to a dose of 4.0×10^{13} cm^{-2} (2),
7.0×10^{13} cm^{-2} (3), and 5.0×10^{13} cm^{-2} (4). The peaks at
39.2° and 84.3° are due to the (200) and (400) planes of
the rutile phase, respectively.

tracks without overlap. In Fig. 6a, two white spots are seen near the upper left
and the lower right corners. In order to observe the white spot carefully, the SEM
image was obtained by enlargement (b). Latent tracks were recognized as the
milky color and an amorphous "island" with a radius of 0.9×0.2 nm observed in
the "sea" of the crystallographic orientation. Also note that atoms were aligned far
from the latent track, and the atoms near the latent track were disordered. The

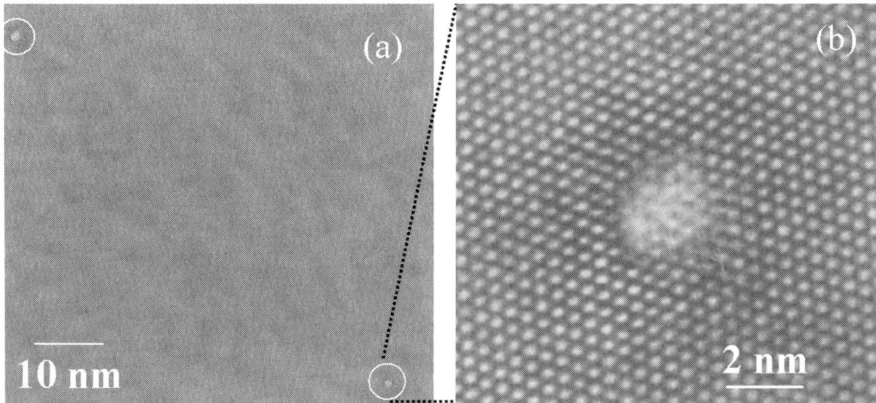

Fig. 6 HERM bright images of a (100) rutile TiO$_2$ single crystal subjected
to 84.5 MeV Cu ions at an accumulated dose of 1.0×10^{10} cm^{-2}.
Scales are denoted in each figure.

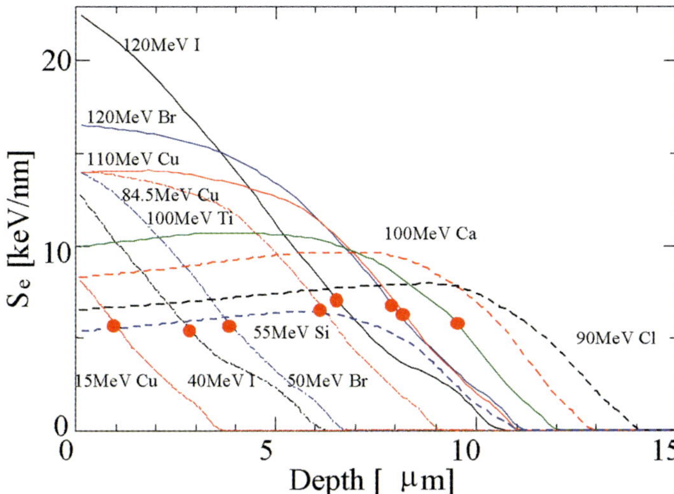

Fig. 7 Calculated electronic stopping power S_e against depth from the sample surface. The closed circles on the curve indicate S_e at the depth where the etching stopped.

disorderly atoms might originate from the stress introduced by passage of a swift heavy ion. The radius of the stressed region was estimated to be about 1.8 nm.

Fig. 7 shows electronic stopping power S_e calculated by SRIM 98 code as a function of the depth from the sample surface [19]. The curves represent the irradiation of 78.8 MeV I, 120 MeV Br, 50 MeV Br, 110 MeV Cu, 84.5 MeV Cu, 15 MeV Cu, and 100 MeV Ti ions. Values of D_s summarized in Tab. 2 are plotted with closed circles in Fig. 7. The point at $S_e=0$ was close to the depth where the ion was stopped. Dotted lines in Fig. 7 imply the region where the etching was observed. At first glance, the saturated D_e (D_s) was shallower than the depth at $S_e=0$. For instance, the D_s and the depth at $S_e=0$ for 78.8 MeV I ion were 4.6 and 8.6 μm, respectively. Furthermore, the first order feature was that the values of S_e at D_s presented with closed circles were always located around the solid line of 6.2 keV/nm irrespective of incident ions or acceleration energies. In other words, a threshold electronic stopping power of 6.2 keV/nm was apparently necessary to commence etching.

Fig. 8 shows the evolution of XRD spectra of rutile (100) single crystal subjected to Ca ion irradiation with various energies at 15.3, 31.9, 50.8, 72.3, 82.0 and 94.7 MeV at an accumulated dose of 8.0×10^{13} cm^{-2}. It seems important that intensity of the peaks at 39.2° and 84.3° was decreased until 50.8 MeV and recovered to the original intensity with increase of the beam energy over 50.8 to 94.7 MeV.

The rutile (100) single crystal subjected to irradiation of 72.3 MeV Ca ion, for example, was apparently not etched with 20% HF. But when the irradiated single crystal was cut perpendicularly to the surface, and subsequently immersed into the 20% HF, a layer in the sample was selectively etched. Fig. 9 shows SEM

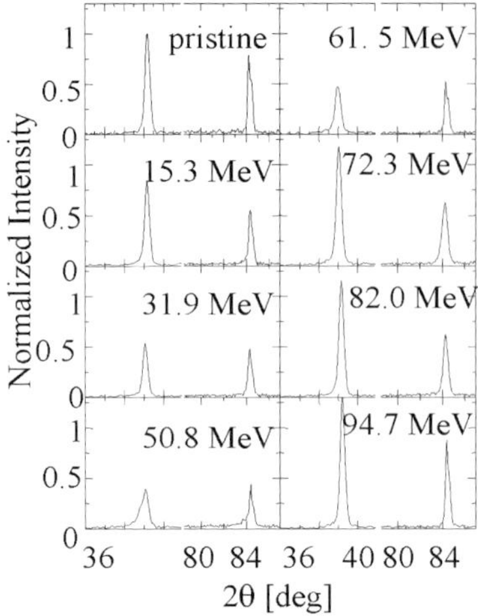

Fig. 8 XRD spectra after irradiation of Ca ions to a dose of 8.0×10^{13} cm^{-2} with various acceleration energies.

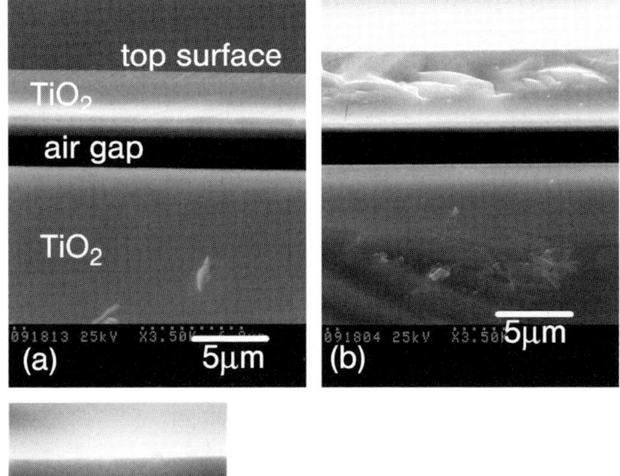

Fig. 9 SEM images of the cross section of the sample etched following the: (a) 72.3, (b) 82.0, and (c) 84.5 MeV Ca ion irradiation to a dose of 3.0×10^{14} cm^{-2}. The ions were irradiated from the top of the image.

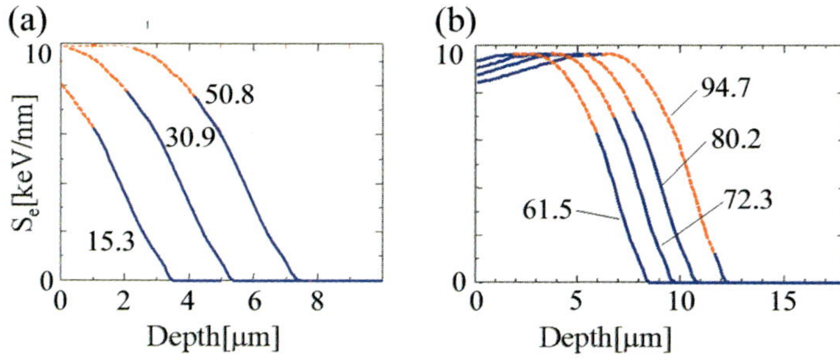

Fig. 10   Calculated values of S_e of Ca and Cl ions as a function of the depth from the sample surface. The curves A and B are for the values of 84.5 MeV Ca ion and 100 MeV Cl ion, respectively. The broken lines on the curves indicate S_e at the region where the etching is observed.

images of the cross section of the sample subjected to (a) 72.3, (b) 82.0, and (c) 84.5 MeV Ca ion irradiation at an accumulated dose of $3.0\times10^{14}\,cm^{-2}$ subsequently cut perpendicularly, then soaked in the 20% HF. The top surfaces where the ion beams were launched from the accelerator are shown with arrows. It is important that the layer of top surface was not etched. An air gap created by etching was observed between the upper and lower layers of rutile single crystal. The thicknesses of the top layers were estimated to be about (a) 4 μm, (b) 5 μm, (c) 5 μm and of air gaps were (a) 3 μm, (b) 2.5 μm, (c) 5 μm.

Electronic stopping power, S_e, against depth was calculated and is depicted in Fig. 10. The 20% HF etching was performed for rutile (100) single crystals irradiated with Ca ions of various energies. The region where the etching was observed was indicated by dotted lines in Fig. 10. Note that etching was commenced from the top surface in the samples irradiated with 15.3, 31.9, or 50.8 MeV Ca ions, while the inside etching was observed in the samples irradiated with 72.3, and 82.0 MeV Ca ions. An inside etching was also observed in the sample irradiated with Cl ions accelerated by 77.0 MeV and higher at an accumulated dose of $1.0\times10^{15}\,cm^{-2}$.

4

Discussion

In the XRD spectra shown in Fig. 5, the 84.5 MeV Cu ion irradiation decreases the intensities of the peaks assigned to the original rutile crystal structure and induces the two peaks at the lower angle side. According to the Bragg law, a peak shift toward a lower angle indicates an increase in lattice constant. Actually, the expansion was observed as the upheaval by ion irradiation as summarized in Tab. 2. Since TiO_2 crystal in rutile phase has the largest density among TiO_2 polymorphs, it seems that the expansion is due to a structural change induced by the ion irradiation. Decrease in intensities of both peaks at 39.2° and 84.3° means the degradation of a crystal structure, that is the formation of amorphous phase.

From these considerations, it was found that the ion irradiation induces the amorphous phase and the expansion of lattice in a rutile TiO$_2$ single crystal. The original peak intensity decreased when the ions were irradiated to a dose of 4.0×10^{12} cm^{-2} (2), but the intensity hardly changed with further irradiation. This does not imply the remaining of the crystal structure that the original peaks were still existed after irradiation, because X-ray intensity from Cu K_α was reduced by half through 50 μm TiO$_2$, which is deeper than the ion penetration depth. In other words, the undamaged layer below the damaged layer could be detected by XRD measurement.

The HREM bright image of the sample surface irradiated with 84.5 MeV Cu ions shows the damage induced by one ion as seen in Fig. 6. The milky color island where crystalline structure is not observed is considered to correspond to the amorphous region induced by the passage of one ion. Recalling the previous discussion, it was assumed that the stressed region surrounding the milky color island corresponded to the region where the lattice constant becomes larger.

As mentioned above, the passage of the swift heavy-ions in the rutile TiO$_2$ single crystal induced the increase in a lattice constant as well as the amorphous phase. In order to clarify whether the etched region was the amorphous region or the stressed region due to the increase of lattice constant etched part or both, the following calculation was performed. We evaluated the area of the amorphous region. The area of amorphous region can be calculated by the Poisson law [20]

$$F_\mathrm{d} = 1 - \exp(-A\phi) \tag{1}$$

where F_d is the ratio of the amorphous area against the sample surface area, A area of the amorphous region induced by one ion, and ϕ the ion dose. As shown in Fig. 6, the amorphous region introduced by one 84.5 MeV Cu ion was estimated by the equation $\pi \times 0.9^2$ nm^2. The solid curve in Fig. 11 shows the change in F_d as a function of dose ϕ in the case of the irradiation of 84.5 MeV Cu ions. As seen in Fig. 11, dosage of more than 3.0×10^{14} cm^{-2} is required to make the amorphous area cover the entire sample surface. However, recalling Fig. 4, etched depth was saturated at an accumulated dose of 2.0×10^{13} cm^{-2}, which is one order of magnitude smaller than the calculated dose by which the amorphous area covers the entire sample surface. Moreover, the etching can be observed even in the sample irradiated at an accumulated dose of 7.0×10^{12} cm^{-2} (Fig. 4) where only 20% of the surface becomes amorphous according to the calculated result. These results indicate that etching cannot be explained with amorphous model. It is another candidate that the stressed lattice region surrounding the amorphous region with a radius of 1.8 nm was also etched. The radius of the stressed lattice region was around 1.8 nm. The broken lines in Fig. 11 indicate the ratio of both the amorphous and the transitional areas against the sample surface area obtained by applying the area of the amorphous and the transitional regions, that is $A = \pi \times 1.8^2$ nm^2, in Eq. (1). The result indicates that the amorphous and the stressed lattice regions cover 50% of the sample surface with the irradiation dose of 7.0×10^{12} cm^{-2}, where the surface etching is observed as mentioned above.

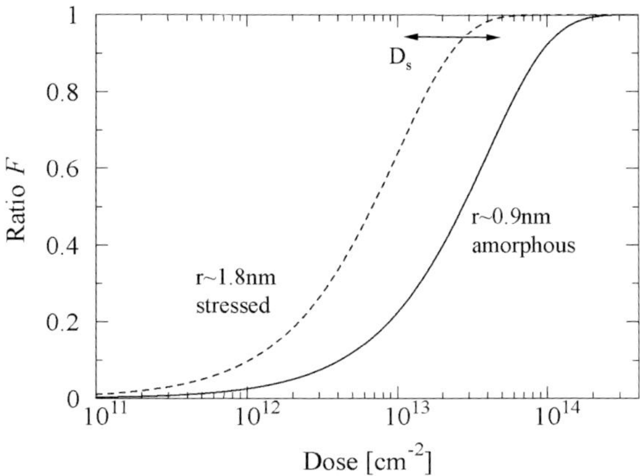

Fig. 11 Change in the ratio of the amorphous area against the sample surface area (solid curve) and that of the amorphous and the transitional area against the sample surface area (dotted lines) with respect to the irradiated dose of 84.5 MeV Cu ions.

Fig. 5 (5) and (6) show the XRD spectra of the sample irradiated 84.5 MeV Cu ions to a dose of 3.0×10^{12} cm^{-2}, which is classified in phase (1) in Fig. 4, before (5) and after (6) etching by the 20% HF. The spectrum before etching shows the peaks at 38.5° and 83.0° assigned to the stressed lattice region. These two peaks disappeared and the XRD spectrum was returned to the pristine spectrum by etching. From the results, it was deduced that the stressed lattice region was also etched.

Experimental results mentioned next also support our models, which proposed the stressed lattice as well as the amorphous regions were etched. The etching rate of the sample irradiated with 84.5 MeV Cu ions to a dose of 7.0×10^{12} cm^{-2} is slower than that of 5.0×10^{13} cm^{-2} as shown in Fig. 3a. Furthermore, the etching does not seem to be uniform in the case that the irradiation dose is 7.0×10^{12} cm^{-2}, which results in the large error bars for these data shown in Fig. 3a. These observed results indicate that the etching rate of the stressed lattice region is slower than that of the amorphous region. It was assumed that non-uniform and slow etching of the sample with irradiated dose of 7.0×10^{12} cm^{-2} was originated from the co-existence of the two phases with different etching ratios.

In phase (2) of Fig. 4, steps were created between irradiated and masked parts, as observed after etching. However, the etched depth was shallower than D_s. The value of D_s was the largest obtained by a certain ion irradiation condition and is achieved by irradiating with enough ions to cause damage all over the irradiated region. The shallower depth in phase (2) indicates that the damage in rutile TiO_2 single crystal due to a swift heavy ion is conical rather than cylindrical. This is because, if the damage were cylindrical, the damage would be uniform from the top to the bottom and the etching to the depth of D_s would be observed even in phase (2).

Kats et al. reported that a slow electron gives higher energy density to a target than a fast electron [21]. It has also been reported that a faster ion makes a smaller amorphous track in ion-irradiation onto yttrium iron garnet when two ions with different speeds have almost equivalent S_e value [22]. These reports indicate that the speed of irradiated particles is also an important factor to induce damage in targets. The reason why the surface was not etched in samples irradiated with 72.3–100 MeV Ca or 33.6–100 MeV Cl ions might be the speeds of these ions. In other words, the speeds of Ca ions accelerated with 72.3 MeV or higher or Cl ions accelerated with 33.6 MeV or higher are too fast to create the etchable region. As the ions go into a sample, the ions slow down, and then the ions create the etchable region.

In order to further discuss the effect of ion speeds, we have analyzed the radial distribution of energy deposited around the path of the ion reported earlier [23], in which ion speeds are taken into account for the evaluation of the energy distribution. This distribution, which is found from the Rutherford formula and experimental compensation, is depicted below.

$$D(t) = D_1(t)(1 + K(t)) \tag{2}$$

where

$$D_1(t) = \frac{Ne^4 Z^{*2}}{amc^2\beta^2 t} \left(\frac{\left(1 - \dfrac{t+\theta}{T+\theta}\right)^{1/a}}{t+\theta} \right) \text{(CGS)} \tag{3}$$

$$K(t) = A\left(\frac{t-B}{C}\right) \exp -\left(\frac{t-B}{C}\right) \tag{4}$$

$D(t)$ is the density of deposited energy in a coaxial cylindrical shell between radius t and $t+dt$ from the path of an ion in case that an ion that has incident relative velocity $\beta = c/v$ and effective charge Z^* is irradiated into the sample that contains N electron per cm^{-3}. e, m, and c are the electron charge, the mass of electron, and the speed of light, respectively.

$$\theta = 4.17 \times 10^{-8} \text{ [g cm}^{-2}] \tag{5}$$

$$T = kW^a \tag{6}$$

where

$$k = 6 \times 10^{-6} \text{ [g cm}^{-2} \text{ keV}^{-a}] \tag{7}$$

$$W = 2mc^2\beta^2/(1 - \beta^2) \tag{8}$$

Effective charge is represented as

$$Z^* = Z[1 - \exp(-125\beta Z^{-2/3})] \tag{9}$$

Z is the atomic number of the irradiated ion.

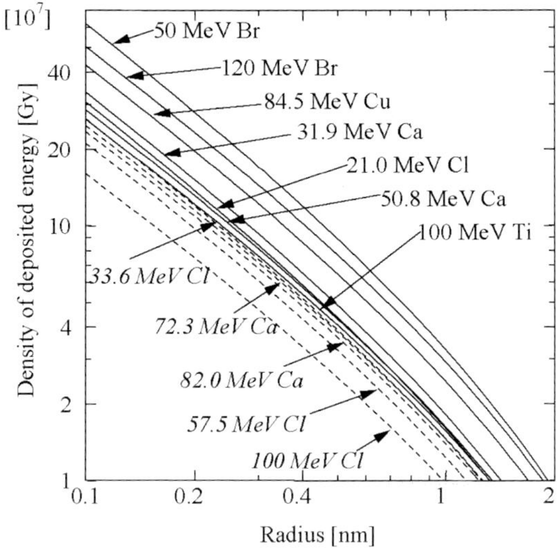

Fig. 12 The radial distribution of deposited energy at the sample surface applied to rutile single crystal. The distribution was reported by Waligorski et al. [23].

Fig. 12 shows the radial distribution of energy deposited on a TiO_2 surface around the path of ions used in this experiment. The solid curves are for the ion irradiation conditions that make rutile TiO_2 single crystal etchable from surface, while the dotted curves are for those that do not. As seen in Fig. 12, the deposited energy in the case that the etching from the sample surface occurs always comes higher than that the etching does not occur. This result indicates that the value of energy deposited by an ion is the most important factor to decide whether the irradiated TiO_2 single crystal becomes etchable or not.

5
Conclusions

We have investigated the structural change in rutile TiO_2 single crystal induced by swift heavy-ion irradiation. It has been clarified that a rutile TiO_2 single crystal is not etched by hydrofluoric acid, however, the amorphous region and the stressed lattice region generated by the irradiation of swift heavy ions become etchable by hydrofluoric acid. When the ion whose value of electronic stopping power was decreased with depth of rutile TiO_2 single crystal and the depth that could be etched was limited to the point where the values of electronic stopping power was decayed to 6.2 keV/nm. It was also found that when the ion whose maximum value of electronic stopping power was located at inside of TiO_2, inside could be selectively etched keeping the surface rutile phase and we can obtain the hollow structure. Radial distribution of energy deposited around the path of the ion showed that there was a threshold whether surface etching can be observed or not. Since the surface and aspect shape of rutile TiO_2 single crystals after etching was very flat in the order of nanometer, it was believed to enable to fabricate three-dimensional nanofabrication by use of this etching technique.

This method also makes it possible to fabricate rutile TiO_2 plates thinner than a few microns with nano-order flatness, which is difficult by conventional methods. Our method will be available for the processing of solar cells, photonic catalysts, or photonic crystals, which require nanofabrication technique.

Acknowledgments

We would like to express our appreciation to T. Sekiguchi of the Kagami Memorial Laboratory for Materials Science and Technology, Waseda University for his help in XRD measurements and Y. Nagasawa, Waseda University for his help in our experiment. This work was financially supported by the Budget for Nuclear Research of the Ministry of Education, Culture, Sports, Science and Technology, based on the screening and counseling by the Atomic Energy Commission and partly supported by a Grant in-Aid for Scientific Research from the Ministry of Education, Culture, Sports, Science and Technology (12450132).

References

1 A. Fujishima and K. Honda, *Nature* **1972**, *238*, 37.

2 A. Mills and S. Le Hunte, *J. Photochem. Photobiol. A* **1997**, *108*, 1–35.

3 R. K. Karn and O. N. Srivastava, *Int. J. Hydrogen Energy* **1999**, *24*, 27.

4 C. Tosello, F. Rossi, S. Ronchin, R. Rolli, G. C. Righini, F. Pozzi, S. Pelli, M. Fossi, E. Moser, M. Montagna, M. Ferrari, C. Duverger, A. Chiappini, C. De Bernardi, *J. Non-Cryst. Solids* **2001**, *284*, 230.

5 T. C. Lu, L. B. Lin, S. Y. Wu, J. Chen, Y. Y. Ahang, *Nucl. Instrum. Meth. B* **2002**, *191*, 236.

6 E. Yablonovitch, *Phys. Rev. Lett.* **1987**, *58*, 2059.

7 S. Y. Lin, V. M. Hietala, L. Wang, and E. D. Jones, *Opt. Lett.* **1996**, *21*, 1771.

8 H. Kosaka, T. Kawashima, A. Tomita, M. Notomi, T. Tamamura, T. Sato, and S. Kawakami, *J. Lightwave Technol.* **1999**, *17*, 2032.

9 A. Mekis, J. C. Chen, I. Kurland, S. Fan, P. R. Villeneuve, and J. D. Joannopoulos, *Phys. Rev. Lett.* **1996**, *77*, 3787.

10 P. R. Villeneuve, S. Fan, J. D. Joannopoulos, K. Y. Lim, G. S. Petrich, L. A. Kolodziejski, and R. Reif, *Appl. Phys. Lett.* **1995**, *67*, 167.

11 C. O. Jung, K. K. Chi, B. G. Hwang, J. T. Moon, M. Y. Lee, J. G. Lee, *Thin Solid Films* **1999**, *341*, 112.

12 Y. Kondo, T. Nagashima, S. Takenobu, N. Sugimoto, S. Ito, *Proc. OFC* **2002**, *11*.

13 M. Lanata, M. Cherchi, A. Zappettini, S. M. Pietralunga, and M. Martinelli, *Opt. Mater.* **2001**, *17*, 11.

14 D. Mergel, *Thin Solid Films* **2001**, *397*, 216.

15 S. Yamazaki, N. Hata, T. Yoshida, H. Oheda, A. Matsuda, H. Okushi, K. Tanaka, *J. Phys. C* **1981**, *42*, 4–297.

16 A. Shishido, I. B. Diviliansky, I. C. Khoo, T. S. Mayer, S. Nishimura, G. L. Egan, and T. E. Mallouk, *Appl. Phys. Lett.* **2001**, *79*, 3332.

17 K. Awazu, S. Ishii, K. Shima, S. Roorda, and J. L. Brebner, *Phys. Rev. B* **2000**, *62*, 3689.

18 C. Trautmann, W. Brüchle, R. Spohr, J. Vetter, and N. Angert, *Nucl. Instrum. Meth. B* **1996**, *111*, 70.

19 J. F. Ziegler, J. P. Biersack, and U. Littmark, *The Stopping Power and Ranges of Ions in Matter, Vol. I*, Pergamon, New York **1985**.

20 T. A. Tombrello, *Nucl. Instrum. Meth. B* **1984**, *2*, 555.

21 R. Kats and E. J. Kobetich, *Phys. Rev.* **1969**, *186*, 344.

22 A. Meftah, F. Brisard, J. M. Costantini, M. Hage-Ali, J. P. Stoquert, F. Studer, and M. Toulemonde, *Phys. Rev. B* **1993**, *48*, 920.

23 M. P. R. Waligorski, R. N. Hamm, and R. Kats, *Nucl. Tracks Radiat. Meas.* **1986**, *11*, 309.

III
Applications

Nanoparticles-Based Chemical Gas Sensors for Outdoor Air Quality Monitoring

Marie-Isabelle Baraton and Lhadi Merhari

1
Objectives

The general public awareness of the consequences of urban air pollution on health has led decision-makers to define stringent regulations for air quality monitoring (AQM) and to set the thresholds of acceptable pollutant concentrations at very low levels. In the European Union, every country has been instructed to establish a network of AQM stations in its main cities and to inform citizens about the air quality on a daily basis. These bulky stations based on complex equipment often including differential optical absorption (DOA) spectroscopy allow precise concentration measurements of different kinds of gaseous pollutants in air. However, one of the major problems that cities are facing is the overall cost of the AQM stations, which may reach up to several millions Euros for each unit. Only large and rich cities can afford them but, still, in an insufficient number for establishing a dense network and for obtaining full coverage. Moreover, the lengthy air sampling and data processing do not allow "real time" dissemination of the information to the public.

Our objective is to provide alternative AQM systems based on cost-effective semiconductor gas sensors. By significantly reducing the cost of the stations, it will become possible to implement dense networks in every large city in every country. In addition, due to the tiny size of semiconductor sensors, it will become possible to integrate the different sensors in small sensing units, thus transforming the bulky expensive AQM stations into cost-effective portable devices. But, to this end, the performance of the semiconductor gas sensors have to be enhanced specially in terms of high sensitivity to gases and low cross-sensitivity to humidity.

In two projects funded by the European Commission under the BRITE-EURAM III and IST programs (Contracts No. BRPR-CT95-0002 and IST-1999-12615), our Consortia worked on the improvement of the semiconductor gas sensor characteristics by using semiconducting nanoparticles for CO, NO, NO_2, and ozone detection. This chapter summarizes the different stages in our progress towards the improvement of the nanoparticle sensors and the achievements of the Consortia to date.

The Nano–Micro Interface: Bridging the Micro and Nano Worlds
Edited by Hans-Jörg Fecht and Matthias Werner
Copyright © 2004 WILEY-VCH Verlag GmbH & Co. KGaA, Weinheim
ISBN: 3-527-30978-0

2
Current Status of Semiconductor Sensors

The most popular semiconductor gas sensors are commercialized by the Japanese company Figaro Engineering Incorporated. Figaro gas sensors are solid-state devices composed of sintered metal oxides (mainly tin oxide, SnO_2). As for all resistive gas sensors, they detect gases through variations of the electrical conductivity when reducing or oxidizing gases are adsorbed on the semiconductor surface. Owing to their low cost, the semiconductor sensors are very popular for indoor air quality (IAQ) control. However, they are not suitable for outdoor air quality (OAQ) monitoring due to their cross-sensitivity to humidity and their high detection thresholds [1].

To illustrate the improvements that had to be made, Tab. 1 gives the maximum authorized concentrations of pollutants that have been defined by the European Environment Agency (EEA) for the European Union [2]. Obviously, any kind of instrument aimed at measuring outdoor pollutant concentrations must be capable of reliably detecting these concentrations. In particular, gas sensors must have detection thresholds well below these authorized concentrations so that most of the pollution episodes can be measured accurately. Tab. 1 compares the detection thresholds we targeted to meet the EU directives with the typical sensitivities of commercial electrochemical and semiconductor sensors as available in 2000. It is clear that none of the commercial sensors had sufficiently low detection thresholds for any of the most common air pollutants, namely CO, NO_2, NO, and O_3.

Tab. 1 Comparison between the gas detection thresholds required to detect the maximum authorized concentrations of pollutants in air and the gas detection thresholds of commercial sensors.

Polluting gases	CO [ppm]	NO₂ [ppb]	NO [ppb]	O₃ [ppb]
Maximum authorized concentrations in air (EU directives)	43 ½ hour average	105 ½ hour average	800 ½ hour average	60 ½ hour average
Target for the detection threshold	3	50	100	20
Commercial sensors: Electrochemical	5	600	900	200
Semiconductor (typical data available in 2000)	100			

3
New Paradigms for the Advancement of Semiconductor Sensors

3.1
Advantage of Using Nanoparticles

In resistive sensors, the grain or crystallite size is one of the most important factors affecting the sensing properties. It has been demonstrated that when the particle radius becomes comparable to the depth of the space-charge layer, the space-charge region can develop in the whole crystallite, thus leading to a drastic resistance increase [3]. For example, the sensitivity of sensors based tin oxide nanoparticles dramatically increases when the particle size is reduced down to 6 nm. Below this critical grain size, the sensor sensitivity rapidly decreases [3]. Because it has been calculated that the Debye length of SnO_2 is $D = 3$ nm at $250\,°C$ [4], it seems that the highest sensitivity is reached when the particle diameter corresponds to $2D$. In addition, the high surface-to-bulk ratio of nanoparticles allows a larger density of molecules to adsorb on the surface, thus leading to a larger effect on the electrical conductivity. We have indeed experimentally shown that the use of nanosized semiconductor particles in the fabrication of chemical gas sensors via thick film technology greatly enhances the sensor sensitivity [5, 6].

3.2
Control of the Physical and Chemical Properties of Nanoparticles

Although several materials, including indium oxide and tungsten oxide, were tested [7], our efforts essentially focused on tin oxide. At first, an empirical approach led to the conclusion that the optimum balance between excellent sensor sensitivity and reasonable nanoparticle production yield (that is reasonable production costs) was reached for tin oxide particles having a diameter around 15 nm. Simultaneously, it clearly appeared that the nanoparticles had to be very systematically controlled to exactly define the synthesis parameters, thus ensuring the reproducibility of the sensor characteristics over different batches. The particle size and shape, the particle size distribution, the crystalline state, the stoichiometry, and the chemical composition were studied by transmission electron microscopy (TEM), X-ray diffraction (XRD) spectrometry, energy dispersive X-ray (EDX) analysis, and Fourier transform infrared (FTIR) spectroscopy. In addition, the control of the surface chemistry of the nanoparticles was shown to be a key issue in the improvement of the sensor reliability, reproducibility, and performance. The surface chemistry was investigated by Fourier transform infrared spectroscopy, which was shown to be an extremely relevant technique to obtain a thorough understanding of the surface phenomena at the origin of the gas detection mechanism [8]. This fundamental approach was considered as a critical step to refine the sensor optimization by tailoring both the surface chemical composition and reactivity of the nanoparticles during and eventually after their synthesis [9].

3.3
Optimization of the Screen-Printing Process

While the particle size reduction is an essential factor for sensitivity increase, it rapidly appeared that this factor alone was not sufficient as the grain size had also to be retained during the entire sensor fabrication process. The usual method to fabricate thick-film gas sensors consists in dispersing the semiconducting particles in a suitable solvent to obtain an ink with appropriate viscosity. This ink is then printed on a substrate (generally an alumina tile) bearing electrodes with a heating element on the back side. The printed substrate is then dried and fired at controlled temperatures to avoid both shrinkage and film cracking [5]. This standard procedure used for microparticles had to be modified for nanoparticles in order to obtain a crack-free sensitive layer and to avoid grain growth. It was additionally concluded from experiments that the suppression of agglomerates should lead to improved adhesion and improved film quality where cracks due to inhomogeneous heat transfer cannot appear. The screen-printing process was therefore carefully adapted taking into account the specific and complex behavior of nanoparticles [5].

4
Results

4.1
Characterization of Nanoparticles

The SnO_2 nanoparticles were obtained by evaporating pellets made of compressed micropowder with the pulsed radiation of a Nd:YAG-laser, followed by nucleation of nanoparticles in a controlled aggregation gas [10]. The chamber was kept under pure oxygen to avoid any contamination. The advantage of this synthesis method is that there is no contamination by non-dissociated precursors and there is no surface contaminant (powder yield: 5–10 g/h at 15 nm particle diameter).

The nanoparticles were analyzed using an X-ray diffractometer to determine both composition and particle size. The XRD spectrum revealed that the material was primarily tin dioxide (rutile phase) with some tin monoxide phase. The calculation of the mean diameter from the Scherrer formula gave a value of about 15 nm for tin dioxide particles. The TEM analysis showed that the tin oxide nanoparticles were arranged in chains and that necks existed between some adjacent nanoparticles. A non-negligible degree of nanoparticle agglomeration was observed and had to be taken into account in the screen-printing process.

The infrared transmission spectrum of the tin oxide nanoparticles dispersed in potassium bromide and pressed into a pellet (Fig. 1a) showed two main bands at 630 and 565 cm^{-1} corresponding to the vibrations of the rutile structure in good agreement with literature data [11, 12].

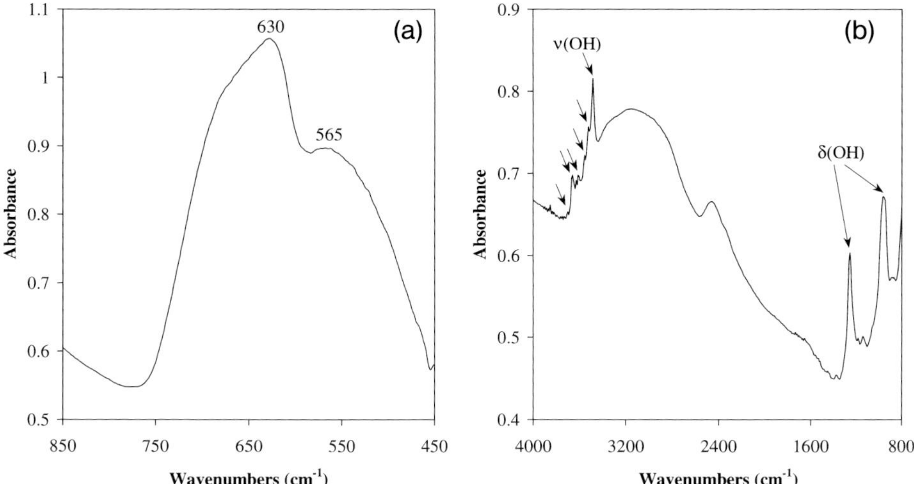

Fig. 1 Infrared spectrum of SnO$_2$ nanoparticles: (a) absorption bands of the bulk modes; (b) absorption bands of the surface species (the arrows in the 4000–3200 cm^{-1} range indicate v(OH) absorption bands corresponding to different types of OH surface groups).

4.2
Surface Chemistry of Nanoparticles

The reproducibility of the surface chemical composition and of the surface reactivity of nanoparticles has been shown to be critical in all applications [13]. This is particularly critical in the case of gas sensors where the surface reactions are at the origin of the gas detection mechanism.

The surface chemistry of nanoparticles can be conveniently investigated by FTIR spectroscopy under specific conditions [13]. The infrared spectrum of the chemical groups at the very surface of the SnO$_2$ nanoparticles is presented in Fig. 1b. All the absorption bands in the 4000–3000 cm^{-1} region are due to the stretching vibrations of the OH groups bonded to tin atoms of the first atomic layer [14]. The corresponding bending vibration δ(OH) absorbs in the 1500–800 cm^{-1} region. The large number of v(OH) bands indicates that several types of OH groups exist at the nanoparticle surface, mainly due to different coordination states of the tin atoms within the surface and to the different environment of these atoms essentially caused by the presence of defects. The different v(OH) absorption frequencies of these OH surface groups indicate their different acid–base nature. Therefore, a reproducibility study of these v(OH) bands is critical to ensure a perfect reproducibility of the surface composition and of the surface chemistry of the nanoparticles.

4.3
Rapid Screening of the Sensing Potential of the Nanoparticles

As the deposition technique is relatively time-consuming, a methodology was established to rapidly screen the semiconducting nanoparticles in terms of gas sensing potentiality before embarking in the fabrication of the devices. This rapid screening is based on the analysis of the absorption of the free carriers in the whole infrared range (Drude–Zener theory) [15]. Indeed, the variations of the background infrared absorption of a semiconductor sample reflect the variations of the free carrier density, that is the variations of the electrical conductivity. In previous works, we demonstrated that the variations of the infrared energy transmitted by the semiconductor nanoparticles versus gas exposures could be directly related to the electrical response of the real sensor [8, 16–18].

As an example, Fig. 2 shows the infrared spectrum of the tin oxide nanoparticles recorded at 300 °C under oxygen (spectrum a). When carbon monoxide is adsorbed on the surface in presence of oxygen, an increase of the background absorption is observed, caused by an increase of the free carrier density corresponding to an electrical conductivity increase (spectrum b). This increase of the electron density is due to the oxidation of carbon monoxide into carbon dioxide by reaction with ionosorbed oxygen at the tin oxide surface [19, 20]. In this reaction, electrons, which are the free carriers in this n-type semiconductor, are released into the conduction band as follows:

$$2\,CO + O_2^- \quad \rightarrow \quad 2\,CO_2 + e^-$$

$$CO + O^- \quad \rightarrow \quad CO_2 + e^-$$

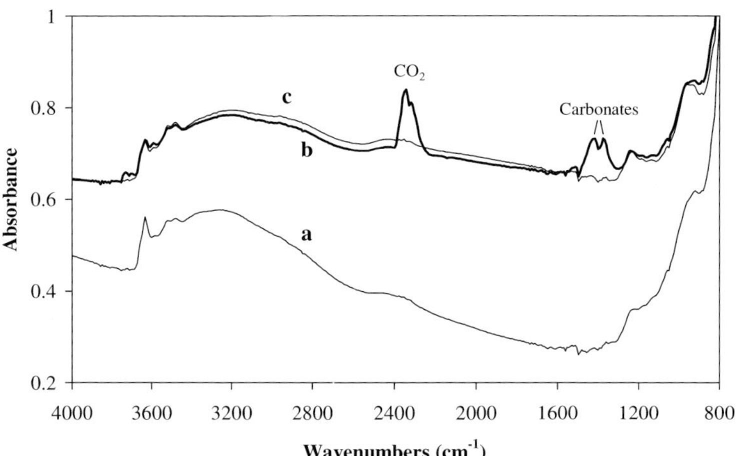

Fig. 2 Infrared spectrum of the SnO$_2$ nanoparticles at 300 °C: (a) under 50 mbar oxygen; (b) after addition of 10 mbar CO in presence of oxygen; (c) after evacuation.

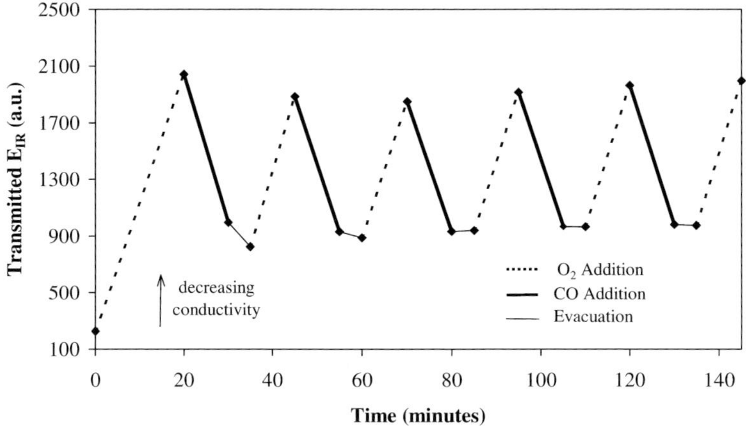

Fig. 3 Infrared energy (E_{IR}) transmitted by the SnO$_2$ nanoparticles at 300 °C versus gas exposures. An increase of E_{IR} indicates a decrease of the electrical conductivity.

The absorption band of gaseous carbon dioxide is indeed noted on the infrared spectrum at 2348 cm^{-1} (spectrum b). Carbonate surface groups are subsequently formed by adsorption of the resulting CO$_2$ on basic surface sites without any change in the electrical conductivity:

$$CO_2 + O^{2-} \quad \rightarrow \quad CO_3^{2-}$$

The absorption bands of the carbonate species are observed in the 1600–1200 cm^{-1} region (spectrum b). Both carbon dioxide and carbonate groups are eliminated by evacuation (spectrum c).

In Fig. 3, the variations of the infrared energy (E_{IR}) transmitted by the SnO$_2$ nanoparticles are reported versus gas exposures. An increase of the transmitted infrared energy corresponds to a decrease of the electrical conductivity and reproducible variations of E_{IR} are observed under carbon monoxide additions. As mentioned above, such a curve has been shown to directly correspond to the sensor electrical response [16, 18, 21].

This type of infrared analysis of semiconducting nanoparticles appeared to be extremely useful to get a relatively quick evaluation of the sensing potentiality of the materials as soon as they are synthesized. By optimization closed-loops, it was then possible to rapidly adjust the synthesis parameters for optimized sensing properties and to select the best batches of nanoparticles to be further processed.

4.4
First Optimization Stage of the Screen-Printing Process

During the first optimization stage of the screen-printing process, a homogeneous dispersion of the nanoparticles in the solvent was achieved by determining the ap-

Tab. 2 Comparison between the gas detection thresholds required to detect the maximum authorized concentrations of pollutants in air, the gas detection thresholds of commercial sensors and the gas detection thresholds of our sensors after the first optimization stage.

Polluting gases	CO [ppm]	NO$_2$ [ppb]	NO [ppb]	O$_3$ [ppb]
Target for the detection threshold	3	50	100	20
Commercial sensors: Electrochemical Semiconductor *(typical data available in 2000)*	5 100	600 –	900 –	200 –
Our sensors 1st optimization round *(data obtained in 2000)*	30	500	800	200

propriate concentration and by applying ultrasonication before screen-printing. Then, it was observed that the sintering temperature had a strong influence on the grain growth and also on the sensor sensitivity. It was found that up to 450 °C the grain growth was not significant but rapidly increased above this temperature. The sintering temperature was set below this critical value (around 400 °C for SnO$_2$ films), and may be varied depending on the metal oxide. When appropriate, a catalyst was added to the metal oxide to increase both sensitivity and selectivity.

Tab. 2 compares the sensitivity that our sensor prototypes could achieve at the end of this first optimization stage (in 2000) with the targeted sensitivities and the sensitivities of the electrochemical and semiconductor sensors commercially available at that time. It was clear that neither the commercial sensors nor our prototypes met our targets for outdoor AQM.

However, our prototypes exhibited a detection threshold to ozone, NO, and NO$_2$ comparable to that of the electrochemical cells, which are known to be very sensitive but much more expensive than semiconductor sensors. As for our CO sensor prototype, it appeared more sensitive than its Figaro counterpart.

4.5
Second Optimization Stage of the Screen-Printing Process

To further optimize our devices, we had to come up with a new concept of fabrication better accommodating the nanoparticles. The goal was to obtain homogeneous stacked layers of nanoparticles. As explained above, it was thought that the suppression of agglomerates should lead to improved adhesion and improved film quality. Additionally, a narrower particle size distribution should lead to a higher surface to bulk ratio and therefore to a higher sensitivity to gases.

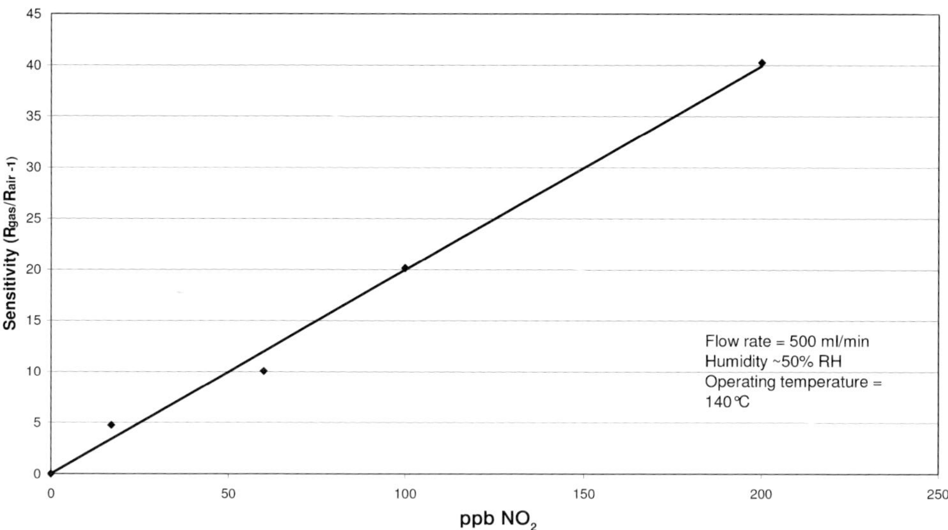

Fig. 4 Sensitivity of our sensor prototypes to NO$_2$ at 140 °C under 50% relative humidity (RH) [22].

Because industrial considerations forbad the manipulation of nanoparticles one by one, we developed a low-cost layer-by-layer deposition method via a wet route where the nanoparticles would preferably pile up in a regular 3D network thanks to their identical size.

The layer-by-layer deposition method consisted in dispersing the nanoparticles in an appropriate solvent to obtain a very diluted colloid. After sedimentation during a few hours, a drop of the supernatant was taken and deposited onto the alumina tile. The thin layer of nanoparticles was allowed to dry before a subsequent layer was deposited. The procedure was repeated up to one hundred times then the sensor was heat-treated below 450 °C. The final film was almost transparent under white light.

The sensitivity to NO$_2$ and to ozone of our sensor prototypes based on 15 nm tin oxide particles is shown in Figs. 4 and 5 respectively. The sensitivity is expressed as the ratio of the resistance of the sensors under synthetic air and under synthetic air mixed with various concentrations of polluting gases. The optimized operating temperatures were determined to be 140 °C for NO$_2$ detection and 120 °C for ozone detection. The threshold detection limit was about 15 ppb for both gases and the response time was of the order of 1 min.

Tab. 3 gives the lowest gas concentrations that our sensor prototypes could reliably detect at the end of the second optimization stage (achieved in 2002). The detection thresholds are compared to the targets previously defined and to the typical detection thresholds of commercial sensors as available in 2002.

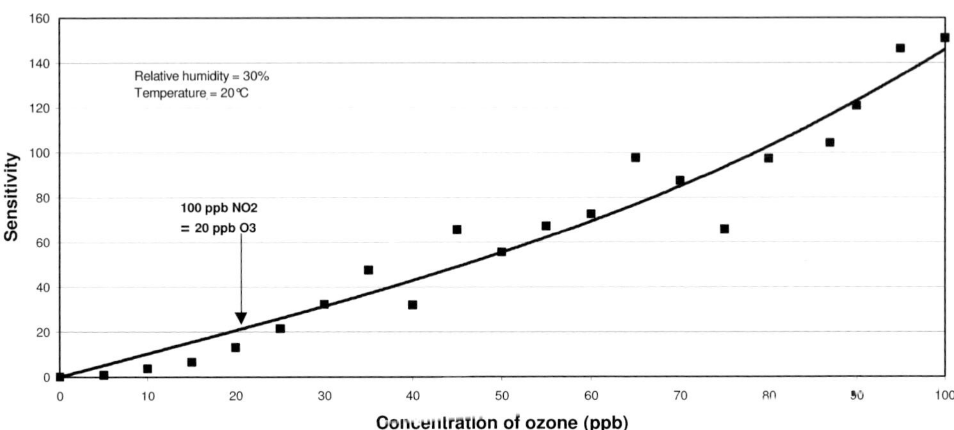

Fig. 5 Sensitivity of our sensor prototypes to ozone at 120 °C under 30% relative humidity (RH). The arrow indicates that the sensitivity of this sensor to 20 ppb ozone is identical to its sensitivity to 100 ppb NO$_2$ [22].

Tab. 3 Comparison between the gas detection thresholds required to detect the maximum authorized concentrations of pollutants in air, the gas detection thresholds of commercial sensors and the gas detection thresholds of our sensors after the second optimization stage.

Polluting gases	CO [ppm]	NO$_2$ [ppb]	NO [ppb]	O$_3$ [ppb]
Target for the detection threshold	3	50	100	20
Commercial sensors				
Electrochemical	1	100	500	50
Semiconductor	5	–	–	–
(typical data available in 2002)				
Our sensors	3	15	100	15
2nd optimization round				
(data obtained in 2002)				

The obvious conclusion is that our sensors can compete with commercial devices. But more important, our sensors can fully meet the required targets for outdoor AQM whereas commercial electrochemical and semiconductor sensors still cannot.

**5
Outlook**

Our Consortium showed that nanoparticles-based semiconductor sensors exhibit higher sensitivities to air pollutants, lower detection thresholds, lower operating temperatures. The device optimization is not a straightforward procedure and requires controlled surface chemistry of nanoparticles, homogeneous dispersion of nanoparticles, deposition of homogeneous layers, a low level of dopant if any, and mild firing conditions.

Although our sensors are chemically and electrically stable, the long-term stability over extended periods of time (several months) has still to be checked. The response time of our sensor prototypes is typically of the order of one minute at the lowest gas concentrations. However, we have come up with an electronic design that will allow the measurement of gas concentrations every 10–20 s. Independently from but simultaneously with the optimization of the sensing layer, the cross-sensitivity to humidity, which is a critical issue for semiconductor sensors, was addressed. The resulting whole sensing unit, which is small enough to be a portable device, will be able to reliably operate under a relative humidity (RH) in the range 10–100% and an outdoor temperature in the range 5–60 °C [22].

In a second step, these portable gas sensing units associated with global positioning systems (GPS) will communicate with a central computer via a wireless network based on the GSM protocol. These microstations, installed on mobile carriers such as city buses, will constitute a dynamic network covering the city and complementing the existing AQM stations. Through the Internet, it becomes therefore possible to not only inform in quasi-real-time citizens on air quality status but also help decision-makers to more efficiently manage road traffic and provide scientists with additional data to refine mathematical models of pollution clouds in cities. With the booming development of telecommunication networks and the recent integration of GSM and GPS technologies in commercially available devices, it appears that this second step of our project designed by the Consortium as early as the beginning of 1999 has now simply become the modification of existing wireless systems for our specific application.

The sensing element represents the most challenging component in our concept of mobile AQM microstations and our proposed dynamic network is relevant only if semiconductor sensors can successfully and reliably detect low gas concentrations. We have now shown that these low-cost gas sensors can meet the criteria in terms of detection thresholds for polluting gases in air. Our concept can now proceed with the complete development of the AQM microstations, which could be implemented easily, rapidly, and in a sufficient number in any city of any country at low cost.

Acknowledgments

The authors are indebted to their partners in the SMOGLESS and INTAIRNET Consortia and more specifically to Drs H. Ferkel and J. F. Castagnet (IWW, Technische Universität Clausthal, Clausthal-Zellerfeld, Germany) for the synthesis of nanoparticles, to Drs G. S. V. Coles and T. Starke (University of Wales at Swansea, UK) for sensor fabrication and their electrical characterization, to Drs V. Burganos and S. Zarkanitis (ICE-HT/FORTH, Patras, Greece) for their study of the cross-sensitivity to humidity, to Drs M. Taylor and L. Witrant (Oldham-France S. A., Arras, France), Drs M. Willett and M. Jones (City Technology Ltd, Portsmouth, UK) for sensor evaluation under industrial conditions.

The financial support of the European Commission under the BRITE-EURAM III and IST programs is gratefully acknowledged.

References

1 R. S. Morrison in *Semiconductor Sensors* (Ed. S. M. Sze), John Wiley & Sons, New York, USA, **1994**, Chapter 8, 383–413.

2 Directive 2000/69/CE of the European Parliament and of the Council of 16/11/ 2000; Directive 2001/81/CE of the European Parliament and of the Council of 23/10/2001 (http://europa.eu.int/eur-lex).

3 Y. Shimizu and M. Egashira, *MRS Bulletin* **1999**, *24*, 18–24.

4 H. Ogawa, M. Nishikawa and A. Abe, *J. Appl. Phys.* **1982**, *53*, 4448–4455.

5 G. Williams and G. S. V. Coles, *J. Mater. Chem.* **1998**, *8*, 1657–1664.

6 M.-I. Baraton and L. Merhari, *Mater. Trans.* **2001**, *42*, 1616–1622.

7 M.-I. Baraton, L. Merhari, H. Ferkel and J. F. Castagnet, *Mater. Sci. Eng. C* **2002**, *19*, 315–321.

8 M.-I. Baraton and L. Merhari in *MRS Symp. Proc. Series* Vol. 581 (Eds. S. Komarneni, J. C. Parker, H. Hahn), MRS, Warrendale, USA, **2000**, 559–564.

9 M.-I. Baraton and L. Merhari, *Nano-Structured Materials* **1998**, *10*, 699–713.

10 W. Riehemann in *MRS Symp. Proc. Series* Vol. 501 (Eds. K. E. Gonsalves, M.-I. Baraton et al.), MRS, Warrendale PA, USA, **1998**, 3–14.

11 M. Ocaña, C. J. Serna, J. V. Garcia-Ramos and E. Matijević, *Solid State Ionics* **1993**, *63–65*, 170–177.

12 M. Ocaña and C. J. Serna, *Spectrochim. Acta* **1991**, *47A*, 765–774.

13 M.-I. Baraton in *Handbook of Nanostructured Materials and Nanotechnology* (Ed. H. S. Nalwa), Academic Press, San Diego CA, USA, **1999**, 89–153.

14 E. W. Thornton and P. G. Harrison, *J. Chem. Soc. Faraday Trans.* **1975**, *71*, 461–473.

15 N. J. Harrick, *Phys. Rev.* **1962**, *125*, 1165–1170.

16 M.-I. Baraton, *Scripta Mater.* **2001**, *44*, 1643–1648.

17 M.-I. Baraton in *Nanostructured Films and Coatings* (Eds. G.-M. Chow et al.), NATO-ARW Series, Kluwer, Dordrecht, The Netherlands, **2000**, 187–201.

18 M.-I. Baraton in *Nanocrystalline Metals and Oxides: Selected Properties and Applications* (Eds. P. Knauth and J. Schoonman), Kluwer, Boston, USA, **2002**, 165–187.

19 V. E. Henrich and P. A. Cox, *The Surface Science of Metal Oxides*, Cambridge University Press, UK, **1994**.

20 P. K. Clifford, Mechanisms of Gas Detection by Metal Oxide Surfaces, PhD Thesis, Carnegie Mellon University, Pittsburg PA, USA, **1981**.

21 SMOGLESS Final Report (Contract No.: BRPR-CT95-0002), unpublished results, **1999**.

22 INTAIRNET Second Year Report (Contract No. IST-1999-12615), unpublished results, **2002**.

Amorphous Electrically Conducting Materials for Transducer Applications

Alex Dommann, Marco Cucinelli, Matthias Werner, and Marc-Aurele Nicolet

1
Introduction

Appropriate physical, electrical, and chemical characteristics as well as reproducible synthesis and long-term stability of the compound are prerequisites for the use of new materials in microelectromechanical systems (MEMS). Bulk metallic glasses have, in general, properties such as high elastic deformation, high fracture strength, and good corrosion resistance at temperatures up to the glass transition and above [1]. Amorphous films composed of tantalum, silicon, and nitrogen exhibit great similarities with bulk metallic glasses and are promising candidates for their use in microstructured systems [2].

2
Mictamict Alloys

Ta–Si–N is representative of a group of materials called "mictamict", a word combined from the Greek "mictos" for mixed and "amictos" for unmixed or pure. That group also includes mixtures of other early transition metals, silicon, and nitrogen (such as Ti–Si–N, Mo–Si–N, W–Si–N) or alloys such as WBN, TiSiO, or RuSiO. Common to all of these materials is that they are mixtures of mutually immiscible or poorly miscible binary compounds (such as TaN and Si_3N_4 in the case of Ta–Si–N) that have one common element of a size smaller than that of the other two. Once formed in an amorphous state by virtue of the method of synthesis used, the amorphous structure effectively resists crystallization. This is because crystallization cannot occur polymorphically in such systems but requires a primary nucleation of one or both of the binary phases. Diffusional transport is required for these processes. However, a significant mobility of only one of the three species will not necessarily initiate crystallization. These conditions enhance the kinetic stability of the amorphous phase. The absence of a simple ternary compound in the system further favors high metastability. High crystallization temperatures are thus a characteristic feature of these mictamict amorphous alloys. Combined with their special physical or chemical properties, these films offer new potential applications, including MEMS structures.

The Nano–Micro Interface: Bridging the Micro and Nano Worlds
Edited by Hans-Jörg Fecht and Matthias Werner
Copyright © 2004 WILEY-VCH Verlag GmbH & Co. KGaA, Weinheim
ISBN: 3-527-30978-0

3
Thin Films

Mictamict alloys have yet to be synthesized in bulk form. Thin films, however, have been obtained by a variety of techniques, including reactive duel-electron-beam co-evaporation, reactive sputtering, metal–organic chemical vapor deposition, and exposure of silicide films to energetic reactive atmospheres. Once formed, the amorphous structure that evolves appears to be largely independent of the mode of deposition as long as nucleation and growth of the equilibrium phases are suppressed. A random distribution of species and minimal atomic mobility during the formation process are main prerequisites, that is, the synthesis must take place away from thermal equilibrium conditions. Equal structural amorphousness does not, however, invariably ensure identical physical characteristics.

4
Properties of Ta–Si–N Films

With a Young's modulus of about 200 GPa and a coefficient of linear thermal expansions of about 6×10^{-6}/K, the mechanical properties of Ta–Si–N are similar to those of silicon (220 GPa and 4.2×10^{-6}/K). Reactive sputtering is the technique most commonly used to form these thin films. High compressive stresses of hundreds of megapascals are typical of such films deposited on substrates near room temperature.

Experiments have shown that these stresses can be reduced by thermal annealing. Heating to about 400 °C decreases the stress while the films remain structurally amorphous. By optimization of the duration of annealing and by anticipating the possible subsequent differential contraction effect between the film and substrate on subsequent cooling, the residual stress reduces nearly a hundred times, or even turned into the tensile direction.

Finally, the amorphous microstructure is commonly known to hamper or impede fatigue effects often observed in polycrystalline materials. The grain boundaries in polycrystalline materials are also often seen to be the first points of chemical attack. The absence of such boundaries in amorphous mictamict alloys up to elevated temperatures constitutes another beneficial feature of these materials. The binary nitrides of the early transition metals and of silicon do oxidize. Correspondingly, the mictamict nitrides do oxidize as well, albeit at reduced rates as was demonstrated with wet oxidation of TiN and Ti–Si–N films.

5
MEMS of Ta–Si–N Films

Sputtered Ta–Si–N films were investigated with a view to their application as saws, sensors, and actuators [3]. In a first step, deposition conditions and film properties such as stoichiometry, crystallization temperature, surface roughness and, in particular, the amorphous structure as a function of nitride concentration were analyzed. In a second step, free-standing Ta–Si–N surface microstructures were made by an all-metal production process. As shown in Fig. 1, this was done by the so-called "sacrificial layer technology". This technology has become a well-established surface micromachining technique for making complex three-dimensional thin-film microstructures on any type of process-compatible substrate. Al was used as sacrificial layer for making free-standing Ta–Si–N. Both layers can be sputter deposited at low temperatures, and are thus well-suited for post-processing on prefabricated electronic components.

6
Surface Micromachining of Ta–Si–N Microbeams

As mentioned above, ternary films turn out to be promising candidates for application in MEMS, due to their interesting physical, electrical, and chemical properties.

An all-metal combination has been chosen for making free-standing amorphous Ta–Si–N films (1–2 μm thick) with Al (2–4 μm) as the sacrificial layer. Both layers can be sputter deposited at low temperature (below 300 °C) offering the possibility of post-processing on a substrate with prefabricated electronic components (Fig. 1). Patterning of Ta–Si–N was performed in an SF_6/O_2 plasma using a photoresist mask. (Further fluorine-based dry etching processes are described elsewhere [4].) The lateral etching of the sacrificial Al in a standard H_3PO_4/NHO_3 mixture was highly selective, so the etchant did not attack the Ta–Si–N film. An as-made microbridge is shown in Fig. 2. The beams are 2 μm thick, 300 μm long, and 80 μm wide. The underetched holes and the tethers are 10 and 20 μm wide, re-

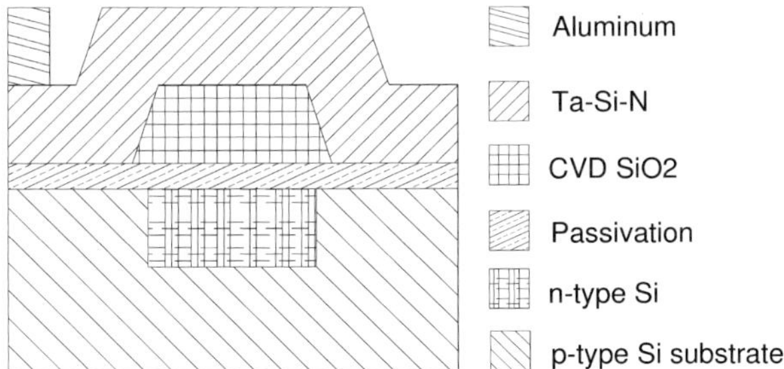

Fig. 1 Schematic cross section of a micromachined Ta–Si–N surface.

Fig. 2 SEM of a side view of a free-standing Ta–Si–N beam (30 µm wide, 2 µm thick)

spectively. The slight buckling of the free-standing double-clamped beam indicates that the Ta–Si–N film is under compressive strain, in agreement with experimental data by Reid [5]. This compressive stress of some hundreds of megapascals has been reduced by post-annealing the bridges at 450 °C for 30 min.

Potential applications for such amorphous metal microbridges are switching devices such as MEMS relays: polysilicon is normally used for such devices [6]. From the mechanical point of view, Ta–Si–N and polysilicon exhibit similar mechanical properties, the former having a somewhat longer lifetime. However, from the electrical point of view, Ta–Si–N has better characteristics than polysilicon. Indeed, the electrical conductivity of Ta–Si–N exceeds that of polysilicon making Ta–Si–N an even more interesting contact material for MEMS compared with polysilicon.

The transfer function of the Ta–Si–N microbridge has been measured under vacuum by optical interferometry. The output signal of the latter has been fed into a gain-phase analyzer. Resonance frequencies of 113 and 140 kHz were measured for the 350 µm and 300 µm length beams, respectively. These devices exhibit a relatively high quality factor of about 1000 [6]. These findings reveal that Ta–Si–N structures are promising candidates for use in MEMS.

7
X-Ray Analysis of Ta–Si–N Films

Crystallinity analyses were carried out using a Philips PW 1050 computer-controlled X-ray diffractometer with filtered Cu Kα radiation. X-ray diffraction patterns of the films deposited at different N_2 partial pressures were measured. To get a well-defined background signal all measurements were performed on films deposited on (100) Si wafers. All samples were X-ray amorphous. Cross-sectional TEM measurements revealed [7] that in the range 5–10% N_2 with respect to the partial pressure of argon, a change from TEM-amorphous to X-ray-amorphous growth occurs.

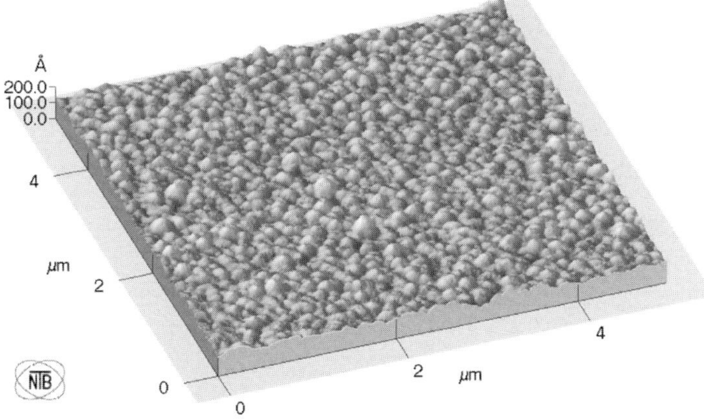

Fig. 3 An atomic force microscope (AFM) image of a 2 μm thick sputtered
Ta–Si–N film on a Si substrate (average roughness of Si = 1 nm).

Rutherford backscattering spectrometry measurements of these Ta–Si–N films
yielded the following atomic compositions (with an error of about 10%) of
$Ta_{49}Si_{14}N_{37}$ for 5% N_2 and $Ta_{44}Si_{14}N_{42}$ for 10% N_2. The crystallization temperature of the Ta–Si–N films as determined by X-ray diffraction is around 800 °C, for
an annealing time of about 1 h. This result confirms the high thermal stability of
this mictamict alloy [2].

Contact atomic force microscopy measurement of a sputtered Ta–Si–N surface is
shown in Fig. 3. An AutoProbe LS Scanning Probe Microscope by Park Scientific
Instruments was used. The surface is extremely smooth having a rms roughness
of less than 5 nm.

The intrinsic strain of the beams was measured by the X-ray rocking curve
method. The strain on the substrate wafer induced by amorphous Ta–Si–N layers
can be measured by high resolution X-ray diffractometry. Our MPD1880/HR-Phi-
lips diffractometer is equipped with a very intense beam source and a Bartels
monochromator. This is a powerful, non-destructive method with a high strain
sensitivity for the investigation of thin films of the order of up to some micrometers, which is in the typical thickness range needed for micromachined actuators. By determining the tensor profile a detailed surface lattice strain analysis can
be performed [8]. The stress can then be derived from the lattice strain data by
using Hook's law.

The amorphous Ta–Si–N layer exhibits a quite interesting stress versus annealing temperature behavior (Fig. 4). It is therefore possible to control the stress by
setting the annealing temperature. This allows, for example, tailoring of the material for its use as a seal in fluidic or gas microstructures, as shown in Fig. 5.

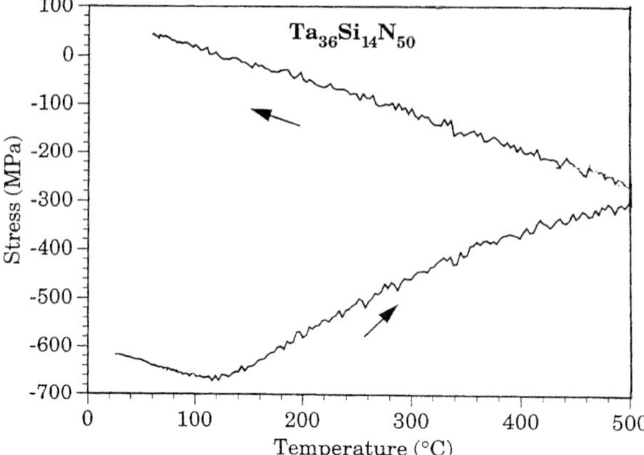

Fig. 4 Stress versus annealing temperature for 400 nm Ta$_{36}$Si$_{14}$N$_{50}$ on Si (111) substrate.

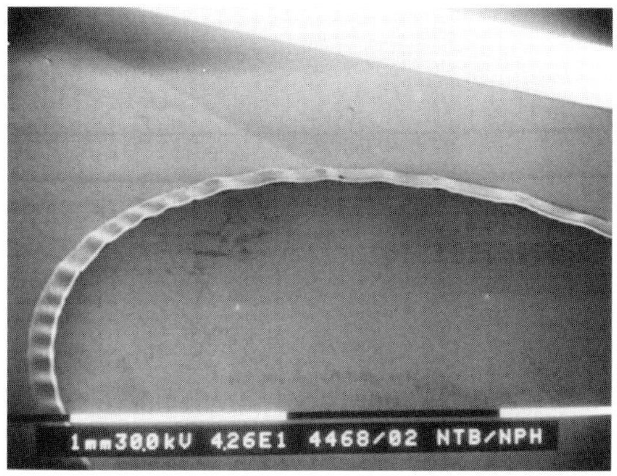

Fig. 5 SEM of a free-standing Ta–Si–N sealing lip (2 μm thick).

8

Ta–Si–N Thin Films as Diffusion Barriers for Cu Metallization

To demonstrate the stability of the Ta–Si–N layers as a promising material for surface microstructured tools we discuss also the barrier properties of these films. The pressure towards ever-increasing levels of integration is pushing aluminum technology to its limits. One alternative to aluminum is copper, which is a fast diffuser in silicon and generates deep levels in the band gap of silicon. Highly effective diffusion barriers are thus essential.

By changing the composition slightly from Ta$_{44}$Si$_{14}$N$_{42}$ to Ta$_{36}$Si$_{14}$N$_{50}$, Ta–Si–N films meet the stringent requirements for diffusion barriers better than any previously known thin films, not only for copper, but also for aluminum and other metals [9]. The ideal diffusion barrier for a metal/silicon contact is electronically transparent and atomically opaque. In addition, the film must be inert, that is, thermodynamically stable with respect to the adjacent materials (metal and silicon or silicide). The main drawback of the widely used TiN as a barrier material is its polycrystalline, columnar structure, which limits its application range. Conceptually, pure amorphous TiN would alleviate the drawback of the columnar structure, but amorphous TiN is structurally rather unstable.

Thermodynamic calculations have shown Me–Si–N alloys with Me = V, Nb, Ta, Cr, Mo, W to be stable in contact with copper, as long as the metal content of the ternary compound exceeds that of silicon by more than $5/3 = 1.67$ [10]. This result is not that surprising, since copper forms no stable compounds with any of these transition metals or with nitrogen. The latter make up the major part of the elements in the ternary films. One therefore expects that amorphous films of similar compositions would be quite stable against copper. Indeed, no reactions have been observed by cross-sectional transmission electron microscopy between amorphous $Ta_{36}Si_{14}N_{50}$ films and copper.

A further requirement for diffusion barriers is electronic:bulk resistivity of the film and its contact resistivity. For films of 100 nm thickness or less, bulk resistivity below about $1\,m\Omega\,cm$ is adequate. In that resistivity range, even the highest current densities nowadays encountered in integrated circuits will generate a voltage drop across the film that is smaller than $kT/q = 26\,mV$.

The stability of the saturation current of a shallow pn-junction critically tests the performance of a diffusion barrier. Figure 6 shows that a 80 nm thick $Ta_{36}Si_{14}N_{50}$ barrier covered with a copper layer, keeps the saturation current constant after annealing of the diode in vacuum for 30 min at 900 °C [9]. It is worth pointing out that this temperature is only 13% below the melting point of copper,

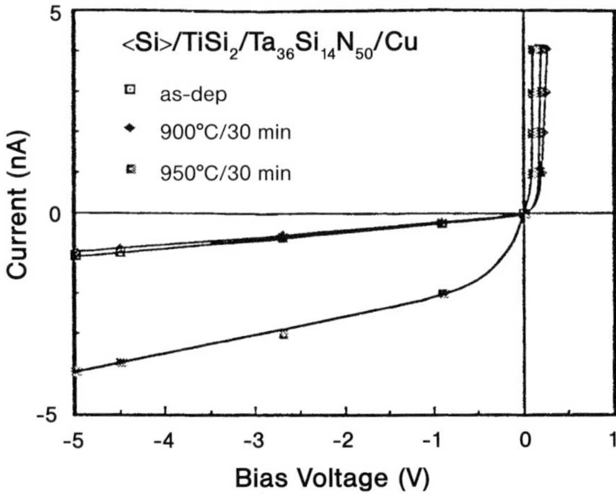

Fig. 6 *I–V* characteristic of a shallow n + p Si junction diode with the $\sigma TiSi_2$ (30 nm)/$Ta_{36}Si_{14}N_{50}$ (80 nm)/Cu (500 nm) metallization before and after annealing at 900 or 950 °C for 30 min [9].

in absolute values. During a heat treatment at this temperature, the copper film changes its topography substantially by solid-phase diffusion. Holes open up in the films, exposing the bare barrier layer, and copper protrusions appear. However, the barrier layer as well as the $TiSi_2$ contacting layer and the shallow junction below it remain unaffected.

The process that initiates the crystallization of the $Ta_{36}Si_{14}N_{50}$ barrier is attributable to the copper-induced crystallization of Ta. In this case there is evidence that a crystallization front may propagate through the barrier layer from its interface with the copper, the layer thereby acting as a sacrificial barrier [11]. Without the barrier, the diode fails at $500\,°C/30$ min. With a gold overlayer diodes withstand $750\,°C/30$ min.

In practice, barriers need to withstand temperature/time stressing of only $550\,°C/30$ min or less. Experience suggests that if the barrier sustains 800 or $900\,°C$ for 30 min, it will do so for a very long time at $550\,°C$. Amorphous ternary nitride layers are the only films so far that have the potential to meet that goal.

In the case of aluminum as the capping layer, the effectiveness of the diffusion barrier is not linked with its inertness of the ternary compound to Al, but rather on their reactivity. Upon annealing, a very thin (a few nanometers) self-sealing layer of AIN forms at the interface, which prevents further solid-state reactions.

Acknowledgments

Thanks also to Nico Onda (NTB) for discussing the chapter.

References

1 W. L. Johnson, *MRS Bull.* **1999**, *October*, 42.

2 P. J. Pokela, J. S. Reid, C. K. Kwok, E. Kolawa and M. A. Nicolet, *J. Appl. Phys.* **1991**, *70*, 2828.

3 J. M. Mrosk, L. Berger, Ch. Ettel, H.-J. Fecht, G. Fischerauer and A. Dommann, *IEEE Trans. Ind. Electron.* **2001**, *48*, 258.

4 G. F. McLane, L. Casas, J. S. Reid, E. Kolawa and M. A. Nicolet, *J. Vac. Sci. Technol. B* **1994**, *12*, 2352–2355.

5 J. S. Reid, Amorphous Ternary Diffusion Barriers for Silicon Metallizations, PhD Dissertation, California Institute of Technology, **1995**.

6 M. A. Grétillat, P. Thiébaud, N. F. de Rooji and C. Linder, in *Proc. IEEE Micro Electro Mech. Syst. Workshop*, Oiso, Japan, **1994**, 97–101.

7 M. A. Grétillat, C. Linder, A. Dommann, G. Staufert, N. F. de Rooij and M.-A. Nicolet, *J. Micromech. Microeng.* **1998**, *8*, 88–90.

8 R. A. Buser and A. Dommann, *Sens. Actuators A* **1994**, *43*, 317–321.

9 E. Kolawa, P. J. Pokela, J. S. Reid, J. S. Chen, R. P. Ruiz and M. A. Nicolet, *IEEE Electron Device Lett.* **1991**, *12*, 321.

10 J. S. Reid, E. Kolawa and M. A. Nicolet, *J. Mater. Res.* **1992**, *7*, 2424.

11 E. Kolawa, J. S. Chen, J. S. Reid, P. J. Pokela and M. A. Nicolet, *J. Appl. Phys. Lett.* **1991**, *70*, 1369.

12 L. E. Halperin, M. Bartur, E. Kolawa and M. A. Nicolet, *IEEE Electron Device Lett.* **1991**, *12*, 309.

Commercial Applications of Diamond-Based Nano- and Microtechnology

Peter Gluche, André Flöter, Stephan Ertl, and Hans-Jörg Fecht

1
Introduction

Diamond possesses outstanding material properties. Especially high mechanical stability in combination with high thermal conductivity and low coefficient of friction, which makes diamond an attractive material for many applications. However, the use of diamond is limited by its cost and the lack of cost-effective shaping methods. In the past seven years, novel CVD processes for diamond growth [1] have been developed and optimized, resulting in the production of polycrystalline and fine-grained diamond films of high mechanical strength [2, 3]. Furthermore, recent progress in plasma shaping of diamond films allows the preparation of diamond parts of complex shape [4]. As examples, the use of diamond for cutting tools and micromechanical parts such as microtoothed wheels is described.

1.1
Properties of Diamond

Tab. 1 shows the main material properties of diamond compared with other materials used for micromechanical parts. Diamond is the hardest of all known materials, which makes it an attractive material for wearing parts. The high fracture strength and Young's modulus make diamond parts very robust and dimensionally stable. The high thermal conductivity, in combination with the low thermal expansion coefficient and the low heat capacity, enable thermally stressed applications such as heat spreaders [5], lenses and laser windows [6], or thermal microactuators [7]. Undoped diamond is electrically insulating. However, diamond can be doped and can therefore also be semiconducting and at high doping concentrations even be metal-like [8]. This effect is used to make diamond microelectronics such as diodes [9], transistors [10], microsensors [11], and diamond MEMS [12].

The Nano–Micro Interface: Bridging the Micro and Nano Worlds
Edited by Hans-Jörg Fecht and Matthias Werner
Copyright © 2004 WILEY-VCH Verlag GmbH & Co. KGaA, Weinheim
ISBN: 3-527-30978-0

Tab. 1 Properties of diamond compared with other materials.

Property		HOD	Diamond	Silicon	Titanium
Specific resistance ρ	(Ωcm)	10^{-3}–10^{16}	10^{-3}–10^{16}	10^{-4}–2×10^{10}	5×10^{-5}
Thermal conductivity λ_{th}	(W/mK)	max. 1500	2000	105–145	22
Specific heat capacity c_p	(J/kgK)	500	500	713	520
Thermal expansion coefficient a_{th}	(10^{-6}/K)	0.8–1.1	1.1	2.6	9
Youngs modulus E_m	(GPa)	max. 850	1143	110–190	115
Fracture strength σ_B	(GPa)	typ. 4.8	10.3	1.4	–
Hardness	(Mohs)	10	10	7	–

1.2
Synthesis of Diamond

In general, there are two methods for diamond synthesis: the high-pressure high-temperature and low-pressure synthesis. The high-pressure synthesis runs in the diamond-regime of the carbon phase diagram in Fig. 1 [13]. Using high pressure and high temperature, diamond crystals of different sizes can be grown from a molten carbon source.

Using metastable conditions at small pressures and rather low temperatures, diamond can also be grown using activated carbon-containing gases, mainly methane, in a mixture with hydrogen. The latter process is based on carbon addition to an already carbonized surface, preventing graphite deposition through simultaneous etching by the hydrogen component. Thus, sp3-bonded carbon is

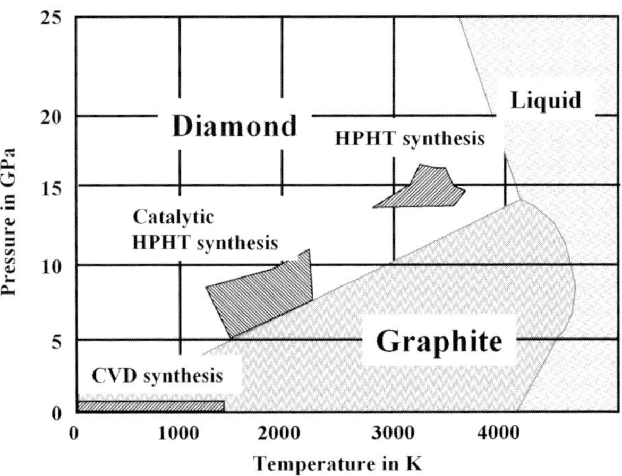

Fig. 1 Phase diagram of carbon showing the pressure and temperature regimes for diamond synthesis.

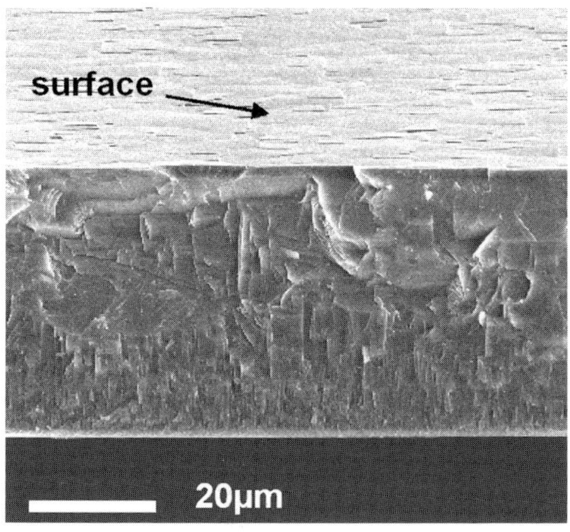

Fig. 2 SEM image of the cross section of an HOD film. The increasing grain size from the substrate towards the growth surface leads to a columnar structure.

more stable against hydrogen, diamond species remain on the surface and lead to growth of diamond crystals. The resulting film quality very much depends on the process parameters. The growth itself is always somewhat perpendicular to the surface, leading to a columnar structure shown in Fig. 2.

The film surface is characterized by the orientation of the microcrystals. This can be clearly seen in Fig. 3, showing the growth surface of three different diamond films. From top to bottom, statistically polycrystalline, textured, and highly oriented diamond (HOD) films are shown. The statistically oriented diamond film does not show any orientation of the crystallites at all. For the textured film, the film surface is composed of the same crystal faces. The HOD film shows an additional azimuthal orientation of the grains. Several grains have coalesced, leading to a low surface roughness.

A widely used method for the deposition of diamond films is the so-called microwave plasma-enhanced CVD method (MWPECVD). Fig. 4 is a sketch of a suitable CVD machine, consisting of a vacuum chamber, a wave-guide, and a microwave generator. The microwave power is coupled to the vacuum chamber, where a standing wave causes a plasma ball if the substrate holder is in a suitable position for microwave resonance. The growth conditions are mainly dependent on the carbon-species concentration, the substrate temperature, and the pressure in the chamber, as described (for example) earlier [14].

Other suitable processes differ from MWPECVD mainly by the methods of carbon and hydrogen molecule activation. As substrates, suitable materials must be chosen to withstand the respective conditions. The most widely used substrates are silicon, carbides, quartz, refractory metals (W, Mo, Ta, Nb, Ti), and ceramics.

Another sophisticated approach is the production of nanocrystalline diamond films. In contrast to conventional CVD growth of polycrystalline diamond films, an inert gas (e.g., Argon) is added during growth. This results in a diamond film

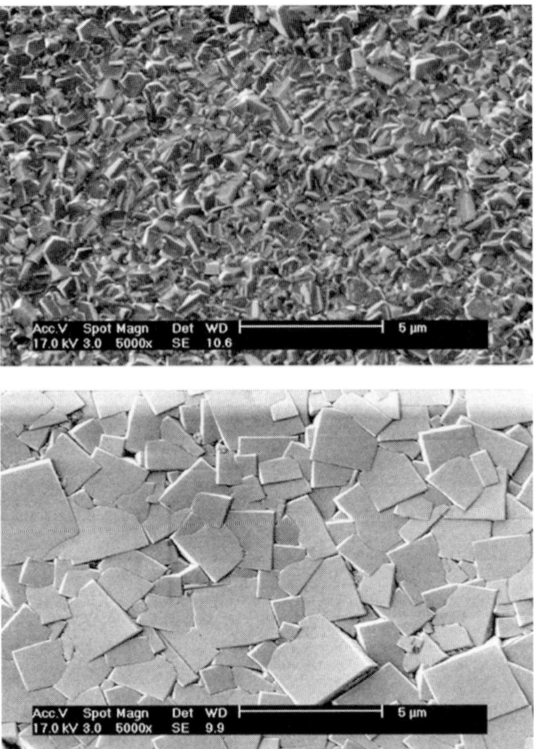

Fig. 3 SEM images of diamond films of different quality (top view): *top:* statistically oriented, no texture; *middle:* statistically oriented, textured (fiber texture); *bottom:* highly oriented diamond (HOD) in which several grains have coalesced.

with grain sizes of 3–5 nm. For these films very good mechanical properties, such as a Young's modulus of 967 GPa, a hardness of 97 Gpa, and a fracture strength of approximately 5 GPa have been reported [3]. However, such films exhibit high internal stress even at low film thicknesses of several micrometers. This leads to substrate deformation and the delamination of the films has been observed [15].

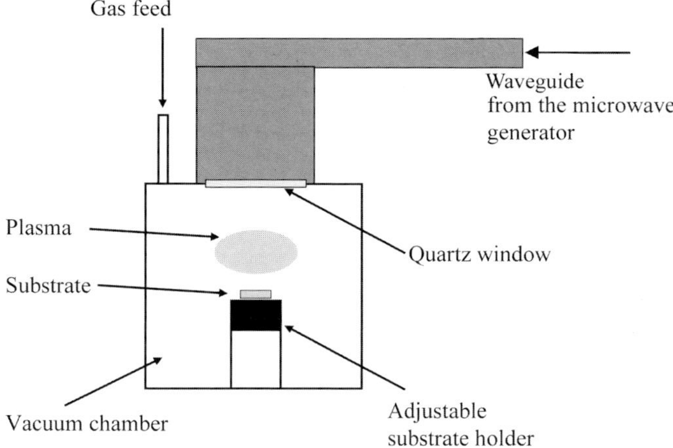

Fig. 4 Diagram of an MWPECVD reactor for diamond growth.

2
Commercial Applications: Cutting Tools and Micromechanical Diamond Parts

Modern surgical methods require sharp microcutting tools. To decrease the patient recovery times, there is a trend towards less invasive surgical techniques, which require less traumatic wounds. Especially in the field of microsurgery, such as eye- and neurosurgery, the demand for ultrasharp cutting tools is continuously increasing [16]. A first major step towards this trend was the introduction of diamond keratomes for eye surgery, made from natural gem stones, about 35 years ago. Since diamond is the hardest of all known materials and has outstanding mechanical properties, it was possible to make microcutting tools having an outstanding and long-lasting sharpness. Although such knives have the highest functionality, only 50% of all cuts are performed with diamond scalpels, mainly due to their high price. To overcome this situation, a novel diamond scalpel is presented, utilizing a thin CVD diamond film on a silicon backing. Furthermore, the shape and the sharpness of the scalpel is no longer the result of manual polishing. A novel plasma process is described resulting in radii of curvatures of less than 5 nm.

These processes could be diversified to the manufacture of micromechanical diamond parts. The excellent mechanical properties in combination with the low coefficient of friction and the low thermal expansion coefficient of diamond may make it possible to operate microdrives without lubricants at high efficiency. The first results of a diamond microgear are presented.

2.1
Diamond Cutting Tools

2.1.1 Application

Conventional precision cutting instruments made of diamond have found their main application in microsurgery, where they are often preferred to steel scalpels for their long lasting sharpness and reusability. The area of ophthalmic surgery offers the largest market potential. Today, approximately 5 million cataract operations are performed in the industrialized nations, making it the most performed surgery worldwide [17, 18]. In Germany a growth rate of about 5% per year is observed due to the aging of the society. However, according to the World Health Organization, the cataract is still the most important cause of blindness worldwide.

Patients suffering from cataract have decreasing vision caused by opacification of the lens. The reason for the development of a cataract is still unclear, but it is strongly correlated with age. To prevent blindness, the opaque lens has to be removed and replaced with an intraocular lens implant (IOL). The surgery is most often performed by opening the anterior chamber of the eye using a metal or diamond scalpel. After removing the scalpel from the wound, the wound channel should be self-sealing, to avoid suturing, which can lead to postoperative astigmatism.

With the introduction of diamond scalpels into the market, surgical techniques have improved towards less traumatic wound geometries, making it possible to offer ambulatory cataract surgery. Nevertheless, the demand for sharper instruments is unbroken and the miniaturization of intraocular lens implants creates a demand for smaller and sharper instruments.

The main cutting parameters for a blade are shown in Fig. 5, where
V = volume
Z = thickness
α_c = apex angle
c_f = coefficient of friction
r_c = radius of curvature of the blade

The sharpness of the blade is mainly determined by the radius of curvature, r_c, at the tip. Since stainless steel is a relative soft material compared with diamond,

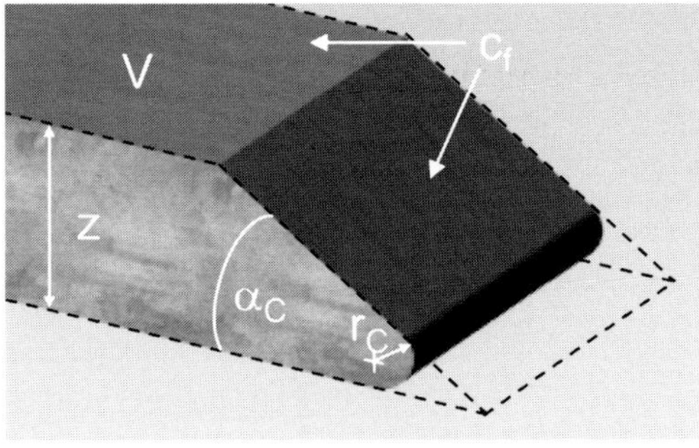

Fig. 5 Diagram of a cutting edge showing the main parameters influencing the cutting behavior.

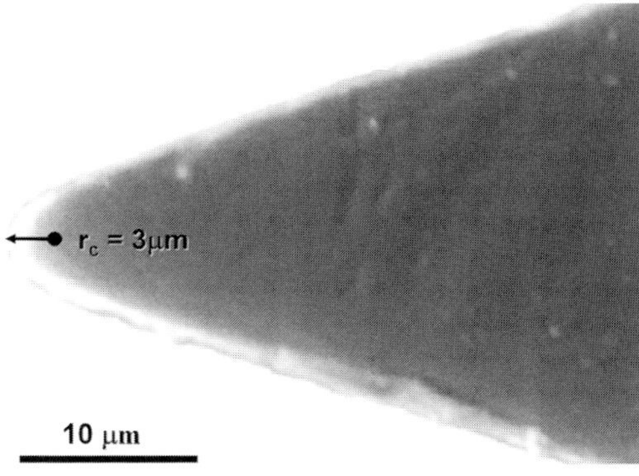

Fig. 6 Micrograph of a conventional metal scalpel. The radius of curvature is about 3 μm.

even a very small radius of curvature will increase rapidly during the cutting of even soft materials, such as a human cornea. As a consequence, most metal blades rapidly loose sharpness and are therefore used only once (disposable products). Furthermore, the residual surface of a corneal cut (and therefore the wound sealing and wound healing) is strongly influenced by the radius of curvature. The cross section of an as-manufactured stainless steel blade is shown in Fig. 6. The radius of curvature is approximately 3 μm, which is expected to increase significantly during the first cut into a human cornea.

Scalpels made from natural diamond are usually manually polished. Such blades show little wear and there is no significant change in radius of curvature on cutting a human cornea. Therefore, diamond scalpels can be used several hundred times, if handled and cleaned carefully. However, diamond can only be polished using diamond powder. The achievable radius of curvature is thus strongly dependent on the polishing process, the powder particle size, and further process parameters, such as the speed of the polishing wheel and the temperature. For most diamond scalpels, the radius of curvature is less than 200 nm. Therefore, with increased sharpness, a decrease of cutting artifacts should be observed [19–21]. On the other hand, due to the mechanical polishing process, the shape of a diamond scalpel is limited and complex shapes such as curved geometries are usually approximated by facets (Fig. 7).

In conclusion, the larger the radius of curvature, r_c, of a blade, the more cutting defects are likely to be observed in the cornea and the higher the residual corneal surface roughness is expected to be. Therefore, a key to further cutting improvement seems to be the decrease of the radius of curvature to a minimum value, which is limited by the radius of one atom. To reduce dulling of the cutting edge, diamond (as the hardest known material) should be chosen.

To exceed the state of the art, a novel plasma-based diamond polishing process has been developed enabling diamond cutting edges to have a radius of less than 5 nm.

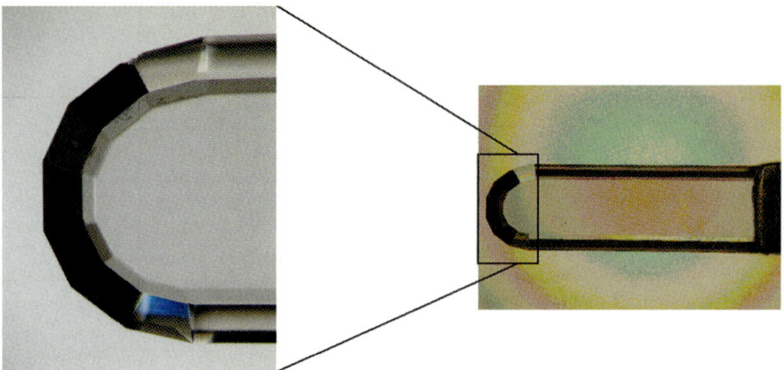

Fig. 7 Natural diamond blade manufactured by manual polishing.
The crescent shape has to be approximated by facets.

2.1.2 **Novel Process: the Diamaze Blade**
In a first step, a thin diamond film is grown on a silicon substrate using an
MWPECVD system. A key factor is diamond nucleation on the silicon surface.
Using a novel bias process, more than 50% of the nucleated grains are oriented
with respect to the silicon substrate. During further outgrowth, this will yield an
HOD film, best described by the low surface roughness of less than 200 nm at
30 μm film thickness and the high mechanical strength reported earlier [2].
Nevertheless, the cross section of such a diamond film shows a columnar structure, since the grains are nucleated at the silicon and increase in size towards the
diamond film surface. Therefore, the material properties are expected to be anisotropic. Fig. 8 shows as grown HOD films on silicon substrates of one and two
inch diameters.

In a second step, the shape of the diamond scalpel is transferred into the silicon substrate using standard photolithography. This allows for the production of
complex shapes at a high lateral resolution of less than 0.5 μm. The resist is used
to transfer the shape of the scalpel in a previously deposited mask by reactive ion
etching (RIE). The shape is then transferred to the silicon substrate using an an-

Fig. 8 HOD films, grown on one and
two inch silicon wafers. The surface is
mirror-like without any treatments after
growth.

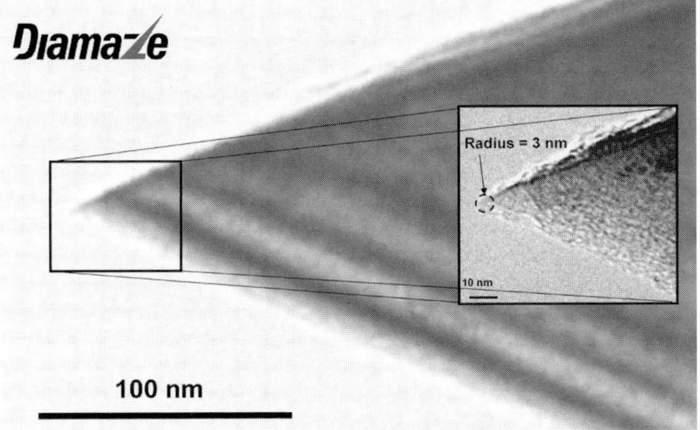

Fig. 9 High-resolution micrograph of a plasma-polished diamond blade. The radius of curvature is less than 5 nm.

isotropic etching process that does not attack the diamond film. Therefore, the etching process stops at the diamond–silicon interface. The result is a relief-structured diamond wafer, where the residual silicon defines the shape of the scalpel.

In the last step, the diamond wafer is plasma polished. The process is very similar to plasma etching. Here, the diamond film is removed atomic layer by atomic layer from the silicon side, resulting in an ultrasharp cutting edge having a radius of curvature of less than 5 nm [22]. The process is self-adjusting and therefore insensitive to over etching. Using this process, a radius of curvature of less than 5 nm can be reproduced as can be seen from the micrograph in Fig. 9. Thus, the

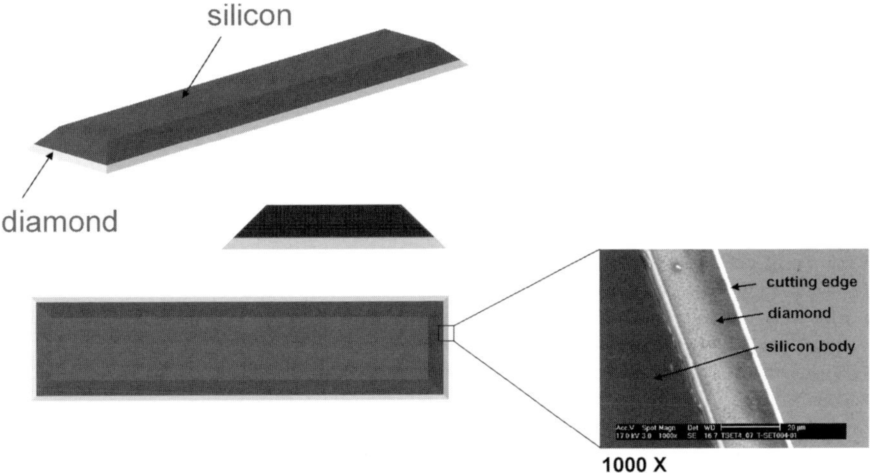

Fig. 10 Diagram of the Diamaze blade. The insert shows the top view of the diamond film and cutting edge.

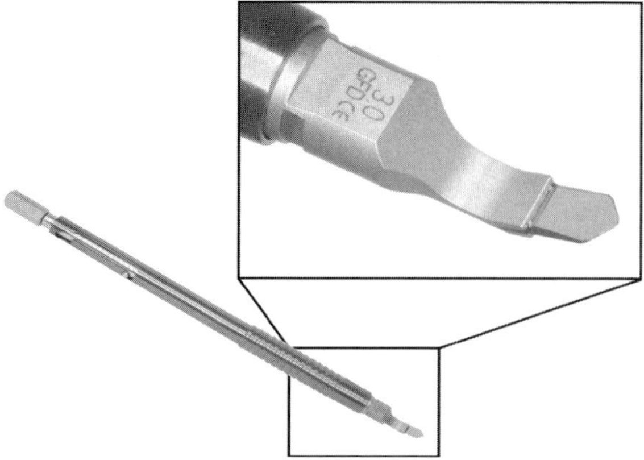

Fig. 11 Diamaze blade mounted into carrier and titanium scalpel handle.

cutting edge represents almost the physical limit, utilizing diamond as the hardest material and a radius of curvature, r_c, which is close to the minimum value. Fig. 10 shows a diagram of the blade. Finally, the blade is mounted into a titanium carrier and screwed to the titanium handle (Fig. 11).

2.1.3 Advantages: Socio-economic Impact

To investigate the cutting performance of the new blade, a comparative cutting study has been performed using porcine eyes. Conventional metal blades, natural diamond blades, and the Diamaze blades have been used to cut into scleral and corneal tissue. After cutting, the residual surface has been prepared and investigated by SEM.

Fig. 12 shows the micrographs of the residual cutting surfaces. As clearly seen from the pictures, the metal blade and the natural diamond blade show cutting artifacts, which are likely to be correlated with the radius of curvature and supposed to have negative influence on the wound sealing and wound healing. On the other hand, the Diamaze blade shows almost no cutting artifacts and very smooth surfaces. This might be a further significant step towards ambulatory treatments at lower risk for the patient as well as the surgeon. Thus, a significant cost reduction can be expected when using the nanostructured diamond blade for this type of surgery.

2.1.4 Future Aspects

The new manufacturing methods seem to be well suited for production of ultrasharp diamond cutting tools. Nevertheless, there is still a bottleneck in the diamond production. The process for making HOD films of high mechanical strength is still performed at low growth rates below 1 µm per hour and scaling is still limited to small wafer diameters. Furthermore, almost all polycrystalline diamond films show a columnar grain structure as can be seen from the cross sec-

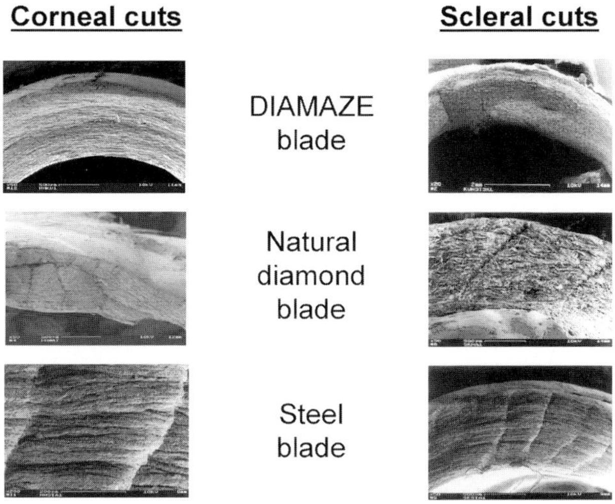

Corneal cuts

Scleral cuts

DIAMAZE
blade

Natural
diamond
blade

Steel
blade

Fig. 12 Comparative study of corneal and scleral cuts, using a conventional metal blade, a natural diamond blade, and the novel Diamaze blade. Almost no cutting artifacts are observed using the Diamaze blade.

tion of a cleaved HOD film in Fig. 2. After plasma structuring of such diamond films, one will observe the columnar structure at the etched side walls, because of different etch rates of the grains and the grain boundaries.

Further improvements are expected, when using a nanocrystalline diamond film instead of a polycrystalline diamond film. It is also expected that structuring of such a film will lead to smoother surfaces. Nevertheless, the mechanical properties of such films have to be investigated in more detail. Especially the strength of the grain boundaries is of interest for the wear resistance of cutting tools.

2.2
Diamond Micromechanical Parts

2.2.1 Application

Using the same process basics, the production of micromechanical diamond parts is possible. The main advantage is from the excellent mechanical properties and the low coefficient of friction. Thus, lubricant-free operation of diamond microgears at high speed, low wear, low moment of inertia, high efficiency, and high reliability might be possible. To evaluate the potential of this application, diamond microgears have been designed, made, and characterized.

2.2.2 Design

Two diamond-toothed wheel designs having a diameter of 550 µm and 8 (M55Z8) or 10 teeth (M55Z10) have been chosen. To keep the final drive ratio as independent as possible of clearance tolerances, an involute gear design was used. The geometrical data are shown in Tab. 2.

Tab. 2 Geometrical data for the diamond toothed wheels.

Diameter [mm]	M55Z8	M55Z10
Tip circle [mm]	0.56	0.70
Pitch circle [mm]	0.44	0.55
Root circle [mm]	0.33	0.46
Center bore [mm]	0.23	0.23

2.2.3 Production of Diamond Micromechanical Parts: Diamond-Toothed Wheels

The small size of the diamond-toothed wheels leads to problems in mounting and adjusting them one to another, not known in standard gear production. To enable a sufficient vertical overlap for precise movement, their thickness was chosen to be 150 µm. Even with very high-aspect-ratio etching processes, a lateral under etch of the mask cannot be entirely suppressed. However, the optimization of the plasma processes allowed the production of microstructured diamond-toothed wheels having sidewall angles of $90°\pm2°$ at a thickness of 150 µm. As in all plasma structured diamond films, the columnar grain structure is revealed, resulting

Fig. 13 Micrograph of the M55Z8 and M55Z10 diamond toothed wheel. The metal part is a steel needle.

Fig. 14 SEM image of a diamond gear wheel. The columnar structure of the plasma etched side walls and the verticality can be seen clearly.

in a surface roughness of 500 nm maximum. A further reduction by the use of a fine grained or a nanodiamond film can be expected.

Fig. 13 shows a picture of an M55Z8 and M55Z10 diamond gearwheel. Fig. 14 shows an SEM image of a diamond gear wheel. From the picture, the verticality of the process and the residual surface roughness caused by the columnar grain structure can be seen clearly.

2.2.4 Diamond Microgear

From the toothed wheels, a microgear was assembled using an aluminum base plate with bores of 0.2 mm diameter for the center axis of each wheel. A nickel-wire of 200 µm diameter was used as the axis material. The vertical adjustment of the wheels was done by gluing copings on top of the axis. For actuation, a micro-motor was used, which was specified to have a maximum speed of 100 000 rpm and a maximum torque of 7.5 µNm. The original output wheel was replaced by a

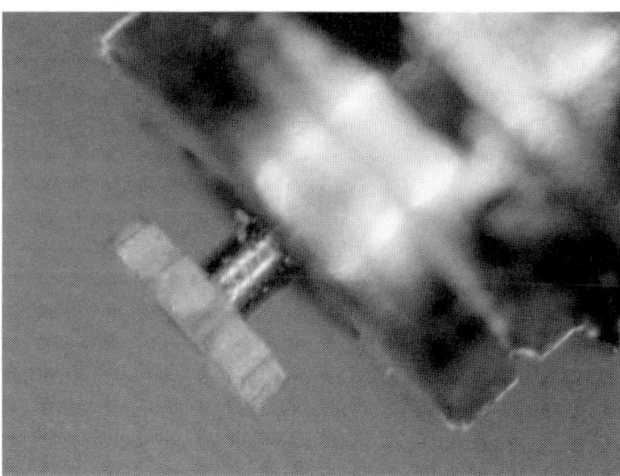

Fig. 15 Micromotor with diamond driving wheel.

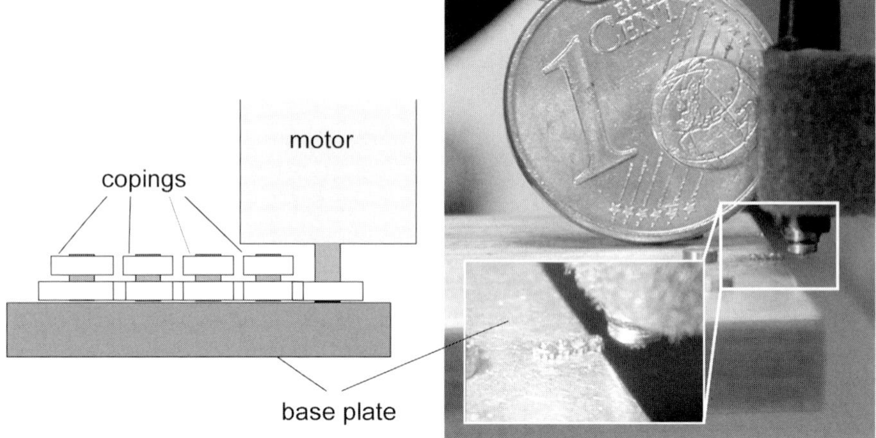

Fig. 16 Left, sketch of the diamond gear; right, picture of the assembled diamond gear prior to testing.

diamond wheel as shown in Fig. 15. A schematic cross section and a photograph of the linear gear set-up are shown in Fig. 16.

The diamond gear could be successfully driven in this test set-up with a maximum speed of 35 000 rpm. After more than 200 min at 20 000 rpm, the gear showed a parasitic destructive breakdown due to failure of the motor, which was caused by metal particles from the base plate and the axes. Fig. 17 shows the base plate and a Ni-axis before and after the operation of the gear. Heavy strong wear

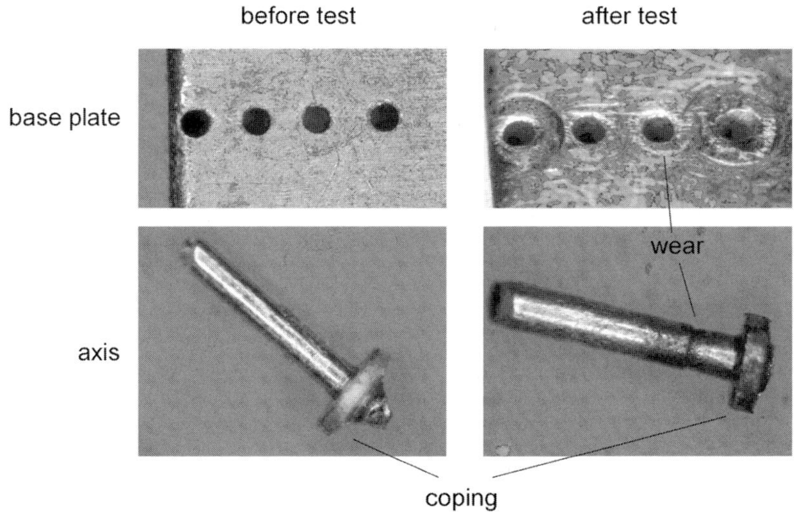

Fig. 17 Wear characteristics on the base plate and the Ni axes after operation of the diamond gear.

can be observed after the operation, which is caused by the still-too-rough diamond sidewall surfaces. The jamming of the gear caused by wear particles, resulted not in breakdown of the wheels, but of the motor. Wear of the diamond wheels could not be observed.

3
Summary

In this chapter, a novel production method, based on CVD diamond deposition and plasma structuring techniques, has been presented. The production of ultrasharp diamond cutting edges with radii of curvature of less than 5 nm has successfully been demonstrated. This process was used to make ultrasharp diamond scalpels for eye and neurosurgery. A comparative study involving metal scalpels, natural diamond scalpels, and the novel Diamaze scalpel has been performed using porcine corneas. Here, significantly reduced cutting artifacts on the residual cornea surfaces could be observed with decreasing radius of curvature. This might be an important step towards ambulatory surgical cataract treatments. Patients and surgeons will also benefit, because a sharper cutting edge will lead to better wound sealing and healing. This means lower risk for the surgeon and the patient. Further improvements might be achieved by the substitution of polycrystalline diamond films by nanocrystalline diamond films. In order to commercialize this step, several problems (such as the high internal stress) have to be solved.

It was possible to diversify the process for the creation of micromechanical diamond parts. Here, 150 μm thick diamond films have successfully been structured using an RIE method. A verticality of less than $2°$ and a residual surface roughness of less than 500 nm have been demonstrated. This is already close to the results achieved in silicon using the Bosch process.

For a first test set-up, two different types of diamond-toothed microwheels have been made. In order to test the capability to operate a diamond microgear without lubricants, a test gear consisting of five toothed wheels has been set up and run up to 35 000 rpm without lubricants. However, a destructive breakdown occurred after 200 min at 20 000 rpm, which was caused by wear of the aluminum base plate and the nickel axes. The diamond parts, however, did not show any wear. It is likely that the failure was caused by the still-too-rough sidewall faces of the diamond parts, acting as micromilling cutters. This surface roughness is thought to be caused by the columnar grain structure of the polycrystalline diamond film. Therefore, in order to improve the sidewall surface roughness, advantage can again be taken of a significant reduction in the diamond grain size. This might be achieved by using nanocrystalline diamond films.

References

1 A. Flöter, H. Güttler, G. Schulz, D. Steinbach, C. Lutz-Elsner, R. Zachai, A. Bergmaier, G. Dollinger, *Diamond Relat. Mater.* **1998**, *7*, 283–288.

2 P. Gluche, M. Adamschik, A. Vescan, W. Ebert, F. Szücs, H. J. Fecht, A. Flöter, R. Zachai and E. Kohn, *Diamond Relat. Mater.* **1998**, *7*, 779–782.

3 H. D. Espinosa, B. Peng, K.-H. Kim, B. C. Prorok, N. Moldovan, X. C. Xiao, J. E. Gerbi, J. Birrell, O. Auciello, J. A. Carlisle, D. M. Gruen, and D. C. Mancini, *Proc. Mater. Res. Soc. Symp.* **2003**, 741.

4 E. Kohn, P. Gluche, M. Adamschik, *Diamond Relat. Mater.* **1999**, 8, 934–940.

5 E. Wörner, C. Wild, W. Müller-Sebert, R. Locher and P. Koidl, *Diamond Relat. Mater.* **1996**, 5, 688–692.

6 E. Woerner, C. Wild, W. Mueller-Sebert, P. Koidl, *Diamond Relat. Mater.* **2001**, *10*, 557–560.

7 P. Gluche, R. Leuner, A. Vescan, W. Ebert, C. Rembe, S. Aus der Wiesche, E. P. Hofer and E. Kohn, *Microsyst. Technol.* **1998**, 5, 38–43.

8 M. Werner, W. Kohly, R. Locher, D. S. Holmes, S. Klose, H. J. Fecht, *Diamond Relat. Mater.* **1997**, 6, 308–313.

9 A. Vescan, I. Daumiller, P. Gluche, W. Ebert and E. Kohn, *Diamond Relat. Mater.* **1998**, 7, 581–584.

10 A. Aleksov, A. Denisenko, and E. Kohn, *Solid-State Electronics* **2000**, 44, 369–375.

11 M. Werner, P. Gluche, M. Adamschik, E. Kohn, H. J. Fecht, *Proc. IEEE Int. Symp. on Industrial Electronics (ISIE '98)*, 7–10 July **1998**, Pretoria, South Africa, 147–152.

12 E. Kohn, M. Adamschik, P. Schmid, S. Ertl, and A. Flöter, *Diamond Relat. Mater.* **2001**, *10*, 1684–1691.

13 F. P. Bundy, *J. Geophys. Res.* **1980**, *B12*, 6930.

14 B. Dischler and C. Wild (eds), *Low Pressure Synthetic Diamond: Manufacturing and Applications*, Springer Series in Materials Processing, Springer, **1998**, Chapter 1–8.

15 J. A. Carlisle and O. Auciello, *Electrochem. Soc. Interface*, Springer, **2003**, 28–31.

16 F. K. Jacobi, B. Dick, R. M. Bohle, *J. Cataract Refract. Surg.* **1998**, 24, 498–502.

17 Gesundheitsberichterstattung des Bundes, Anzahl der aus dem Krankenhaus entlassenen vollstationären Patienten einschließlich Sterbe-, ohne Stundenfälle (Jahre, Alter, Geschlecht, Diagnose/Behandlungsanlass nach IDC9), Online in the Internet: http://www.gbe-bund.de, Date: 07.01.2003.

18 World Health Organisation, Blindness and visual disability, part II of VII: Major causes worldwide, Online in the Internet: http://www.who.int/inf-fs/en/fact143.html; Date: 07.01.2003.

19 E. Cabernard, *Klein. Mbl. Augenheilkunde* **1975**, *167*, 134–136.

20 E. J. Galbary, *Ophth. Surg.* **1984**, *15*, 203–205.

21 J. G. Vincent, J. Doting, *Eur. J. Cardiothoracic Surg.* **1989**, 3, 373–375.

22 Ch. Dieker, W. Jäger, S. Ertl, S. Strobel, *Dreiländertagung Elektronenmikroskopie*, Innsbruck, **2001**.

Bio-Inspired Anti-reflective Surfaces by Imprinting Processes

Thomas Sawitowski, Norbert Beyer, and Frank Schulz

1
Introduction

Nanotechnology is the most promising new technology arising in the past 3–7 years [1–4]. It is a striking for improving or even creating new materials and material properties by simply changing the size and the aspect ratio of a given material [3]. On the other hand nanotechnology is not a technology that can be classified based on traditional concepts in disciplines such as physics, chemistry, biology, or engineering. Nanotechnology is an overall technology involving experts from every discipline covering aspect from all these fields somehow related to materials. A second very important aspect is that nanotechnology always deals with surface and interface effects. This stems from the fact that nanomaterials always exhibit a huge surface-area-to-volume ratio due to their small size. This ratio gives a very high energy state of matter, so methods for making nanostructures today are very sophisticated. Furthermore, those structures very often tend to increase in size so they have to be stabilized until they can finally show the unique properties arising from the diminishing dimensions. In this chapter the concept of enhancing desired surface properties by very simple imprinting methods will be described as an example of an interdisciplinary approach in physics, engineering, and chemistry.

2
Aluminum Oxide: Template and Lithographic Tool

For decades it has been known that valve metals such as magnesium, titanium, aluminum, and others form a stable oxide layer upon electrochemical oxidation [5]. These layers are formed by oxidizing the metal in different diluted electrolytes and by subsequent reaction of the activated surface or adsorbed metal ions with water molecules or hydroxide ions.

The most prominent valve metal is aluminum. For economic reasons it is necessary to protect the unnoble metal (standard corrosion potential: -1.676 V at pH <7; -2.310 V at pH >7) [6] against corrosion under ambient conditions. This protection

The Nano–Micro Interface: Bridging the Micro and Nano Worlds
Edited by Hans-Jörg Fecht and Matthias Werner
Copyright © 2004 WILEY-VCH Verlag GmbH & Co. KGaA, Weinheim
ISBN: 3-527-30978-0

is achieved, for example, by the so-called ELOXAL process (ELOXAL stands for electrochemical oxidation of aluminum) [6]. The oxidation is normally carried out in diluted acidic electrolytes typically of sulfuric or chromic acid by applying potentials of the order of a few tens to a few hundreds of volts and a direct current. In this process the aluminum surface is converted into an amorphous aluminum oxide and hydroxide surface, which can best be described as a Boehmite composition of aluminum oxide [7–12]. After oxidation, the aluminum is dipped into a bath of hot water or treated with steam to close the pores in the oxide surface and complete the corrosion protection [13, 14].

The very first step in the oxidation process is the formation of a thin dense layer of oxide on the metal surface. The thickness of this layer depends on the applied potential. A steep drop in potential occurs, equal to a field strength of up to 10^{10} V/m, across this few-nanometer-thin oxide film. This high electrical field, together with some initial surface perturbation (coming from the natural surface roughness or from grain boundaries, for example), is the reason for the first pores to be formed [15, 16]. In this electrical high-field regime at slight perturbations the oxide crystal lattice is deformed and the electrolytes dissolve more rapidly the oxide causing pores to be formed [17]. While the electrical field determines oxide formation and dissolution, the pore geometry can be controlled by the electrical field and thus by the potential applied in the process of anodizing aluminum. In the end a structure such as that given in Fig. 1 is reached.

Pores are ordered parallel to each other and perpendicular to the substrate surface. At the bottom of the pores a thin oxide layer, the so-called barrier layer, remains. As a rule of thumb one can say that for each volt of anodic potential the pore diameter increases by 1.5 nm. So by applying 50 volts, pores of the order of 80 nm are formed. The pores are packed hexagonally with amorphous Boehmite forming the pore wall in between. Pore densities can reach values up to 10^{11} pores/cm^2 while the porosity always remains the same around 30% because smaller pores are packed more densely than larger ones (Fig. 2).

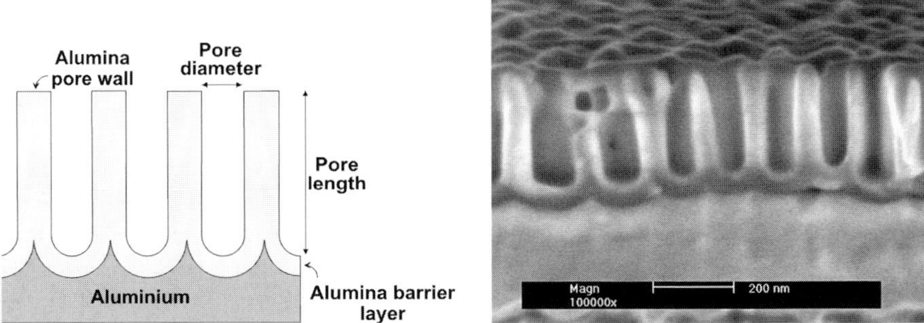

Fig. 1 Left, diagram of a cross-sectional area of nanoporous alumina attached to aluminum; right, cross-sectional SEM image of the nanoporous alumina attached to aluminum. The pores with the underlying barrier layer and the remaining metal can clearly be seen.

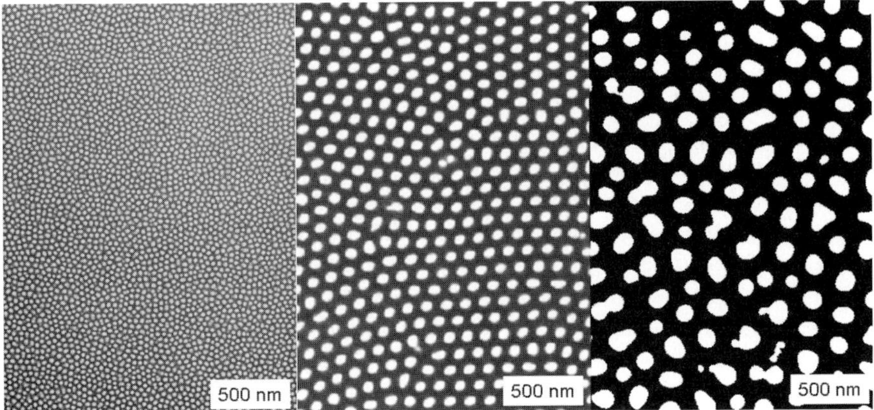

Fig. 2 TEM images of nanoporous alumina. All images are made under the same magnification showing three different pore sizes: on the left the pores are 20 nm in diameter, in the middle 50 nm, and on the right side about 100 nm.

The pore length is more or less controlled by the electrical charge, which is proportional to the time of anodic oxidation. Increasing time leads to an increase in oxide layer thickness until there is an equilibrium reached between oxide formation and porous layer dissolution in the electrolyte. Thicknesses up to some 100 μm are common. Upon heat treatment the oxide layer tends to crystallize. In a differential scanning calorimetry measurement the loss of water around 100 °C from the hydrated oxide can be seen as an endothermic process indicated by an arrow (Fig. 3).

At around 275 °C a broad exothermal peak indicates the crystallization process of the amorphous Boehmite. At around 450 °C, γ-alumina is formed without a change in the pore geometry. These processes are non-reversible as shown by a

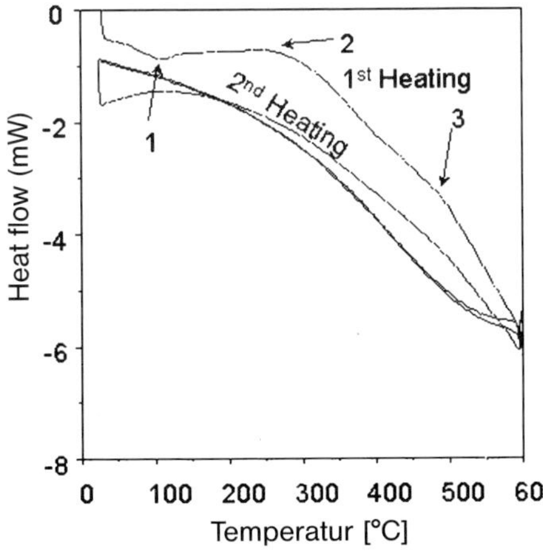

Fig. 3 Differential scanning calorimetry indicates the structural changes undergoing in the alumina when the materials is heat-treated: (1) water is removed, (2) the amorphous Boehmite crystallizes, (3) the Boehmite material undergoes a phase change into γ-alumina [11].

second thermocycle where no signal can be detected. This indicates that the nanoporous oxide layer does not change in structure under elevated temperature (demonstrated up to 1200 °C) and that with a heat-treatment a mechanically even more stable ceramic surface can be created.

With respect to the shape and extension of the aluminum work piece undergoing electrolytic oxidation, there is in principle no limit assuming that no Faraday shielding occurs, for example inside some cavities. Only the geometry of the starting aluminum determines the final shape of the porous surface modified tool. For all these reasons anodic oxidized aluminum is a very useful material in nanoscience [18], since the synthesis of nanosized materials can be controlled using this nanosized test-tube. The adjustable one-dimensional structure of nanoporous alumina offers in addition unique possibilities for wire-like nanomaterials to be formed [18–33].

For example, when filling small particles of metals or semiconductors (also called colloids and clusters) of about 10 nm, wire-like structures can be built and have been intensively investigated for their unique electronic and optical properties [34, 35]. A strong anisotropy has been observed, caused by the one-dimensional nature of this template. The same holds for pores filled with solid metals, such as gold or silver [31, 32]. Those metallic nanorods show angle-dependant optical properties related to the plasmon resonance excited along the short and the long axis. For this reason gold-filled alumina templates appear to be blue looking parallel to the pores and red looking perpendicular to the long pore axis.

The enormous pore density of up to 10^{11} pores/cm^2 can be seen as an extreme dense array of bits. Each pore can represent on its own or filled with some detectable material a "1" in a data storage application. When filling such pores with magnetic materials the magnetic hysteresis becomes a function of aspect ratio and degree of pore filling [36–38]. Therefore, highly ordered arrays of alumina have been made and filled with magnetic material with the final goal of a high-density magnetic data storage device being created [37].

But there are many hurdles to overcome in using nanoporous alumina as a data storage device. One is the low order of the pore structure. It can be improved by pre-structuring the alumina surface [39, 40]. At every pre-determined point on the surface a pore is formed. In this case even square or triangular pores can be made in alumina. Another disadvantage stems from the fact that the surface of a storage device needs to be absolutely smooth since the read/write head would fly over the disk at speeds of meters per second. This could be overcome if silicon wafers with the same pore density as the alumina, could be used. For these reasons porous alumina has been evaluated as a lithographic mask to transfer the pattern from the alumina into the wafer using reactive ion etching processes (RIE) [41, 42]. After coating silicon wafers with aluminum and subsequent oxidation of the metal a nanoporous mask on silicon can be made. By using RIE methods, the pore pattern could be transferred onto the silicon wafer and thus lead to a high-density data storage device.

Those methods are still very sophisticated, involving at least five steps such as sputtering of aluminum, oxidizing of the metal, RIE of the remaining barrier layer at the bottom of the pores, RIE of the silicon, and removal of the remaining oxide. For the envisioned application of storing data at high densities this might be feasible. If one intends to structure surfaces for rather lower-cost applications such as display coverings, such highly sophisticated methods are not applicable. In these cases reliable and inexpensive technologies such as injection molding or casting of polymers have to be used and methods for nanostructuring those surfaces have to be adapted to those techniques.

If one combines the idea of filling the pores with some different material or using the pore structure as a template, the idea of using nanoporous aluminum as a tool for structuring surfaces by applying mechanical forces is born [43]. Properties that change when changing surface structures of polymers are surface wetting, light transmission, friction, wear, and many more. One of the most challenging applications is definitely the improvement of light transmission for transparent polymers.

3
Reflection of Light

Whenever an electromagnetic wave penetrates into a material with a different refractive index, part of the wave is reflected back at an angle equal to the angle of incident [44]. This is a well-known phenomenon from optical physics and can be described mathematically by the law of reflection. The light that propagates into the medium is refracted towards (higher refractive index) or away (lower refractive index) from the normal plane, due to the change in refractive index (Fig. 4).

The degree of light reflection mainly depends on the material properties determined by the refractive index. The refractive index, n, is a complex quantity. It can be described as the sum of the real, n_R, and the imaginary part, k [45]

$$n = n_R + ik \tag{1}$$

The imaginary part is very often called the damping coefficient or the absorption coefficient. When the absorption can be neglected, the complex refractive index is reduced to its real part only.

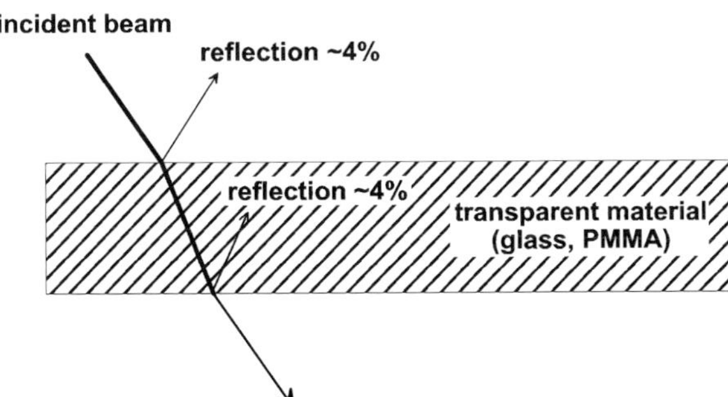

Fig. 4 Light beam refracted penetrating into a medium with higher refractive index. Part of the beam is reflected at each interface due to the changing optical density.

Tab. 1 Optical properties of transparent media.

Medium	Refractive index	Reflectivity [%]
Air (20 °C, 1.013,25 hPa)	1.0002734	–
Water	1.33	2.0
Ethanol	1.336	2.1
Glass	1.5–1.55	4.4
Diamond	2.4	16.9
PMMA	1.49	3.9
Polycarbonate (PC)	1.58	5.0
Polystyrene (PS)	1.59	5.2
PMMA-PS copolymer	1.56	4.8

Taking the Fresnel equations, the reflectivity of a surface can be calculated based on the complex refractive index. When a non-polarized light wave hits perpendicular onto a surface the reflectivity, the ratio of the intensity of the reflected light, I_r, and the incident light, I_i, is equal to [44]

$$R = \frac{I_r}{I_i} = \frac{\left(n_{\text{medium 2}} - n_{\text{medium 1}}\right)^2 + k^2}{\left(n_{\text{medium 2}} + n_{\text{medium 1}}\right)^2 + k^2} \tag{2}$$

Assuming $k \ll n_{\text{medium 1}} - n_{\text{medium 2}}$ (no significant absorption occurs) Eq. (2) can be simplified to

$$R = \frac{\left(n_{\text{medium 2}} - n_{\text{medium 1}}\right)^2}{\left(n_{\text{medium 2}} + n_{\text{medium 1}}\right)^2} \tag{3}$$

For PMMA, for example, with a refractive index of approx. 1.49 this results in a reflectivity of 3.8% at each interface. In Tab. 1 the values for refractive indices and the calculated reflectivity for some transparent polymers (medium 1 = air) are given [46–49].

For highly refractive materials the reflectivity is higher. For diamonds and other gemstones, for example, the high refractive index causes a high reflectivity, which is the reason for their brightness.

For the material scientist it is challenging to reduce the reflectivity of highly refractive polymers such as polycarbonate or polystyrene as well as to improve transparency and contrast for other transparent materials such as glass, PMMA, or copolymers. Different methods are already used or under development today. Besides plasma coating [50], the most promising technology is to nanostructure the polymer surface to improve light transmission [51–55].

4
Anti-reflective Coatings and Surface Structures

The most common method to change surface reflectivity is simply to enhance surface roughness. Perturbations in the size of the wavelength that hits the surface causes scattering of the wave. Depending on the size regime, this can be described, for example, by Mie theories [56] or by the law of Rayleigh [44]. The scattered light gives rise to the impression that the surface is non-reflective (to be more precise, it is a non-glossy surface).

If one thinks about improved light transmission through transparent media, the method of roughening the surface cannot be applied since scattering not only reduce light reflection but also drastically reduce directed light transmission accompanied with randomly propagating light causing disturbing fogging. In addition the scattered light leads to a huge decrease in contrast making it even more difficult to obtain clear imaging through such structures. Therefore, other methods such as the aforementioned coatings are applicable and will be described in more detail.

4.1
Plasma Coating

The most advanced way of achieving a highly transparent surface is plasma coating [50]. Thin transparent coatings of metal oxides are applied to a given substrate, each of which matches the substrate optical properties for a given window in the optical region. The thickness is determined by the optical wavelength and is equal to ¼ of the wavelength for which reflectance should be blocked. The advantages of this process is that even complex shaped geometries can be coated at reasonable cost. However, plasma coating is an expensive, low-throughput, and sophisticated ultra-high vacuum process, and sometimes it is necessary to apply as many as 10–15 different layers, which is a real obstacle to mass production of low-cost consumer goods. Also it is very difficult to adjust completely the absorption characteristics of the coating to the whole optical range of about 400–800 nm. This trade-off between costs and demands leads to the acceptance of slight fluctuations that give rise to colors occurring on the coated substrate especially when tilting the substrate to the incident light. Here the change in penetration depth causes colors to occur due to absorption and interference phenomena. Although plasma technology is very well developed, major inherent drawbacks such as interference colors and expensive processing are still not overcome. It is worthwhile to consider the feasibility of alternative methods of achieving anti-reflective properties on polymers and glass.

4.2
Porous Ceramics

A second approach to a non-reflective surface is to apply porous coatings with a precisely designed pore structure on a given substrate (Fig. 5) [57].

At the outermost part of the porous layer, air is the dominant medium. This means the refractive index is dominated by the refractive index of air. At the innermost part of the layer, the substrate is the dominant material and the refractive index is determined by the refractive index of the base material (when the coating is adapted to the optical properties of the base material). Between these end points, the material composition changes from air into substrate. The same holds for the refractive index. To be more precise, the refractive index can be calculated assuming an effective medium of air and substrate material at any given height of the surface structure [44].

This change in refractive index causes light to penetrate into the substrate almost without being scattered or reflected. Such coatings are made of silica and applied to glass substrates. Using a sol-gel system the glass is coated with the partially organic mixture. When firing the coating, the organic component is removed, leaving a porous film behind. The main advantage is the ability to coat complex shaped substrates. The main disadvantages are the need for a precise adjustment of the refractive index of the coating to the substrate and the high temperature process is not applicable to polymers. So again there are limitations with respect to substrates on which this method cannot be used. In particular, polymeric materials are very often used today as display coverings where the need for improved light transmission is an important factor. Any technique that targets nano-

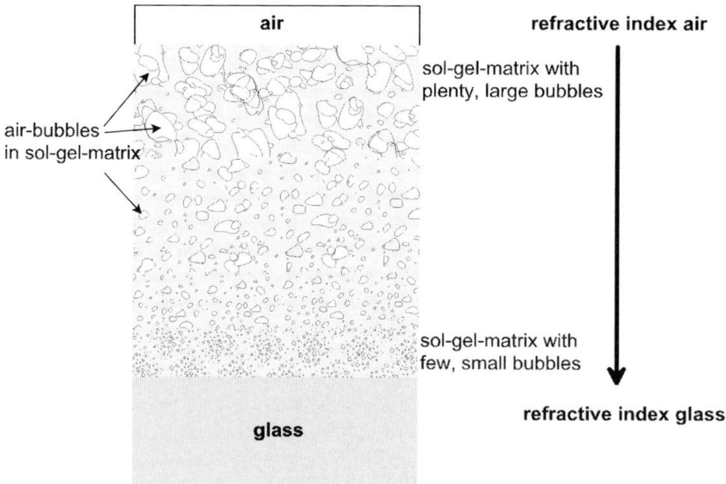

Fig. 5 Anti-reflective nanoporous sol-gel coating. The continuous change in porosity causes a continuous change in the refractive index thus reducing the intensity of the light being reflected.

structured polymeric surfaces has been compatible with one of the most important polymer processing technologies such as injection molding, casting, or extruding.

4.3
Moth-Eye Structures

Owing to the above-mentioned difficulties with present technologies, other approaches to the development of an an anti-reflective surface have been consider-ed. Those approaches have been inspired by nature. For creatures active in the night, light transmission through the lens of the eye is very important for two rea-sons. First, light is information and during the night there is limited information about the environment; if light is reflected from the surface of the eye the infor-mation content is reduced still further. Second, light reflected from the surface of the eye can give away its position and expose the creature to danger from preda-tors. Moths have anti-reflective eyes. Moth-eyes reflect less light that is expected by the laws of physics. On closer inspection little perturbations on the surface of the eye can be seen: hexagons of the order of 200–250 nm in diameter and height (Fig. 6).

These little half-circles (not to be confused with the well-known hexagonal fa-cets of the multi-lens structure of insect eyes) made of the same substance as the eye surface are the cause of the reduced light reflection and are a perfect of biona-notechnology. What causes these perturbations to increase light transmission?

Based on the theory given above, light is scattered when the surface roughness exceeds ½ λ (λ = wavelength. The half-spheres on the moth-eye are just below this limit so they obviously do not scatter light. A diagram of a cross-section through the moth-eye structure is shown in Fig. 7.

Again, at the outermost part of the nanostructure, air is the dominant medium. This means the refractive index is represented by the refractive index of air. At the

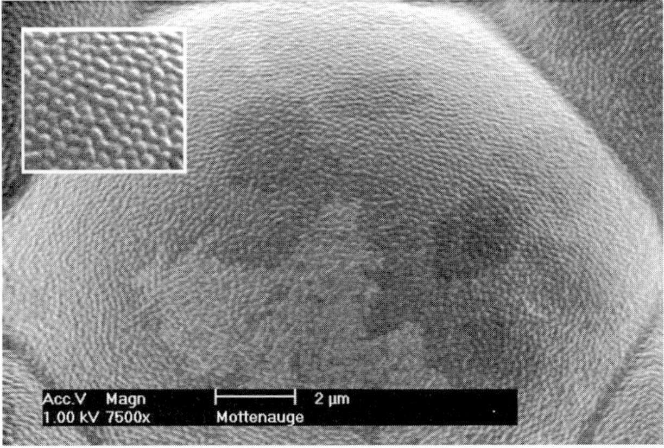

Fig. 6   SEM image of a moth eye facet. On top of the facet small semi-circles of about 200–250 nm can be seen.

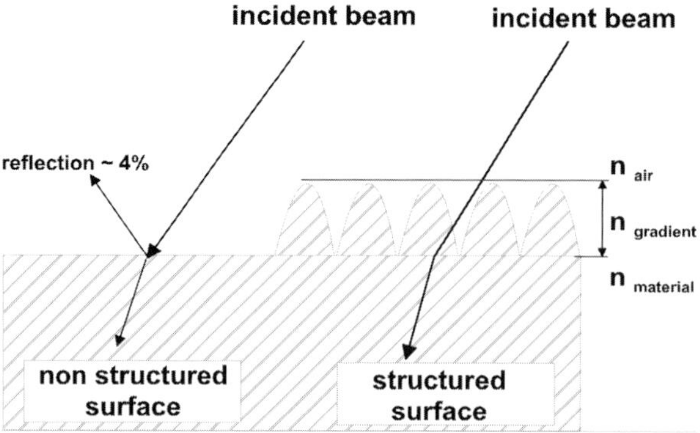

Fig. 7 Comparison of light hitting on a planar or nanostructured PMMA surface.

inner part of the nanostructure the substrate is the dominant material and the refractive index is given by the refractive index of the base material. This change causes light to penetrate into the substrate almost without being scattered or reflected. The challenge for physicists, chemists, and engineers is simply to copy the eye of a moth. Different approaches have been tested so far and only few are suitable for routine production of structured polymers.

4.3.1 LIGA Technology

When trying to structure polymeric surfaces in the sub-500 nm region using high throughput technologies, there is a demanding need for a kind of master structure that can be replicated into polymers easily and reproducibly. For this reason the initial goal in anti-reflective technology is to create a master for replicas with a feature size of 150–300 nm. In modern silicon technology this is achieved by electron beam or UV lithography. These high-cost methods are not suitable for large areas such as display coverings as well as non-planar substrates. An alternative is the so-called LIGA (German: Lithographie, Galvanoformung, Abformung) [53]. In this process a photoresist on a planar surface is structured using optical techniques such as holographic interference of laser light. This surface is copied into a metal and this metal is used as a tool for further replications.

By this method highly periodic arrays of lines or half-circles in photoresists can be made down to a size mainly depending on the wavelength of the laser light, the properties of the photoresist, and the quality of the optical system. Typical values are of the order of 250 nm in lateral dimension. This structure can then be copied by electroplating, for example, nickel on top. When placing this nickel foil into an injection-molding machine one can make polymeric parts with a nanostructured surface directly from the injection molding process (Fig. 8).

When building a roller out of the nickel foil planar surfaces can be structured up to the size of the roller while the length is determined by the extension of the roller.

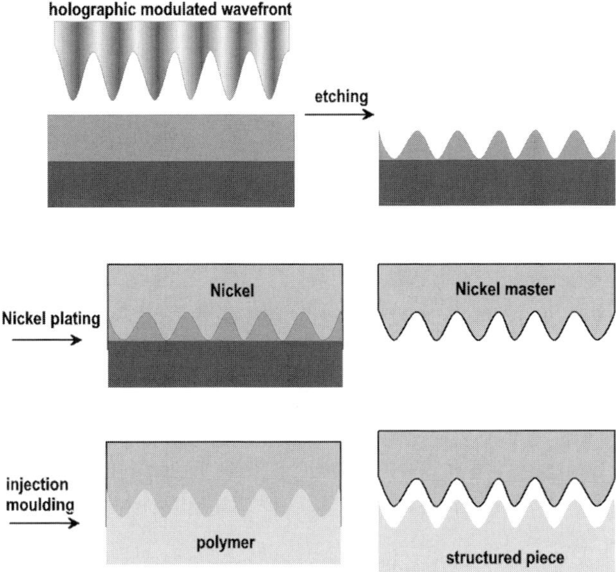

Fig. 8 Diagram of the LIGA process. A holographically structured polymer is galvanically replicated into a metal such as nickel. This replica can then be used as a tool in injection molding machines to replicate the initial pattern many times again in polymers or polymeric coatings.

The limitation of this technology is reached when the size of the master is increased even more. Since the initial step involves optical lithography the size of the optical systems has to increase as well when a larger master should be made. When increasing the optical components, errors such as astigmatism or aberration increase drastically as well. The method fails completely when the substrate for the holographic exposure is not flat. Taking the physical properties of those surface structures into consideration it is in addition of interest to create feature sizes less than 150 nm, which are difficult or even impossible using interference of visible laser light.

So there is still a need for a simple and easy way to create a moth-eye surface structure to improve light transmission in large areas or for complex shaped surfaces very common, for example, for automotive display coverings.

4.3.2 Nanoporous Alumina Tools (AlCoStruct)

The most intriguing feature of alumina is the adjustable pore geometry. Taking into account the cost of making a 250 nm structure by LIGA technology, using nanoporous alumina easily brings feature sizes down to 50 nm if desired on even large and complex shaped structures. So the benefits of using alumina templates are:

- pore formation by a self-assembling electrochemical process,
- adjustable diameter of the surface structure from 50 to 500 nm,

- adjustable height of the surface structures from a few nanometers up to several hundreds of nanometers,
- inert and abrasion-resistant ceramic surface as a master,
- surface can be chemically modified if desired, and
- complex 3D geometries can easily be structured.

Coming from the physics of anti-reflective surface structures it is of benefit to adjust surface feature sizes to different materials since these materials differ in refractive index. For PMMA, for example, an optimal light transmission enhancement is expected to occur with sizes of 100–200 nm. When increasing feature size, scattering at lower wavelengths can occur while smaller sizes will not improve transmission at higher wavelengths. In Fig. 9 an SEM image shows a 280 nm surface structure made by injection molding.

The aspect ratio can be estimated to be about 1. The features are closely packed, giving rise to a gradient in refractive index, which is the reason for improved anti-reflective properties. This surface structure is spectroscopically compared with a second of only 120 nm in size and a blank sample as a reference. The transmission spectra are given in Fig. 10.

The blank sample does show light transmission for about 91% of the complete spectral range. With the 120 nm surface structure, this light transmission can be enhanced by 6% again over the whole spectral range. In contrast to these results, the larger structure leads to an increase in transmission above 650 nm but to a slightly reduced transparency for shorter wavelengths. This is a good example to demonstrate the impact of the feature size (120 nm versus 280 nm) on the anti-reflective effect. The light transmission for the PMMA with a 120 nm surface structure is increased by 5–6% over the spectral range as predicted by theory. The remaining 2–3% are absorbed in the material.

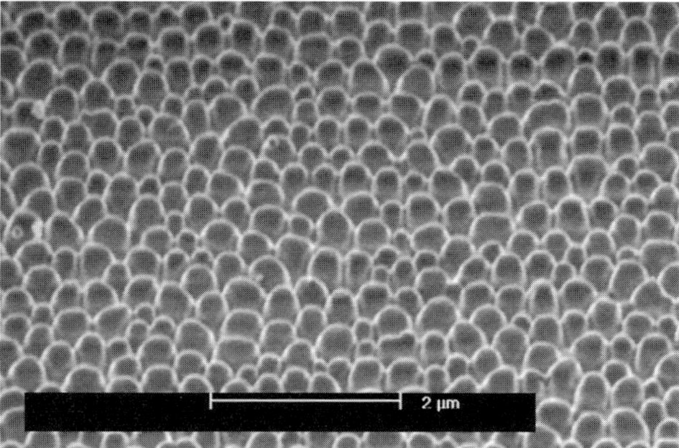

Fig. 9 Optimized PMMA surface made by injection molding.
The feature size is about 280 nm in diameter with an aspect ratio of 1.

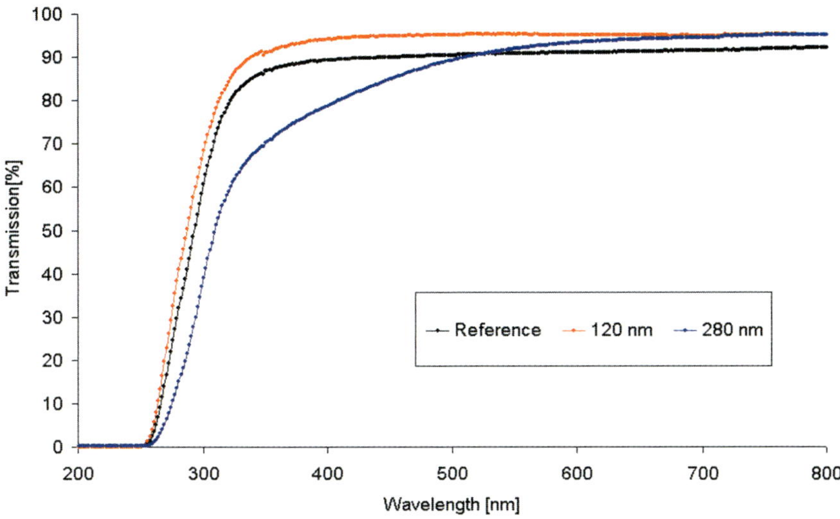

Fig. 10 The transmission spectra of specimens made by injection molding indicate the differences in transparency for different surface structures.

This effect can be visualized when comparing a structured specimen with a reference sample. In Fig. 11 stamps are covered with two PMMA specimens illuminated by diffuse light coming from a neon tube. The stamp covered with the structured PMMA plate can easily be visualized. Owing to the diffuse light reflection the other stamp under the reference can hardly be seen.

The same observation holds for other transparent polymers such as PC, PS, or different copolymers. Again the structure has to be adapted, which is no problem at all using the nanoporous alumina technology.

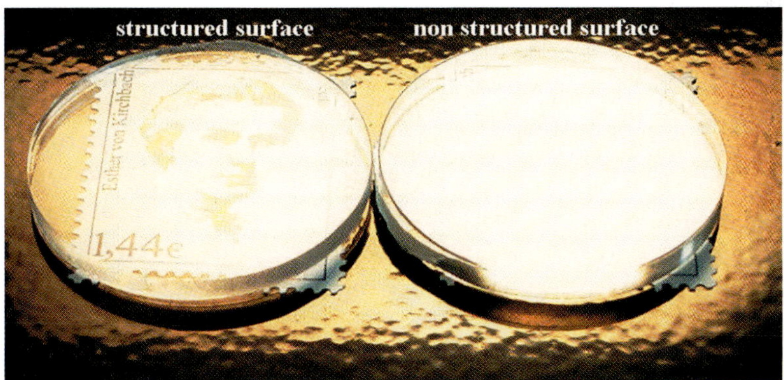

Fig. 11 Two stamps covered by a nanostructured and a plain surface. The enhancement in light transmission in the case of the nanostructured surface can be seen on the right side.

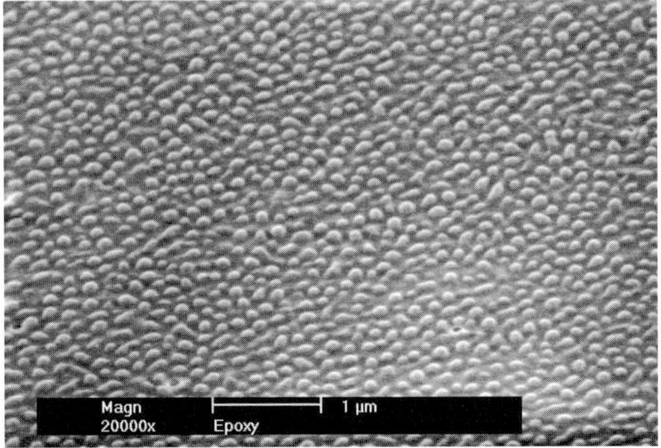

Fig. 12 Replicated pore structure on a two-component epoxy system made by casting onto a master surface. The feature size is about 120 nm in diameter with an aspect ratio of 0.8.

Besides using rather small tools for injection molding it is also possible to cast or imprint the surface structure into a desired material on a larger length scale. Casting can be applied to any kind of curable system. For example epoxy resins or polydimethysiloxane can be cast onto the porous surface leading to a perfect replication. In Fig. 12 a SEM image of a two-component epoxy-system surface made by casting can be seen.

The diameter of the pillars in this example is about 120 nm with an aspect ratio of 0.8. When making these replicas in polydimethysiloxane they can again be used as templates to structure a curable resin, for example, yielding a porous surface.

When taking a seedless roller, endless structuring of surfaces becomes feasible. Sol-gel or UV-curable coatings can be structured and cured immediately after leaving the roller. For solar energy applications every percentage point of light passing through the cover of solar cells counts, and therefore an anti-reflective surface of the cover of those cells is very interesting. Direct structuring of glass is not possible up to now, but coating the surfaces with a sol-gel coating, transferring the structure of the nanoporous alumina, and curing the gel afterwards leads to the desired surface structure. The nature of the sol-gel system has to be carefully adjusted, but this technology is well investigated.

5
Surface Wetting

As mentioned before, not only optical properties but also for example surface wetting can be influenced by a designed surface structure. The influences of surface roughness or surface porosity were described at the beginning of the 20th century

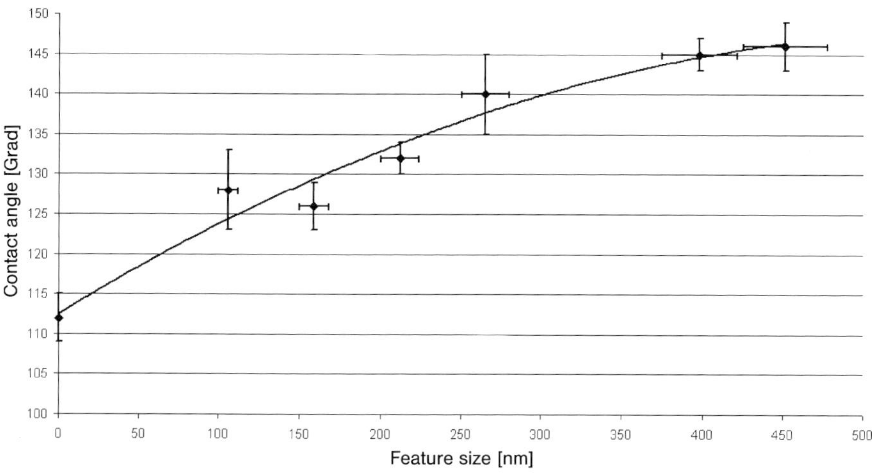

Fig. 13 Contact angle of water as a function of surface feature size made by imprinting into PTFE.

by Young's equation [58, 59] and adding factors for surface roughness [60, 61]. These early investigations of the influence of surface perturbations on the wetting behavior of liquids were broaden by the discovery of Barthlott that the dirt on certain plants is easily removed by water [62–67]. The so-called lotus effect is based on a specially designed micro- and nanostructured hydrophobic leaf surface that repels water very efficiently. Dust and dirt particles are only loosely bound to the low energy surface and removed easily by water droplets collecting those particles from the plant surface. The contact area of a micron sized particle on a lotus leaf surface is reduced to about 2% causing this weakening in the interaction. Low energetic surfaces such as PTFE can be structured in the nanometer range using the nanoporous alumina and the influence of these surface structures on surface wetting has been investigated. Increasing the diameter from 100 nm to 450 nm leads to an increase of the contact angle against water from 110° to 145° (Fig. 13).

This increase can best be explained by reducing the chance that the drop of water reaches the "ground" between the pillars, when those structures increase in diameter and height. In this way the surface becomes superhydrophobic, as shown by the contact angle and the contact angle hysteresis. The hysteresis of the contact angle can be as low as 2° for certain feature sizes. In Fig. 14 on the left a droplet of water on a plain PTFE surface can be seen.

The value of the contact angle determined on the plain substrate is about 110°. On the nanostructured PTFE surface it rises to more than 145°. This contact angle can be increased even more when combining the specially designed nanostructure with a machined microstructure underneath [43].

Fig. 14 Left, a drop of water sitting on a plain PTFE surface with a contact angle of 110°. Right, drop of water on a structured PTFE surface (450 nm) with a contact angle of about 145°.

6
Conclusions

Optical properties of matter are determined by surfaces and interfaces. Surface structures play a key role in the understanding of the interaction of electromagnetic radiation with these surfaces. The so-called lotus leaf effect shows that engineers, chemists, and physicists can learn from nature. The surface structure of certain insect eyes that eveloved by natural selection was the inspiration for a novel method of making an anti-reflective surface. The necessary continuous gradient in refractive index normally made by sophisticated plasma technologies can also be achieved by nanostructuring surfaces in a very special way. Present technological developments utilize the formation of a porous silica layer on glass substrates when an organo-modified sol-gel coating is fired at elevated temperature. Removing the component leaves a porous surface whose refractive index exhibits a gradient matching the optical properties of air on one side and glass on the other. The major drawback stems from the firing process: heat sensitive materials such as polymers cannot be coated using this technology. The so-called "moth-eye-structure" is a bio-inspired alternative solution to this problem consisting of little half-circles each with a diameter of 200 nm, increasing the efficiency of light transmission. They are closely packed on the surface allowing the refractive index to continuously change from the value for air to the substrate material value. Copying this structure can be done either by sophisticated holographic approaches or by utilizing nanoporous alumina known for decades as a template material. There is no additional material that has to be carefully coated onto the substrate: the material itself becomes anti-reflective. In addition the structuring technology can be used to change surface-wetting behavior significantly. For hydrophobic surfaces a large increase in contact angle can be observed for sub-500 nm surface structures. This can be compared with the lotus effect, namely that certain natural and artificial surfaces repel dirt and water significantly. The final goal is to combine the different effects of anti-reflectivity and surface wetting to give a real improvement in display covering technology.

Nanotechnology is an enabling technology. Different disciplines are overlapping in the field of small sized structures, manipulating almost atom by atom to build up new structures, to create new or improved properties of tomorrow's materials. The strongest impact of nanotechnology can be expected from all areas related to surface and interface effects since in the region of diminishing dimensions surfaces become very important and the control over surface chemistry and structure is the key to future applications of nanotechnology.

References

1 V. ALTSTÄDT, W. BLECK, W. DRACHMA, A. DAMMANNS, A. GRUNWALD, H. HARIG, H. HOFMANN, W. A. KAYSSER, C. J. LANGENBACH, J. MÜSSIG, K.-M. NIGGE, R. RENZ, T. SAWITOWSKI, G. SCHMID, V. THOLE, *Neue Materialien für Innovative Produkte. Entwicklungstrends und gesellschaftliche Relevanz, Europäische Akademie zur Erforschung von Folgen wissenschaftlich-technischer Entwicklungen Bad Neuenahr-Ahrweiler GmbH*, Springer, Heidelberg, **1999**.

2 R. W. SIEGEL, *Nanophase Materials, Encyclopedia of Applied Physics*, **1994**, *11*, 173.

3 J. H. FENDLER (Ed.), *Nanoparticles and Nanostructured Films*, Wiley-VCH, Weinheim, **1998**.

4 G. A. OZIN, *Nanochemistry: Advanced Materials* **1992**, *4*, 612.

5 M. M. LOHRENGEL, *Mater. Sci. Eng.* **1993**, *6*, 241.

6 A. F. HOLLEMANN, E. WIBERG, *Lehrbuch der Anorganischen Chemie*, 101 edition, Walter de Gruyter, Berlin, **1995**.

7 J. W. DIGGLE, T. C. DOWNIE, C. W. GOULDING, *Chem. Rev.* **1969**, *69*, 365.

8 W. GAISSER, *Galvanotechnik* **1981**, *72*, 244.

9 G. E. THOMPSON, G. C. WOOD, *Treatise Mater. Sci. Technol.* **1983**, *23*, 205.

10 K. WEFERS, C. MISRA, *Oxides and Hydroxides of Aluminium*, Alcoa Laboratories, **1987**.

11 R. KNIEP, P. LAMPARTER, S. STEEB, *Adv. Mater.* **1989**, *1*, 975.

12 H. GINSBERG, K. WEFERS, *Metallkunde* **1963**, *17*, 202.

13 DROSEL et al., *Aluminium-Taschenbuch*, Vol. 2, 15th ed., Aluminium-Verlag Düsseldorf, **1999**.

14 F. KELLER, M. S. HUNTER, D. L. ROBINSON, *J. Electrochem. Soc.* **1953**, *100*, 411.

15 T. P. HOAR, J. YAHALOM, *J. Electrochem. Soc.* **1963**, *110*, 614.

16 D. D. MACDONALD, *J. Electrochem. Soc.* **1993**, *140*, L27.

17 J. P. O'SULLIVAN, G. C. WOOD, *Proc. Roy. Soc. London* **1970**, *317*, 511.

18 C. MARTIN, *Chem. Mater.* **1996**, *8*, 1739.

19 C. M. ZELENSKI, G. L. HORNYAK, P. K. DORHOUT, *Nanostructured Materials* **1997**, *9*, 173.

20 P. HOYER, H. MASUDA, *J. Mater. Sci.* **1996**, *15*, 1228.

21 H. MASUDA, K. FUKUDA, *Science* **1995**, *248*, 1466.

22 P. HOYER, N. BABA, H. MASUDA, *Appl. Phys. Lett.* **1995**, *66*, 2700.

23 C. A. FOSS, G. L. HORNYAK, J. A. STOCKERT, C. R. MARTIN, *J. Phys. Chem.* **1994**, *98*, 2963.

24 R. V. PARTHASARATHY, C. R. MARTIN, *Nature* **1994**, *369*, 298.

25 J. D. KLEIN, R. D. HERRICK, II, D. PALMER, M. J. SAILOR, *Chem. Mater.* **1993**, *5*, 902.

26 P. HOYER, H. MASUDA, *Jap. J. Appl. Phys.* **1992**, *31*, L1775.

27 C. J. BRUMLIK, C. R. MARTIN, *J. Am. Chem. Soc.* **1991**, *113*, 2174.

28 D. ALMAWLAWI, N. COOMBS, M. MOSKOVITS, *J. Appl. Phys.* **1991**, *70*, 4421.

29 C. R. MARTIN, *J. Am. Chem. Soc.* **1990**, *112*, 8976.

30 G. L. HORNYAK, Dissertation, Fort Collins, **1997**.

31 G. L. HORNYAK, C. J. PATRISSI, C. R. MARTIN, *J. Phys. Chem. B* **1997**, *101*, 1548–1555.

32 C. A. Foss, G. L. Hornyak, J. A. Stockert, C. R. Martin, *J. Phys. Chem.* **1992**, *96*, 7497.

33 T. Hanaoka, H.-P. Kormann, M. Kröll, T. Sawitowski, G. Schmid, *Appl. Organometall. Chem.* **1998**, *12*, 367–373.

34 T. Sawitowski, Thesis, University of Essen, Germany, **1999**.

35 T. Hanaoka, A. Heilmann, H.-P. Kormann, M. Kröll, T. Sawitowski, G. Schmid et al., *Eur. J. Inorg. Chem.* **1998**, 807–812.

36 T. G. Sorop, K. Nielsch, P. Göring, M. Kröll, W. Blau, R. B. Wehrspohn, U. Gösele, L. J. de Jongh, Submitted to *J. Magn. Magn. Mater.* **2003**.

37 N. Tsuya, Y. Saito, H. Nakamura, S. Hayano, A. Furugohri, K. Ohta, Y. Wakui, T. Tokushima, *J. Magn. Magn. Mater.* **1986**, *54–57*, 1681.

38 J. C. Lodder, L. Cheng-Zhang, *IEEE Trans. Magn.* **1989**, *25*, 4171.

39 H. Masuda, H. Asoh, M. Watanabe, K. Nishio, M. Nakao, and T. Tamamura, *Adv. Mater.* **2001**, *3*, 189.

40 H. Masuda, H. Yamada, M. Satoh, H. Asoh, M. Nakao, T. Tamamura, *Appl. Phys. Lett.* **1997**, *79*, 2770.

41 D. Crouse, Yu-Hwa Lo, A. E. Miller, M. Crouse, *Appl. Phys. Lett.* **2000**, *76*, 49–51.

42 C. Y. Liu, A. Datta, N. W. Liu, C. Y. Peng, and Y. I. Wang, *Appl. Phys. Lett.* **2004**, *84*(14), 2509–2511.

43 M. Levering, Thesis, Essen, **2003**.

44 C. Gerthsen, H. O. Kneser, H. Vogel, *Physik*, 15 edn., Springer, Berlin, **1996**.

45 R. E. Hummel, *Electronic Properties of Materials*, 2nd edn., Springer, Berlin **1992**.

46 L. Weber, W. Ehrfeld, *Kunststoffe* **1998**, *10*, 88–92.

47 H. Reichenauer, G. Meyer, *Fachkunde der Plastverarbeitung*, Deutscher Verlag der Grundstoffindustrie, Leipzig, **1971**.

48 B. Carlowitz, *Kunststofftabellen*, Carl Hanser, München, **1995**.

49 H. Dominghaus, *Die Kunststoffe und ihre Eigenschaften*, VDI Verlag, Düsseldorf, **1992**.

50 U. Schulz, N. Kaiser, *Beschichtung von Kunststoffen für Optik und Optoelektronik – Laser Opto*, AT-Fachverlag Stuttgart, **1999**.

51 W. Beh, I. Kim, G. Whitesides, *Adv. Mater.* **1999**, *11*, 1038–1042.

52 M. Korn, T. Körfer, P. Roentgen, *J. Vac. Sci. Technol.* **1990**, *B8*, 1404–1410.

53 E. Becker, W. Ehrfeld, *Phys. Bl.* **1988**, *44*, 166–172.

54 T. Lippert, T. Gerber, M. Goto, *Appl. Phys. Lett.* **1999**, *75*, 1018–1021.

55 C. Daniel, F. Mücklich, Z. Liu, *Appl. Surf. Sci.* **2003**, *317*, 208–209.

56 G. Mie, *Ann. Phys.* **1908**, *25*, 377–445.

57 A. Gombert, W. Glaubitt, K. Rose, J. Dreiholz, B. Bläsi, A. Heinzel, D. Sporn, W. Döll, V. Wittwer, *Thin Solid Films* **1999**, *351*, 73–78.

58 J. Israelachvili, *Intermolecular & Surface Forces*, Academic Press, London, **1992**.

59 H.-D. Dörfler, *Grenzflächen- und Kolloidchemie*, VCH, Weinheim, **1994**.

60 R. D. Hazlett, *Trans. Faraday Soc.: J. Coll. Interface Sci.* **1990**, *137*, 527–533.

61 A. B. D. Cassie, S. Baxter, *Trans. Faraday Soc.* **1944**, *40*, 546–551.

62 S. Wagner, C. Nienhuis, W. Barthlott, *Acta Zoologica* **1996**, *77*, 213.

63 W. Barthlott, *Klima- und Umweltforschung an der UNI Bonn*, **1992**, 116–120.

64 W. Barthlott, C. Nienhuis, *Planta* **1997**, *202*, 1–8.

65 L. Rentschler, *Planta* **1971**, *96*, 119–135.

66 W. Schill, R. Schill, W. Barthlott, *Progress in Botany* **1981**, *43*, 27–38.

67 W. Barthlott, *Plants* **1990**, 69–83.

Preparation and Properties of MgO–Ni(Fe) Nanocrystalline Composites

Oldřich Schneeweiss, Naděžda Pizúrová, Yvonna Jirásková, and Tomáš Žák

1
Introduction

Nanocrystalline composites formed by nickel, iron, and cobalt were investigated as materials having good potential for practical applications as catalysts [1–3] or sensors based on magnetic properties and magnetoresistance. The composites were prepared by chemical methods. Magnesium oxide-based composites with transition metals were observed by investigation of Mg alloys with Fe, Ni, or Co. These alloys have been studied as potential materials for hydrogen storage [4]. An additional alloying of the magnesium was studied from the point of view of hydriding behavior. Nickel and cobalt form some intermetallic compounds with magnesium, the solubility of iron is almost zero, and intermetallic compounds have not been reported before. Various techniques have been used to extend a low equilibrium solubility limit, such as ion implantation [5], the coevaporation technique [6], or mechanical alloying [7].

The spark erosion (synthesis) method is an alternative way of preparing Mg–Fe alloys. This technique was used for the preparation of amorphous, nanocrystalline, or crystalline powder materials [8, 9]. The conditions of erosion are characterized by high temperature (above 10^4 K) and pressure (~ 280 MPa) [13] in the plasma channel, where a synthesis of the electrode materials and the environment occurs, and high cooling rate ($\sim 10^8$ K s^{-1}). It allows solubility limits reached by classic alloying to be overcome and new materials to be synthesized. Also, it is possible to modify the composition of the product by varying the parameters of sparks (voltage or time) and/or chemical composition, temperature, and density (pressure) of the gaseous or liquid dielectrics.

In this chapter we describe some results obtained by synthesis of magnesium with nickel–iron alloys with the aim of preparing nanocomposite materials in which ferromagnetic nanoparticles are separated and protected against oxidation by a magnesium oxide shell.

The Nano–Micro Interface: Bridging the Micro and Nano Worlds
Edited by Hans-Jörg Fecht and Matthias Werner
Copyright © 2004 WILEY-VCH Verlag GmbH & Co. KGaA, Weinheim
ISBN: 3-527-30978-0

2
Experimental

The Mg–Fe powder was prepared by spark erosion of electrodes of pure Mg (99.9%) and Ni–Fe alloys with 20 at.-% Fe carried out in hydrogen at atmospheric pressure as dielectric. Heat treatments were carried out in a vacuum better than 10^{-3} Pa or in a pure hydrogen (better than 5 N) atmosphere combined with the vacuum pre-annealing (at $\sim 100\,^\circ$C) to remove gases adsorbed during sample handling in air. It was shown that the Mg–(Fe,Ni) nanopowders prepared in this way [10, 11] are very sensitive to oxidation and on handling in air or in argon of technical (2.5 N) purity and magnesium nanoparticles oxidize to MgO. ^{57}Fe Mössbauer spectra were measured by a standard transmission method. The pure α-Fe foil was used for calibration. Isomer shifts are reported relative to α-Fe at room temperature. The computer processing of the spectra done by the CONFIT package [12] yielded intensities, I, of the components, their hyperfine inductions, B_{hf}, isomer shifts, δ, and quadrupole splittings, σ. X-ray powder diffraction was performed with a Siemens D5005 (Bruker AXS, Germany) using Cu Kα radiation (40 kV, 45 mA) and a diffracted beam monochromator. Qualitative analysis was performed with the Diffrac-Plus software package (Bruker AXS, Germany, version 7.0) and JCPDS PDF-2 database. For quantitative analysis of XRD patterns we used PowderCell for Windows, version 2.3 with structural models based on the ICSD database. Thermogravitometry and differential thermal analysis were completed by quadrupole mass spectrometer for the analysis of released gases (multiple ion detection) during heating ($\Delta T/dt = 1$ K/s) in dynamic Ar atmosphere (~ 75 mL/min). As a reference sample α-Al$_2$O$_3$ was used. Thermomagnetic curves were measured using a vibrating sample magnetometer. The measured sample was in vacuum $\sim 10^{-2}$ Pa and in 50 Oe external magnetic field.

3
Results and Discussion

Basic information about phase composition was derived from Mössbauer spectra. In the as-prepared powder Fe–Ni, Fe^{2+}, Fe^0, and γ-Fe phases were analyzed. An example of the spectrum is shown in Fig. 1 and parameters of the phases and their components are summarized in Tab. 1. The ferromagnetic Fe–Ni phase is represented in the spectrum by four sextets SA1–SA4 with the mean $B_{hf} = 26.76$ T. The parameters of the sextets agree with those obtained for the Ni–Fe electrode material. The next components (doublets DA1, DA2, and singlet LA1) can be ascribed to the Fe^{2+}. They have the hyperfine parameters, which are in good agreement with the values of Fe$_{1-x}$O reported elsewhere [13, 14]. Fe^{2+} probably comes from the substitution of Mg^{2+} in the MgO lattice. The singlet LA2 can be interpreted as Fe^0, that is, superparamagnetic iron clusters (particles). The last component (the singlet LA3) represents the fcc γ-Fe. Its isomer shift $\delta = -0.09$ mm/s agrees well with that of fine γ-Fe precipitates in Cu [15]. It also confirms the observation of

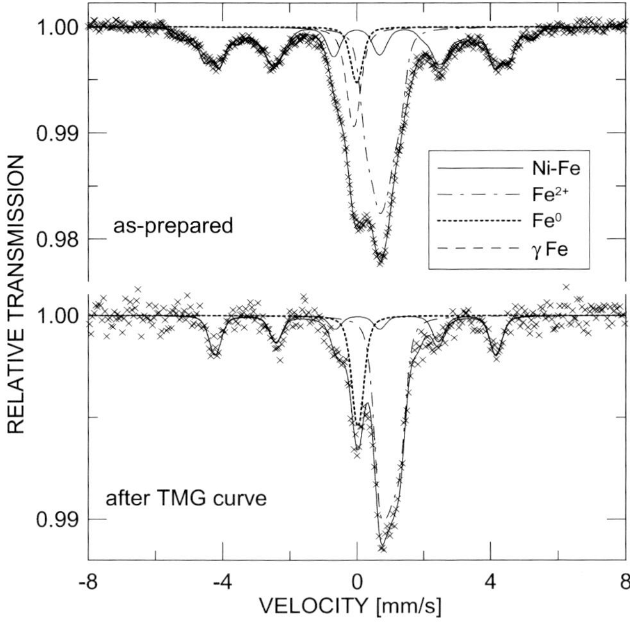

Fig. 1 Mössbauer spectra of the MgO–(Fe,Ni) powders.

Tab. 1 Components and their parameters derived from the Mössbauer spectra analysis.

Component	I	δ [mm/s]	σ [mm/s]	B_{hf} [T]	ΣI	Mean B_{hf} [T]	Phase
As-prepared powder							
SA1	0.05 ± 0.01	$0.03\,0.01$	0.00 ± 0.01	32.34 ± 0.08			
SA2	0.12	0.00	0.00	28.21	0.39	26.76	Ni–Fe
SA3	0.16	0.00	0.00	25.55			
SA4	0.06	−0.03	0.02	21.92			
DA1	0.19	0.62	0.34	–			
DA2	0.15	0.92	0.36	–	0.43	–	Fe^{2+}
LA1	0.09	0.72	–	–			
LA2	0.06	0.00	–	–	0.06	–	Fe^{0}
LA3	0.12	−0.09	–	–	0.12	–	γ-Fe(Ni)
Powder after TGM curve measurement							
SB1	0.24	−0.01	0.02	26.04	0.24	26.04	Ni(Fe)
DB1	0.36	1.13	0.16	–	0.60	–	Fe^{2+}
LB1	0.24	0.67	–	–			
LB2	0.16	0.02	–	–	0.16	–	Fe^{0}

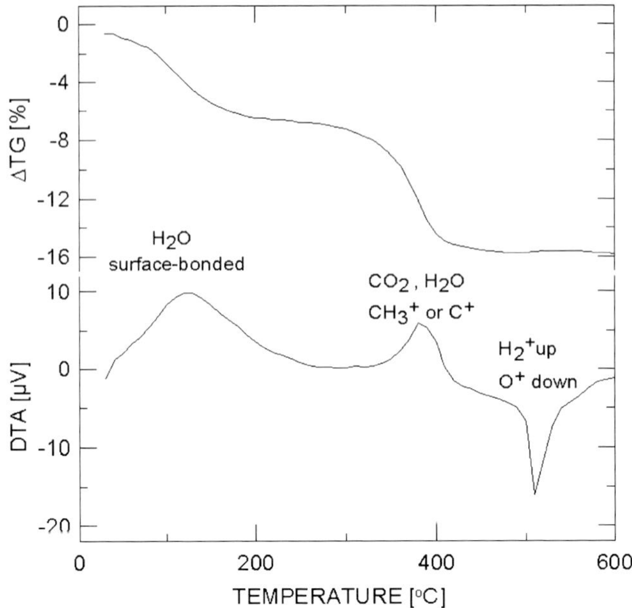

Fig. 2 Thermogravitometric (TG) and differential thermal analysis (DTA) of the as-prepared Mg–(Ni,Fe) powder.

the iron nanometer-size fcc phase in MgO composite films prepared by vacuum codeposition [16, 17].

In the X-ray diffraction patterns of the as-prepared powder, MgO phase was identified. Contents and/or coherent volumes of other phases were below detection limits. It can be explained as a result of miscibility of MgO–NiO and MgO–FeO over the entire molar fraction range because of their very similar structures [18, 19].

The curves of TG and DTA recorded by heating up to $600\,^{\circ}C$ are shown in Fig. 2. The first step/peak on the TG/DTA curves at $\sim 130\,^{\circ}C$ can be ascribed, according to mass spectrometry, to the desorption of water from the surface of the powder particles. Similarly, desorption of structure-bonded water and other gases can explain the step/peak at $380\,^{\circ}C$. Besides that, unexpected fragments of CH_3^+ and/or C^+ were also detected. They could be a result from some catalytic reactions of CO_2 or carbon impurities from the electrodes with hydrogen from water or from hydrogen diluted by spark erosion.

An important exothermic effect at $512\,^{\circ}C$ was observed on the DTA curve. The mass spectrometry has shown an increase in partial pressure of hydrogen (H^+). In this temperature range a small mass increase (see ΔTG curve in Fig. 2) and a fine decrease of oxygen (O^+) were detected. This could be ascribed to progress in oxidation of Mg that was below this temperature protected by hydrogen bonding.

The X-ray diffraction taken on the powder after the TG/DTA measurement (Fig. 3) showed clear peaks of MgO only.

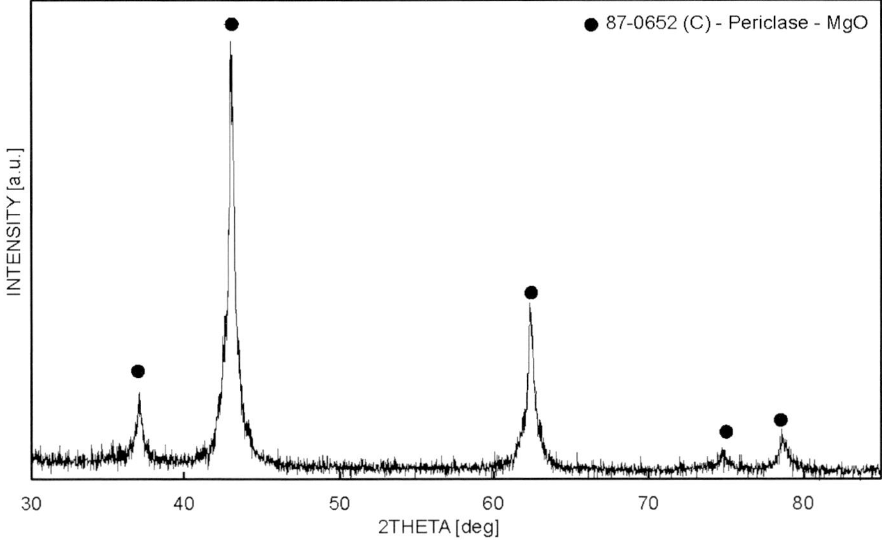

Fig. 3 X-ray diffraction of the MgO–(Fe,Ni) powder after TG/DTA measurement.

The thermomagnetic curves measured at increasing and decreasing temperatures up to 800 °C are shown in Fig. 4. An increase in magnetic moment with temperature up to 350 °C and its slight decrease above this temperature is connected with the Curie temperature of Ni rich Ni–Fe phase. The Curie temperature of the pure Ni is 354 °C [20]. The steps on the curve at temperatures of $\sim$750 °C and 770 °C correspond to the Curie temperatures of Fe rich Fe–Ni and α-Fe phases [20]. An additional fine kink at 250 °C at decreasing temperature could be connected either with the Neel temperature of NiO [20] or with the Curie temperature of $MgNi_2$ (235 °C [21]). The ferromagnetic behavior of the later phase was disputed [21] and neither phase was confirmed by the methods used for the phase analysis in present studies.

Mössbauer spectra taken on the sample after heat treatment up to 500 °C showed negligible changes in the atomic arrangements and phase composition. More pronounced changes were observed in the spectrum taken on the sample used for the TMG curve measurement. The ferromagnetic NiFe phase represented by sextet SB1 (see Tab. 1) corresponds to Ni with low Fe concentration [22, 23]. This phase agrees well with the observed step on the TMG curve at $\sim$350 °C. On the other hand α-Fe and Fe rich Fe–Ni phases were not distinguished in the spectrum although their Curie temperatures on TMG curve witness their presence. The intensity of component corresponding to Fe^{2+} increases in comparison with the as-prepared state. This means that oxidation and/or diffusion of iron in MgO take place during the annealing at decreasing temperature. As consequence, the clusters or nanoparticles of Fe and Fe rich Fe–Ni diminish and Ni–Fe is depleted in iron content. The intensity of the Fe^0 component also increases and the γ-Fe cannot be distinguished there. A simple comparison of the intensities of Fe^0 and γ-Fe in the as-prepared and annealed states indicates that a transition of γ-Fe into Fe^0 occurs.

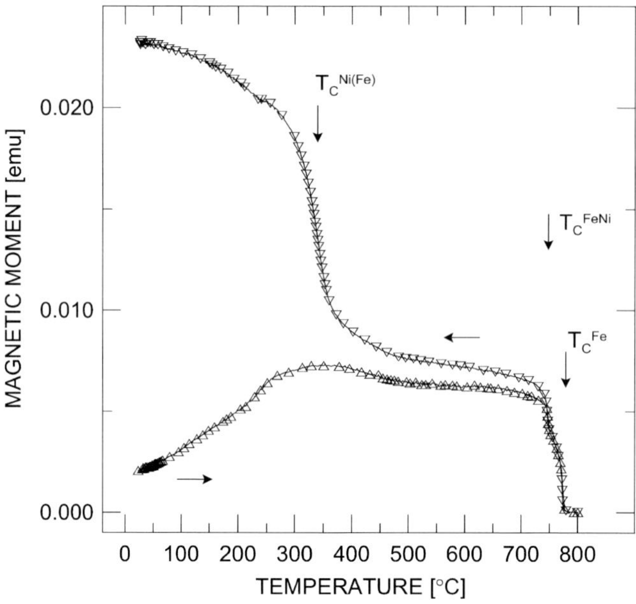

Fig. 4 Thermomagnetic (TMG) curve of the as-prepared MgO–(Ni,Fe) powder.

4
Conclusions

In the as- prepared powder several phases were found. Besides dominating MgO, Ni–Fe, and Fe^{2+}, a small amount of Fe^0 and γ-Fe were identified using Mössbauer spectroscopy and X-ray diffraction.

The TG/DTA completed with mass spectroscopy measurement by heating up to 600 °C show gas desorption during the temperature increase. It also indicates catalytic effects causing appearance of fragments of CH_3^+ and/or C^+. An increase of partial pressure of hydrogen and a slight progress in oxidation was observed at 512 °C.

The thermomagnetic curve measured at heating and cooling of the powder up to 800 °C reflects changes connected with Curie temperatures of the Ni with low Fe content, Fe rich Fe–Ni, and pure α-Fe phases.

The detailed phase analysis shows that the as-prepared powder is stable approximately up to 500 °C. By annealing at higher temperatures hydrogen bonded in the powder phases is released and oxidation takes place both in vacuum and under protective atmosphere. Iron clusters diminish and nanoparticles of Ni–Fe are depleted in iron content after the heat treatment connected with TMG curve measurement.

Acknowledgments

This work was supported by the Academy of Sciences of the Czech Republic (K10101040) and Grant Agency of the Czech Republic (Contract No. 202/01/0668).

References

1 H. Hattori, *Chem. Rev.* **1995**, *95*, 537.

2 B.Q. Xu, J.M. Wei, H.Y. Wang, K.Q. Sun, Q.M. Zhu, *Catalysis Today* **2001**, *68*, 217.

3 Y.H. Hu, E. Ruckenstein, *Catal. Rev. Sci. Eng.* **2002**, *44*, 423.

4 R. Schulz, J. Huot, G. Liang, S. Boily, G. Lalande, M.C. Denis, J.P. Dodelet, *Mater. Sci. Eng.* **1999**, *A267*, 240.

5 H. Reuther, M. Betzel, W. Matz, E. Richter, *Hyperfine Interaction* **1998**, *113*, 391.

6 T. Shinjo, *Hyperfine Interaction* **1986**, *27*, 193.

7 A. Hightower, B. Fultz, R.C. Bowman, Jr, *J. Alloys Compounds* **1997**, *252*, 238.

8 A.E. Berkowitz, J.L. Walter, *Mater. Sci. Eng.* **1982**, *55*, 275.

9 J.L. Walter, *Powder Metallurgy* **1988**, *31*, 267.

10 O. Schneeweiss, Y. Jirásková, T. Žák, J. Šebek, Lin Shao-fan, Yao Xinkan, *Czech. J. Phys.* **2002**, *52*, 167.

11 O. Schneeweiss, Y. Jirásková, J. Šebek, *Phys. Stat. Sol. A* **2002**, *189*, 725.

12 T. Žák in *Mössbauer Spectroscopy in Materials Science* (Eds. M. Miglierini and D. Petridis), Kluwer Academic Publishers, Dordrecht, **1999**, p. 385.

13 C. Gohy, A. Gérard, F. Grandjean, *Phys. Stat. Sol. A* **1982**, *74*, 583.

14 C.A. McCammon, D.C. Price, *Phys. Chem. Minerals* **1985**, *11*, 250.

15 U. Gonser, M. Ron in *Applications of Mössbauer Spectroscopy*, Vol. II (Ed. R.L. Cohen), Academic Press, New York, **1980**, p. 288.

16 N. Tanaka, F. Yoshizaki, K. Katsuda, K. Mihama, *Acta Metall. Mater.* **1992**, *40*, S275.

17 N. Tanaka, F. Yoshizaki, K. Mihama, *Mater. Sci. Eng.* **1996**, *A217*, 311.

18 H.B. Nussler and O. Kubaschewski, *Z. Phys. Chem. Neue Folge* **1980**, *121*, 187.

19 JCPDS-ICDD records from 77-2365 to 77-2369; A.H. Jay, K.W. Andrews, *J. Iron Steel Inst.*, London, **1946**, *152*, 15.

20 D.R. Lide (ed.), *CRC Handbook of Chemistry and Physics*, 80th edn, CRC Press, Boca Raton, **1999**, pp. 12–119.

21 K.H.J. Buschow, *Solid State Commun.* **1975**, *17*, 891.

22 G. Longworth, B. Window, *J. Phys. F.: Metal. Phys.* **1973**, *3*, 832.

23 J.Y. Ping, D.G. Rancourt, R.A. Dunlap, *J. Magn. Magn. Mater.* **1992**, *103*, 285.

Nanocrystalline Oxides Improve the Performances of Polymeric Electrolytes

Silvia Licoccia and Enrico Traversa

1
Introduction

Environmental problems lead to the need for new technologies in the fields of energy production and storage for sustainable development, to reduce the pollutant emissions from fossil fuel combustion [1]. Among the possible systems investigated, fuel cells seem to be very promising as electrochemical power sources for application in portable technology and in electric vehicles, in particular polymeric electrolyte fuel cells (PEFCs) [2]. However, lithium ion batteries are already commercialized for electronic products, although improvement is needed for other applications (with the use of solid electrolytes) [3]. These electrochemical devices are based on polymeric electrolyte membranes, which have also been used for biomedical devices such as cardiac pacemakers and neurostimulators [4].

Polymer electrolytes consist of a polymer matrix into which a salt is dissolved. The ionic transport is generally described as the ionic species being moved through the electrolyte by the motion of the polymer chains. When a solvent is added to the polymer matrix the systems are called polymer gel electrolytes: the solvent improves ionic mobility either favoring segmental motions of the polymer chains or becoming itself the medium for ionic transport [5, 6]. Conductivity may be due to the motion of different ions, such as protons for fuel cells or lithium ions in batteries. The peculiar properties of these materials has made them the subject of numerous investigations, however several technological problems are still to be solved for specific applications. A possible strategy is the use of composites; composite materials have been shown to develop new or multifunctional properties when materials with differing properties are integrated together [7]. For instance, the addition of inorganic fillers is effective in improving the ionic conductivity, mechanical strength, and thermal stability of polymeric electrolyte systems [8]. Moreover, reduction in size at the nanometric level reveals unique physicochemical properties of materials [9]. The uniform dispersion of nanosized particles as fillers in polymers can lead to an ultralarge interfacial area between the constituents per volume of material, resulting in the development of new classes of materials with unique structure [10].

The Nano–Micro Interface: Bridging the Micro and Nano Worlds
Edited by Hans-Jörg Fecht and Matthias Werner
Copyright © 2004 WILEY-VCH Verlag GmbH & Co. KGaA, Weinheim
ISBN: 3-527-30978-0

In this chapter we show how the development of nanocomposite materials prepared by filling polymeric matrixes with nanocrystalline oxides is effective in improving the performances of membranes for direct methanol fuel cells (DMFC), lithium ion polymeric batteries, and electrodes for electrophysiological measurements. The approach of using nanocomposite membranes seems to be quite versatile given that it was successful both for H^+ and Li^+ conductors and with different types of polymers.

2
Results and Discussion

2.1
Direct Methanol Fuel Cells

The commercial material used in PEFCs is Nafion®, a perfluorinated sulfonated polymer, which acts as a proton conductor. For vehicles, the use of methanol instead of hydrogen as a fuel has several practical benefits such as easy transport and storage. The limiting factors for the use of direct methanol fuel cells (DMFCs) are the slow oxidation kinetics of methanol and its crossover through the membrane, which reduce the efficiency of the cell [11, 12]. An increase in operation temperature up to 150 °C is needed to overcome these problems; in fact the kinetics of methanol oxidation would be faster and the poisoning of the anode electrocatalyst would be reduced because of the lower amount of adsorbed methanolic residues. However, given its conduction mechanism due to the presence of water within its structure, Nafion cannot be used at temperatures above 100 °C. In recent years significant efforts have been directed to the development of polymer electrolyte membranes for DMFC alternative to Nafion. For example, Nafion/inorganic composite membranes have been examined and have shown enhanced water retention characteristics, hence allowing their operation in DMFCs at higher temperatures [13, 14]. Our approach to solve the problems was the use of nanocrystalline oxides as fillers for the composite membranes [15, 16].

Nanocrystalline TiO_2 and ZrO_2 powders were prepared by rapid hydrolysis of ethanolic solutions of $Ti(OiPr)_4$ and $Zr(OPr)_4$ respectively [17]. The powders were calcined at 500 °C on the basis of simultaneous thermogravimetric and differential thermal analysis (TG/DTA) results and characterized by X-ray diffraction (XRD) using Scherrer's equation to determine the average crystallite size [18]. The titania powders were found to have the anatase structure with a crystallite size of 12 nm. The zirconia fired at 500 °C showed a mixture of monoclinic and tetragonal structures both with an average crystallite size of 25 nm. Morphology and mean particle diameters were determined by scanning electron microscopy (SEM). For the titania powders, as previously reported [17], the particle size was about 10 nm in good agreement with XRD findings. The SEM observation of zirconia powder showed particle size of about 20 nm.

Nafion-based membranes containing 3, 5, and 10 wt.-% of the prepared oxides were fabricated with the recast technique and tested in a prototype DMFC. The thickness of all membranes was about 100 µm. Membrane electrode assemblies (MEAs) were prepared for fuel cell experiments, which were carried out in a 5 cm^2 single cell (GlobeTech, Inc.). A 2 M aqueous solution of methanol and oxygen was preheated at 85 °C and fed to the cell. The catalyst used for methanol oxidation was 60% PtRu (1 : 1)/Vulcan (E-TEK), while a 30% Pt/Vulcan (E-TEK) was used for oxygen reduction. The Pt loading for both electrodes was 2 mg cm^{-2}.

Introducing the nanosized powders into the Nafion membranes was effective in raising the operation temperature of the cells. All the MEAs fabricated with the composite membranes were capable of operation at 145 °C, while for the pure recast Nafion membranes the maximum operation temperature was 120 °C. Figs. 1 and 2 show the polarization curves and the power density curves, respectively, obtained at 145 °C for the DMFCs using the nanocomposite membranes with the various concentrations of TiO$_2$, and for comparison the polarization curve and power density curve of Nafion measured at 90 °C. The best electrochemical performances were observed for the membranes containing 3–5 wt.-% of titania, while increasing the titania content up to 10 wt.-% caused an increase in cell resistance resulting in worse electrochemical performance. A maximum power density of 350 mW cm^{-2} was reached at a current density of about 1.1 A cm^{-2} for the Nafion-based nanocomposite membrane filled with 5 wt.-% of TiO$_2$.

The addition of nanocrystalline ZrO$_2$ powders was also effective in increasing the operation temperature up to 145 °C, though the electrochemical performances were slightly lower than those observed for titania filling at the same concentrations. Fig. 3 shows the comparison of polarization and power density curves for 3 wt.-% TiO$_2$ and ZrO$_2$ nanocomposite membranes, measured at 145 °C. The max-

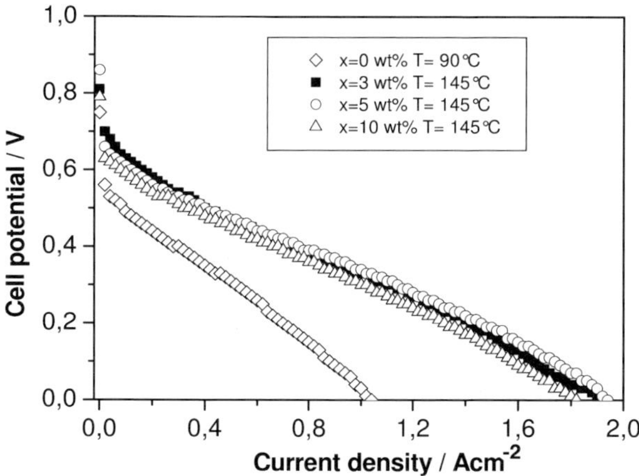

Fig. 1 Polarization curves for the MEAs with Nafion$^®$ membrane measured at 90 °C, and with the various nanocomposite membranes with different TiO$_2$ loadings, measured at 145 °C, in the presence of oxygen feed and 2 M methanol.

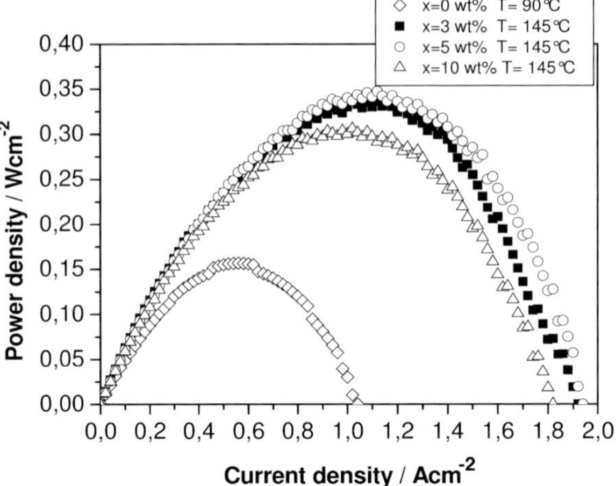

Fig. 2 Power density curves for the MEAs with Nafion® membrane measured at 90 °C, and with the various nanocomposite membranes with different TiO_2 loadings, measured at 145 °C, in the presence of oxygen feed and 2 M methanol.

imum power density for the membrane filled with 3 wt.-% of TiO_2 was 340 mW cm^{-2} at a current density of about 1.1 A cm^{-2}, while for the membrane filled with 3 wt.-% of ZrO_2 a power density of about 320 mW cm^{-2} was measured at a current density of about 1.0 A cm^{-2}.

The improvement of the DMFCs performance upon addition of nanocrystalline oxides might be ascribed to the enhancement of water retention at temperatures

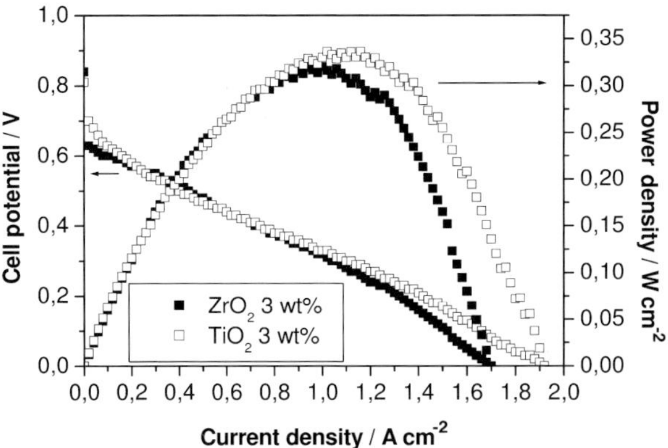

Fig. 3 Comparison of the fuel-cell performance of the MEAs with nanocomposite Nafion®-based membranes with 3 wt.-% of TiO_2 and ZrO_2, measured at 145 °C, in the presence of oxygen feed and 2 M methanol.

above 100 °C, due to water chemisorption on ceramic surfaces. Moreover, ceramic particles may act as barriers that hinder methanol crossover.

2.2
Lithium Ion Polymeric Batteries

Membranes made of polymeric gels containing lithium salts for Li batteries couple the high energy, long life characteristics of the lithium process with the reliability and easy processing of the plastic configuration [19, 20]. Among the most investigated ion-conducting polymer gels are those based on poly(ethylene oxide) (PEO). High molecular weight PEO is a tough and highly crystalline solid with good electrochemical stability [21]. The main problem for the practical application of PEO is its low crystallization temperature, 65 °C [22]; the crystallinity of PEO is accompanied by a decrease in ionic conductivity due to the entanglement of polymer chains that hinder their mobility. Moreover, the decomposition of PEO occurs at about 3.8 V and thus the use of commercially available cathodes is not suitable.

Several strategies have been followed to improve the performances of PEO-based membranes: the use of plasticizers, either salts [23] or low molecular weight organic solvents [24, 25], or the addition of inorganic fillers [8]. Ceramic fillers enhanced the ionic conductivity of PEO-based membranes both below and above its crystallization temperature and improved the performances of polymer electrolyte lithium batteries [26]. The improvement has been demonstrated to be related also to the filler particle size and was higher for smaller particles [27]. One of the mechanisms proposed to explain the observed improvements was the interaction of surface acidic –OH groups with the polymer electrolyte components. Given that zirconia is characterized by the presence of surface Lewis acid groups, we have studied the effect of the addition of nanocrystalline ZrO_2 powders to a PEO-based polymer electrolyte. To study the performance of this electrolyte in Li batteries, a cathode based on $LiFePO_4$ was selected because its working potential is compatible with the PEO decomposition potential [28].

The nanocomposite polymer electrolyte was prepared in acetonitrile using the casting technique from PEO 600000, $LiCF_3SO_3$ and ZrO_2 nanometric powder prepared as described above. The salt/polymer ratio was 1:20 and the concentration of zirconia was 10 wt.-% of the total $PEO_{20}LiCF_3SO_3$ weight. Drying under vacuum at 50 °C achieved membranes of 100 μm average thickness. Electrochemical impedance spectroscopy (EIS) was used to measure the resistance of the electrolyte membranes by sandwiching them between two polished stainless steel electrodes. The metallic-Li complete cell was assembled by sandwiching a lithium foil as anode, the nanocomposite membrane as electrolyte, and a film-type cathode made of 83 wt.-% $LiFePO_4$ active material added with Ag (1 wt.-%), mixed with 12 wt.-% of a carbon conductive additive (MMM Carbon Belgium Super P) and 5 wt.-% of a PVdF binder. The galvanostatic cycling tests were performed at different temperatures using a Maccor battery tester by setting the charge cut-off voltage at 3.8 V and the discharge cut-off voltage at 3.0 V.

Fig. 4 shows the Arrhenius plots of the ionic conductivity measured by EIS on the PEO-based membranes with and without the addition of nanocrystalline zirco-

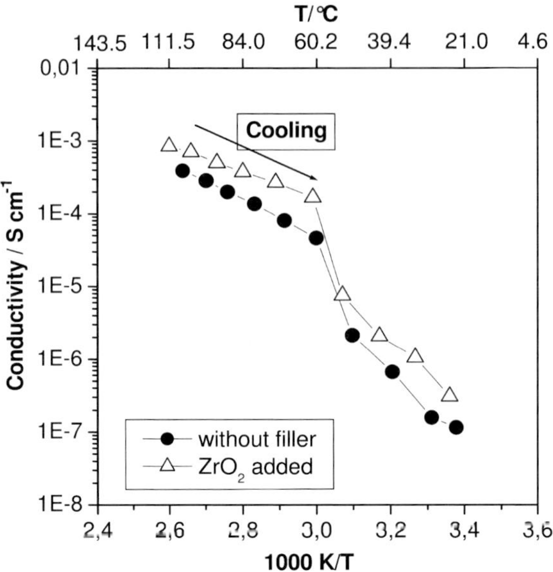

Fig. 4 Arrhenius plots of the conductivity of $PEO_{20}LiCF_3SO_3$ membrane with and without the addition of 10 wt.-% ZrO_2.

nia. The nanocomposite membrane showed a significant enhancement in conductivity values with respect to those measured on the reference ceramic-free PEO membrane, both above and below the PEO crystallization temperature.

The introduction of nanocrystalline zirconia improved the capability and cyclability of PEO-based polymer electrolyte membranes for lithium batteries. Fig. 5 shows the cycling performance of the cell $Li/PEO_{20}LiCF_3SO_3 + 10$ wt.-% $ZrO_2/$ $LiFePO_4 + 1$ wt.-% Ag, operating at 100 °C with different C/rates (the current intensity for 1C corresponds to about 0.25 mA cm^{-2}). The specific capacity measured at C/5 was about 160 mA h g^{-1}, very close to the theoretical value. Even at 1C the specific capacity was as large as 100 mA h g^{-1}. Fig. 6 shows the effect of temperature on the specific capacity of the cell. The measurements were performed at constant C/rate (C/5) decreasing the temperature in 10 °C steps from 100 to 60 °C, varying. No remarkable capacity loss was observed over the entire tested range of temperature. In particular, the specific capacity measured between 80 and 60 °C was about 130 mA h g^{-1}, demonstrating that the cell can work even at 60 °C, below the PEO crystallization temperature. Moreover, since the tested cathode has a working potential of 3.5 V, its compatibility with PEO was excellent.

The introduction of nanocrystalline fillers induced an improvement of performances also in the case of Li$^+$ conductors. The large specific surface area of the oxide is effective in promoting the specific interactions between the filler, the polymeric chains, and the ions deriving from salt dissociation, causing an enhancement in conductivity.

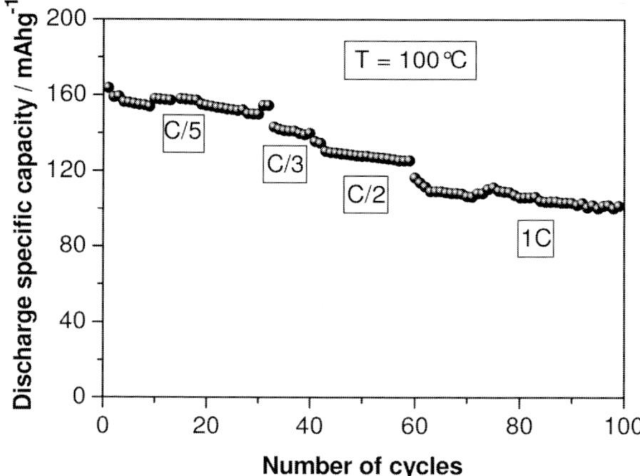

Fig. 5 Specific capacity as a function of the number of cycles at different C/rates, at T = 100 °C of the cell $LiFePO_4$ in $PEO_{20}LiCF_3SO_3$ + 10 wt.-% ZrO_2. Anodic limit: 3.8 V, cathodic limit: 2.4 V.

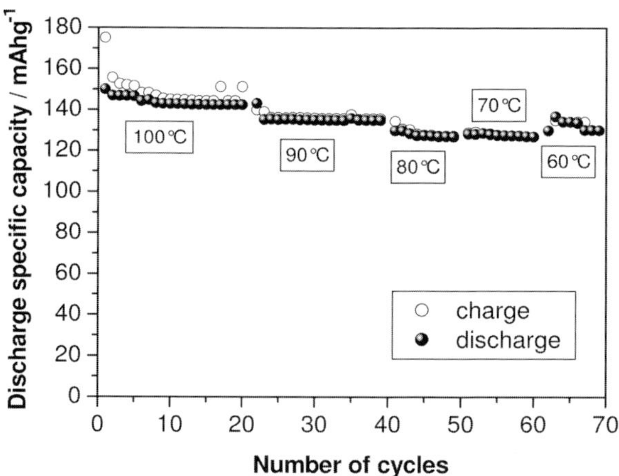

Fig. 6 Specific capacity as a function of the number of cycles at different temperatures of the cell $LiFePO_4$ in $PEO_{20}LiCF_3SO_3$ + 10 wt.-% ZrO_2. Anodic limit: 3.8 V, cathodic limit: 2.4 V.

2.3
Electrophysiological Measurements

Conventional electrophysiological measurements, such as the electroencephalogram (EEG), use scalp electrodes that are indirectly connected with the skin via an electrolyte bridge formed by a liquid gel or paste applied between the electrode and the skin. Furthermore the skin must be accurately cleaned to optimize the measurements. Both skin cleaning and the use of liquids are major drawbacks if the measurements have to be carried out in extreme conditions such as multielectrodic registrations, emergencies, non-cooperative patients, or in the absence of gravity. The initial aim of our work was in fact the development of alternative electrodic material to be used in space. The purpose was to study the risk for functional status of the visual system and higher brain functions of astronauts involved in long space missions [29]. It has been determined that a correlation exists between the composition of cosmic radiation crossing the brain and functional anomalies of the visual system and the mechanism of the interactions are under investigation [30].

The new electrodes must be characterized by high conductivity, chemical stability, complete lack of toxicity, proper mechanical characteristics, and ease of use. All these characteristics must be maintained for long periods of time. The use of polymeric membranes based on the dispersion of $LiClO_4$ solutions in 1,2-diethoxyethane in a PMMA polymeric matrix gave satisfactory results both in electrochemical characteristics and EEG recording [31]. However, a high value (about 100 $k\Omega$) of the skin/electrode interfacial resistance was determined.

To reduce such interfacial resistance, we have modified the composition of the membranes and prepared new gel electrolytes adding water and ethanol as co-solvents, to implement the compatibility with biological fluids and tissues. However, we observed a rapid degradation of the gel membranes with time. To improve chemical stability, and mechanical and electrical properties, the idea was again to use nanocrystalline oxides as fillers in these gel membranes.

SiO_2 (Aldrich, 7 nm), Al_2O_3 (Aldrich, 6 nm), and TiO_2 (synthesized as described above, 12 nm) were chosen as fillers at the concentration of 5 and 10 wt.-%. The preparation of the membranes involved homogeneous dispersion of the selected ceramic powder in the electrolyte solution, formed by $LiClO_4$, 1,2 diethoxyethane, PMMA, and $EtOH/H_2O$ 9:1 [32]. The concentration of [Li^+] was kept constant at 0.8 M. Composition and conductivity values of the studied membranes are reported in Tab. 1. The conductivity values, measured by EIS, were quite high (10^{-3}–10^{-4} sscm^{-1}) for all the membranes prepared. For each specific filler doubling the amount of ceramic from 5 to 10 wt.-% resulted in reducing the conductivity to almost one half. The fast ion transport is promoted by the ceramic/polymer interactions, but at higher concentration lithium ion mobility is inhibited by the excess oxide particles [33].

Fig. 7 shows the effect of the filler addition on the long term stability of the membranes, where the conductivity values as a function of time measured for samples C2 and C4 (containing 5 wt.-% TiO_2 and Al_2O_3, respectively) are reported together with those of a reference ceramic-free membrane. A rapid decay of conductivity values was observed for the ceramic-free sample after few days. The conductivity for the nanocomposite gel was on the other hand stable even after more than two months. No deviations from linearity were observed in the impedance spectra of all samples after different storage periods, demonstrating that the integrity of the samples is maintained.

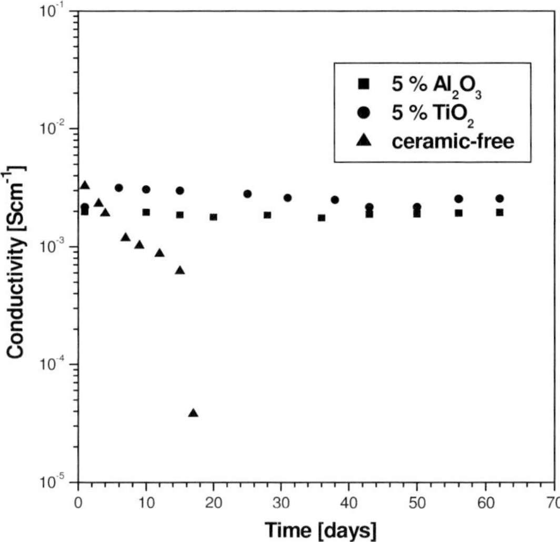

Fig. 7 Time evolution of the conductivity of nanocomposite samples C2 (containing 5 wt.-% TiO$_2$) and C4 (containing 5 wt.-% Al$_2$O$_3$) and that of the reference ceramic-free sample C1. Data obtained at room temperature by impedance spectroscopy.

Tab. 1 Composition and conductivity values of nanocomposite membranes for the recording of bioelectrical signals.

Label	*Electrolyte membrane*	*Conductivity (S cm⁻¹)*
C1	LiClO$_4$–DEE–PMMA	3.28×10^{-3}
C2	LiClO$_4$–DEE–PMMA+5 wt% TiO$_2$	2.70×10^{-3}
C3	LiClO$_4$–DEE–PMMA+10 wt% TiO$_2$	1.48×10^{-3}
C4	LiClO$_4$–DEE–PMMA+5 wt% Al$_2$O$_3$	1.97×10^{-3}
C5	LiClO$_4$–DEE–PMMA+10 wt% Al$_2$O$_3$	9.88×10^{-4}
C6	LiClO$_4$–DEE–PMMA+5 wt% SiO$_2$	2.90×10^{-3}
C7	LiClO$_4$–DEE–PMMA+10 wt% SiO$_2$	1.52×10^{-3}

DEE = 1,2-diethoxyethane.
Molar composition LiClO$_4$: DEE : PMMA (%) = 7.2 : 64.0 : 28.8; [Li$^+$] = 0.8 M.

The nanocomposite gel electrolytes allowed the registration of spontaneous and stimulus-related cerebral bioelectric signals without any skin cleaning and/or preparation. Fig. 8 shows the comparison of the EEG trace recorded using membrane C2 (containing 5 wt.-% titania), C4 (containing 5 wt.-% alumina), and that obtained with the paste used in the conventional clinical practice. The two electrodes were placed on adjacent scalp locations. The new gels allowed the registration of spontaneous electrophysiological cortical signal morphologically comparable to the standard. The measurements performed with silica, however, were not adequate for medical applications.

Fig. 9 shows the visually evoked potential responses for samples C2 and C4. The bottom trace in each frequency spectrum represents the difference between the registration obtained with the composite membranes and the average of several registrations obtained with the paste used in the conventional clinical prac-

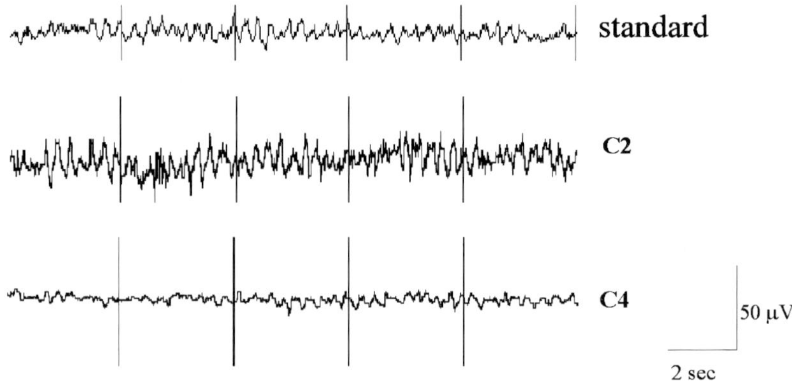

Fig. 8 Comparison of EEG traces recordings obtained with a commercial fluid electrolyte and with the nanocomposite gel membranes C2 (containing 5 wt.-% TiO_2) and C4 (containing 5 wt.-% Al_2O_3).

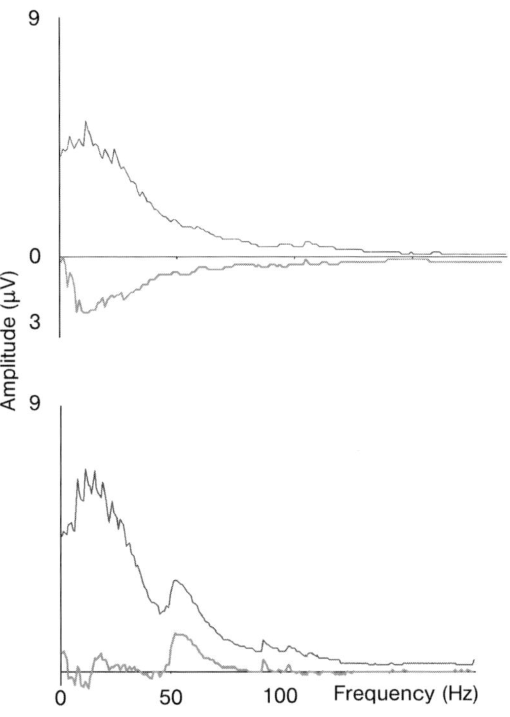

Fig. 9 Responses over the occipital following visual stimuli (VEP) as detected by the nanocomposite gel membranes C2 (containing 5 wt.-% TiO_2) and C4 (containing 5 wt.-% Al_2O_3). The bottom trace in each graph represents the difference between the registration obtained with C2 and C4 and the average of several registrations obtained with the reference commercial fluid electrolyte.

tice. The performance of the new membranes is comparable to the measurements carried out with conventional wet electrolytes.

The addition of the nanocrystalline oxide improved the chemical stability of the gels, reducing the evaporation of the solvent that are chemisorbed on the oxide surface. Also the mechanical properties became more suitable for the given application: the membranes are to be attached to the skin without preparation, and thus have to be plastic but consistent, and should easily adapt to the skin. All these requirements were met by the nanocomposite gel membranes developed.

3
Conclusions

The addition of nanocrystalline oxides was found to improve the performance of a diverse range of polymeric electrolytes, including Nafion for fuel cells applications, PEO for lithium ion polymeric batteries, and PMMA gels for electrophysiological measurements. The improvement of the electrical performance was observed for both protonic and lithium ion conductors. In the case of Nafion the addition of the nanosized ceramic filler allowed operation at higher temperatures, while the opposite was observed in the case of PEO when lower operating temperatures were obtained. Therefore, the mechanisms involved are different: even though the mechanisms have to be fully elucidated, in the former case the increase in the operation temperature is due to the ability of ceramic powders to keep chemisorbed water on the surface, while for PEO the addition of the nanosized filler inhibits crystallization allowing to obtain larger conductivity values. In the case of the PMMA-based gels the ceramic filler improves the retention of the solvents within the polymer matrix, increasing long-term stability of electrical and mechanical characteristics. Therefore, the use of nanocrystalline oxides as fillers for polymeric composites is very versatile and can be applied to a wide variety of systems.

Acknowledgments

The authors wish to express their thanks to all the colleagues and collaborators who contributed to this work. For Direct Methanol Fuel Cells: Prof. P. L. Antonucci (University of Reggio Calabria), Dr. V. Antonucci, Dr. V. Baglio, Dr. A. Di Blasi, Dr. A. S. Aricò (CNR-ITAE, Messina), Dr. F. Serraino Fiory (University of Rome Tor Vergata); for Lithium Ion Polymeric Batteries: Prof. B. Scrosati (University of Rome La Sapienza), Prof. F. Croce (University of Chieti), Dr. A. D'Epifanio, Dr. F. Serraino Fiory (University of Rome Tor Vergata); for membranes for electrophysiological measurements: Prof. W. G. Sannita (University of Genova), Prof. L. Narici, Dr. P. Romagnoli (University of Rome Tor Vergata).

We acknowledge MIUR (Italian Ministry for Education, University and Research) and ASI (Italian Space Agency) for financial support.

References

1 A. B. STAMBOULI, E. TRAVERSA, *Renew. Sustain. Energy Rev.* **2002**, *6*, 297.

2 A. S. ARICÒ, S. SRINIVASAN, V. ANTONUCCI, *Fuel Cells* **2001**, *1*, 1.

3 G. PISTOIA, *Lithium Batteries, New Material, Developments and Perspectives,* Elsevier Science B.V., **1994**.

4 C. F. HOLMES, *Electrochem. Soc. Interface* **1999**, *8*, 32.

5 O. BOHNKE, G. FRAND, M. REZRAZI, C. ROUSSELOT, C. TRUCHE, *Solid State Ionics* **1993**, *66*, 97.

6 *Application of Electroactive Polymers* (Ed. B. SCROSATI), Chapman & Hall, London, UK, **1993**.

7 E. TRAVERSA, *J. Intell. Mater. Syst. Struct.* **1995**, *6*, 860.

8 F. CROCE, G. B. APPETECCHI, I. PERSI, B. SCROSATI, *Nature* **1998**, *394*, 456.

9 Y. M. CHIANG, *J. Electroceram.* **1997**, *1*, 205.

10 R. A. VAIA, E. P. GIANNELIS, *MRS Bull.* **2001**, *26*, 394.

11 X. REN, M. WILSON, S. GOTTESFELD, *J. Electrochem. Soc.* **1996**, *143*, L12.

12 S. WASMUS, A. KUVER, *J. Electroanal. Chem.* **1999**, *461*, 14.

13 A. S. ARICÒ, P. CRETÌ, P. L. ANTONUCCI, V. ANTONUCCI, *Electrochem. Solid-State Lett.* **1998**, *1*, 6.

14 P. L. ANTONUCCI, A. S. ARICÒ, P. CRETÌ, E. RAMUNNI, V. ANTONUCCI, *Solid State Ionics* **1999**, *125*, 431.

15 V. BAGLIO, A. DI BLASI, A. S. ARICO, V. ANTONUCCI, P. L. ANTONUCCI, F. SERRAINO FIORY, S. LICOCCIA, E. TRAVERSA, *Solid State Ionics-2002* (Eds.: P. KNAUTH, J.-M. TARASCON, E. TRAVERSA, H. L. TULLER), Materials Research Society, Warrendale, PA, USA, **2003**, Symp. Proc., Vol. 756, p. 345.

16 V. BAGLIO, A. DI BLASI, A. S. ARICO, V. ANTONUCCI, P. L. ANTONUCCI, F. SERRAINO FIORY, S. LICOCCIA, E. TRAVERSA, *J. New Mater. Electrochem. Sys.* **2004**, in press.

17 E. TRAVERSA, M. L. DI VONA, S. LICOCCIA, M. SACERDOTI, M. C. CAROTTA, L. CREMA, G. MARTINELLI, *J. Sol-Gel Sci. Technol.* **2001**, *22*, 167.

18 S. ENZO, S. POLIZZI, A. BENEDETTI, *Z. Kryst.* **1985**, *170*, 275.

19 P. ROMAGNOLI, G. B. APPETECCHI, F. CROCE, U. HEIDER, R. OESTEN, B. SCROSATI, *Electrochem. Commun.* **1999**, *1*, 83.

20 P. ROMAGNOLI, G. B. APPETECCHI, F. CROCE, L. PERSI, R. MARASSI, B. SCROSATI, *Electrochim. Acta* **1999**, *45*, 23.

21 P. G. BRUCE, *Electrochim. Acta* **1995**, *40*, 2077.

22 R. HUQ, G. C. FARRINGTON, R. KOKSBANG, P. E. TONDER, *Solid State Ionics* **1992**, *57*, 277.

23 L. A. DOMINEY, V. R. KOCH, T. J. BALKELEY, *Electrochim. Acta* **1992**, *37*, 1551.

24 I. E. KELLY, J. R. OWEN, B. C. H. STEELE, *J. Electroanal. Chem.* **1984**, *168*, 467.

25 C. WANG, Q. LIU, Q. CAO, Q. MENG, L. IANG, *Solid State Ionics* **1992**, *53–56*, 1106.

26 F. CROCE, F. SERRAINO FIORY, L. PERSI, B. SCROSATI, *Electrochem. Solid-State Lett.* **2001**, *4*, A121.

27 M. C. BORGHINI, M. MASTRAGOSTINO, S. PASSERINI, B. SCROSATI, *J. Electrochem. Soc.* **1994**, *142*, 2118.

28 F. CROCE, A. D'EPIFANIO, J. HASSOUN, A. DEPTULA, T. OLCZAC, B. SCROSATI, *Electrochem. Solid-State Lett.* **2002**, *5*, A47.

29 L. NARICI, V. BIDOLI, M. CASOLINO, M. P. DE PASCALE, G. FURANO, I. MODENA, A. MORSELLI, P. PICOZZA, E. REALI, R. SPARVOLI, S. LICOCCIA, P. ROMAGNOLI, E. TRAVERSA, W. G. SANNITA, A. LOIZZO, A. GALPER, A. KHODAROVICH, M. G. KOROTKOV, A. POPOV, N. VAVILOV, S. AVDEEV, V. P. SALNITSKII, O. I. SHEVCHENKO, V. P. PETROV, K. A. TRUKHANOV, M. BOEZIO, W. BONVICINI, A. VACCHI, N. ZAMPA, R. BATTISTON, G. MAZZENGA, M. RICCI, P. SPILLANTINI, G. CASTELLINI, P. CARLSON, C. FUGLESANG, *Adv. Space Res.* **2003**, *31*, 141.

30 M. CASOLINO, V. BIDOLI, L. NARICI, M. P. DE PASCALE, P. PICOZZA, E. REALI, R. SPARVOLI, G. MAZZENGA, M. RICCI, P. SPILLANTINI, M. BOEZIO, W. BONVICINI, A. VACCHI, N. ZAMPA, G. CASTELLINI,

W.G. Sannita, P. Carlson, A. Galper,
M.G. Korotkov, A. Popov, N. Vavilov,
S. Avdeev, C. Fuglesang, *Nature* **2003**,
422, 680.

31 P. Romagnoli, M.L. Di Vona, L. Narici,
W.G. Sannita, E. Traversa, S. Licoccia,
J. Electrochem. Soc. **2001**, *148*, J63.

32 P. Romagnoli, M.L. Di Vona,
E. Traversa, L. Narici, W.G. Sannita,
S. Carozzo, M. Trombetta, S. Licoccia,
Solid State Ionics-2002 (Eds.: P. Knauth,
J.-M. Tarascon, E. Traversa,
H.L. Tuller), Materials Research Society,
Warrendale, PA, USA, **2003**, Symp.
Proc., Vol. 756, p. 325.

33 S. Rajendran, O. Mahendran,
R. Kannan, *J. Phys. Chem. Solids* **2002**,
63, 303.

Optimized Electromechanical Properties and Applications of Cellular Polypropylene – a New Voided Space-Charge Electret Material

Michael Wegener and Werner Wirges

1
Introduction

Thin electromechanically active polymer films or systems may be divided in three classes according to the origin of the electromechanical effect.

- In (non-voided) polar and often ferroelectric polymers, the orientation of molecular dipoles causes the symmetry breaking necessary for electromechanical (piezoelectric) properties (Fig. 1). Therefore, the polarization is found on the microscopic scale. The electromechanical activity is based on the change of the dipole density during application of mechanical stresses (direct piezoelectric effect) or electrical fields (inverse piezoelectric effect). Examples for such piezoelectric polymers are polyvinylidene fluoride (PVDF) and its copolymers with trifluoroethylene (P(VDF-TrFE)), tetrafluoroethylene and hexafluoroethylene, several odd-numbered polyamides, as well as some polyureas and polyurethanes [1, 2]. PVDF and P(VDF-TrFE) show piezoelectric transducer coefficients of around 20 pC/N, elastic moduli of around 9 GPa, coupling factors of around 0.2 and thickness–extension resonances typically in the megahertz range, depending on the film thickness.

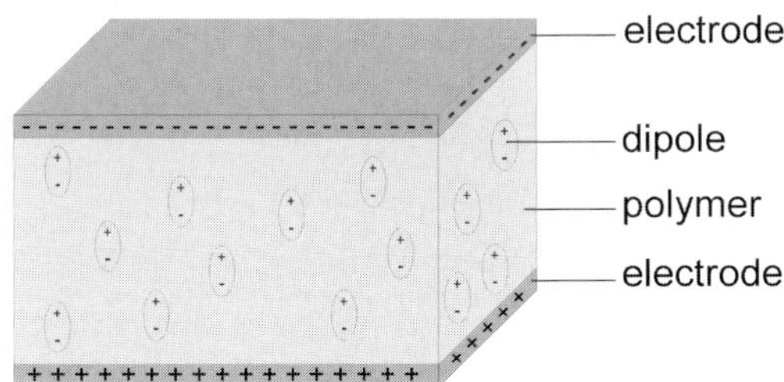

Fig. 1 Symmetry breaking through the orientation of molecular dipoles within domains of non-voided polar piezoelectric polymers.

The Nano–Micro Interface: Bridging the Micro and Nano Worlds
Edited by Hans-Jörg Fecht and Matthias Werner
Copyright © 2004 WILEY-VCH Verlag GmbH & Co. KGaA, Weinheim
ISBN: 3-527-30978-0

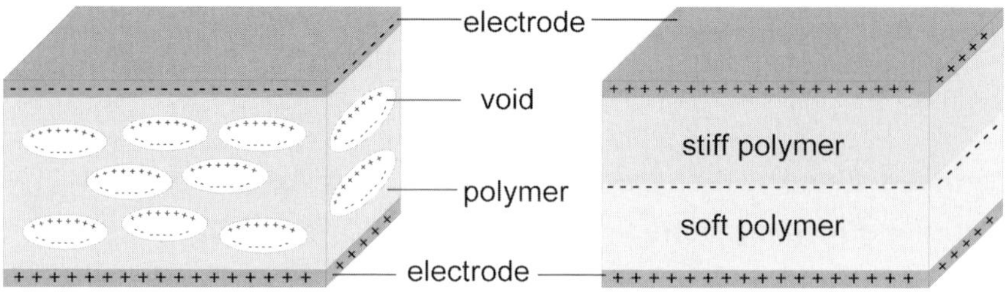

Fig. 2 Symmetry breaking through charge trapping at internal surfaces of:
left, voided space-charge electrets; right, electret-layer systems [3].

- Recently, a new class of cellular (porous or voided) space-charge electrets was developed. These polymers usually contain air-filled voids in their matrix (Fig. 2, left). Up to now the most suitable voided polymers are cellular propylene (PP) [3], porous Teflon®AF [4], porous polyethylene (PE) [5], and porous polytetrafluoroethylene (PTFE) [6, 7]. Detailed information about the preparation and the applications of voided space-charge electrets are found elsewhere [3, 8].
- The development of this new class of cellular materials originates partly from research on electromechanically active systems of two or more space-charge electret films with different mechanical stiffnesses (Fig. 2, right). The working principle of such systems is comparable to that of electret microphones, in which the soft layer is usually an air gap. The symmetry breaking within such a system is achieved with a charge layer deposited onto one or both inner surfaces of the layer system before the films are put together. If the polymer-layer sandwich employed as sensor very high electromechanical activity may be obtained. Because of the preparation and the charging behavior of the individual polymer films, electromechanical properties can only be discussed for the complete layer system, not for the individual polymer films [9]. Such systems are interesting for research [6, 10, 11] and for applications [12].

In this chapter we focus on the description of cellular electromechanically active polymers. We describe the preparation of cellular PP, its investigation, and its optimization as well as implemented and proposed applications in some detail.

2
Investigations on Cellular PP

2.1
Film Preparation

In order to prepare cellular PP films [13], various procedures were developed. In earlier investigations, the void formation was initiated during stretching at elevated temperature or by use of chemical foaming agents [14]. Presently, preparation usually follows the route shown in Fig. 3.

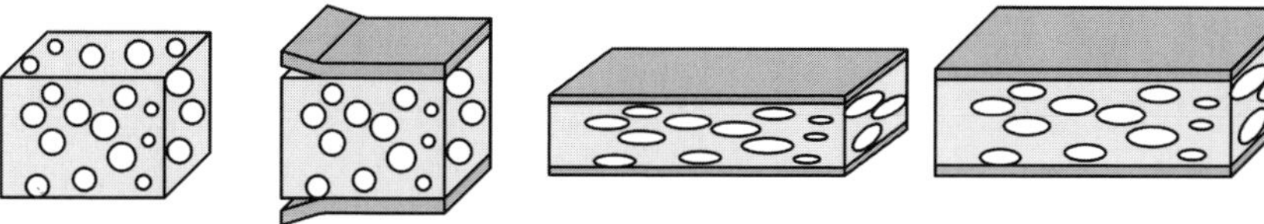

Fig. 3 Preparation of cellular polymer films: (1) extrusion of a polymer/filler composite, (2) co-extrusion of non-cellular cover layers, (3) biaxial stretching, (4) thickness change due to expansion.

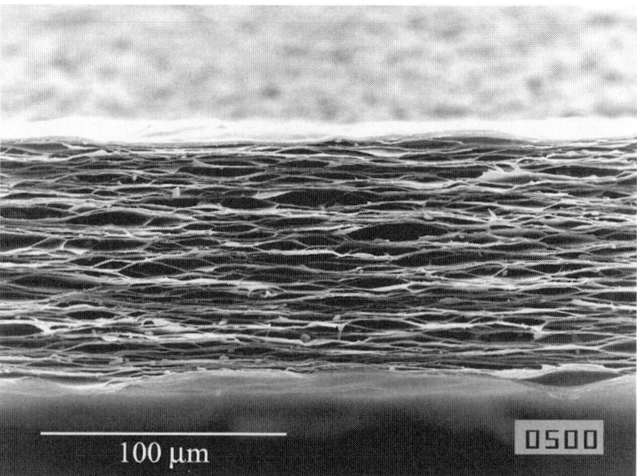

Fig. 4 SEM cross-section micrograph of cellular PP [19].

- A mixture of polypropylene granules and filler particles, typically $CaCO_3$, is extruded. Because of the slight stretching during extrusion cavities form around the filler particles.
- Additional non-cellular polypropylene layers are co-extruded onto both outer film surfaces in order to facilitate the later charging and electroding processes.
- Biaxial stretching leads to void generation based on the cavities around the filler particles within the thin polymer films [15–17].
- In order to adjust the void sizes, films are expanded by means of high-pressure treatments at elevated temperatures, typically at 20 bar and 90 °C [18].

This so-called gas-diffusion expansion (GDE) process leads to a viscous and thus irreversible change of the film thickness by up to 100%, such as from 37–70 µm. Because of the thickness increase, the expanded films are softer than the non-inflated films. The density of 70 µm thick films is typically around 0.33 g/cm^3. As a result, the non-polar cellular PP films contain internal lens-like voids with heights and lateral dimensions typically in the range 1–5 µm and 10–100 µm, respectively (Fig. 4).

2.2
Electro-Active Properties

Cellular PP electret films exhibit useful electromechanical properties. Typical transducer coefficients of 15 and 200 pC/N are found on 37 and 70 µm thick cellular PP films, respectively. The direct and inverse electromechanical properties could be demonstrated with quasistatic [20, 21], dynamic [22, 23], and acoustical measurements [6, 24, 25]. Furthermore, thickness–extension resonances during electrical excitation [26, 27] reveal the elastic modulus c_{33} and the coupling factor k_t. Elastic moduli and resonance frequencies of around 1.3 and 4 MPa and 0.68 and 2 MHz are obtained for the 70 and 37 µm thick films, respectively. From the above-mentioned transducer coefficients, elastic moduli and resonance frequencies for the 37 and 70 µm thick films, it can be concluded that GDE is a suitable technique in order to optimize dynamical sensor properties of cellular PP electrets.

Because of the highly anisotropic structure of cellular PP, only the thickness extension is found to be significant. The transducer coefficient of the length extension is typically around 2 pC/N: two orders of magnitude lower than that of the thickness extension [28]. This large difference leads to advantages if the materials are used in applications. Depending on the desired purpose, the rather small pyroelectric activity may also be an advantage.

2.3
Assessment of the Charging Process

It is now usually assumed that the polarization in cellular PP is generated by electrical breakdown in voids and charge trapping at their inner surfaces during charging at high electric fields. During the last three years, several relationships and phenomena were investigated in order to confirm this assumption.

2.3.1 Dependence on the Charging Field: Threshold Behavior

If electrical charging of the voids is effected by internal breakdown, there should be a threshold field strength for microdischarges in the voids. Fig. 5 shows the surface potential as well as the transducer coefficient of 37 µm thick cellular PP samples measured after charging as a function of the voltage applied to the corona tip during charging. The effective charging field (or the surface potential) increases nearly linearly with increasing corona-tip voltage. However, in order to generate electromechanical activity, the corona-tip voltage must overcome a threshold value of around −12 kV, which corresponds to an effective poling field of around −5 kV. A further increase of the corona-tip voltage leads to higher transducer coefficients until saturation occurs at the highest corona-tip voltages. On the more sensitive cellular PP films with thicknesses of 70 and 100 µm the threshold behavior is more prominent [29].

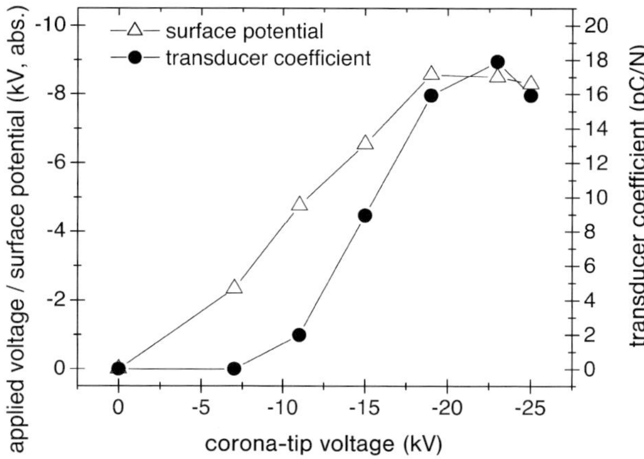

Fig. 5 Surface potential and transducer coefficient of 37 μm thick cellular PP samples as functions of the corona-tip voltage employed during charging.

2.3.2 Independence from the Charging Method

Internal void charging should not strongly depend either on the amount of injected charges or on the polarity of the applied field. In order to investigate this, we performed corona charging [29], where high amounts of charges are injected, and also charging in direct contact [30]. We found the same threshold behavior as well as comparable transducer values on samples charged by means of different charging methods (Fig. 6). The resulting transducer coefficients are also indepen-

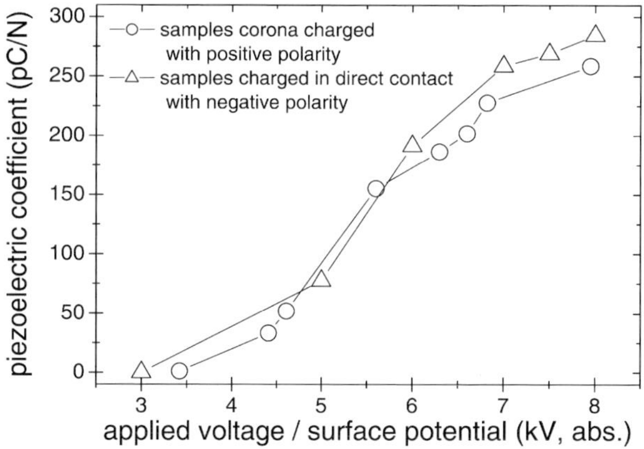

Fig. 6 Threshold for generating electromechanical properties within cellular PP, charged by means of a corona discharge with positive polarity (open cycles, *x* axis: surface potential) and a direct contact with negative polarity (open triangles, *x* axis: applied voltage).

dent of the polarity of the applied electric field, as shown in Fig. 6 for samples corona charged with positive polarity and for samples charged in direct contact with negative polarity. Therefore, we conclude that charge injection is not an essential process in order to achieve internal void charging.

2.3.3 Dependence on the Charging Time

If the internal charging originates from microdischarges charging should be a fast process. In order to check this assumption, different charging times were used. It was found that corona charging by means of electric fields higher than the threshold for charging times longer than 2 s will not enhance the electromechanical properties [29]. A charging time of 2 s is the shortest possible time with the available corona. With electrical charging in direct contact, much shorter charging times can be realized. It was found that the application of short high-voltage pulses (FWHM 45 µs) was sufficient to initiate microdischarges and to achieve electromechanical activity in 37 and 70 µm thick cellular PP films [30].

2.3.4 Switching of Polarization

The internal charging of the voids should be repeatable, therefore a reversal of the polarization should occur when electric fields of opposite polarity are applied. In order to demonstrate this, we performed several charging experiments on a sample of cellular PP with a thickness 52 µm, using different charging fields. After each charging, we measured the transducer properties via piezoelectric thickness resonances. We found a threshold of around ±3 to ±3.5 kV and significant electromechanical activity of around 90 pC/N after charging with a voltage of −7 kV (Fig. 7). Subsequently, charging with opposite polarity and low field, such as an

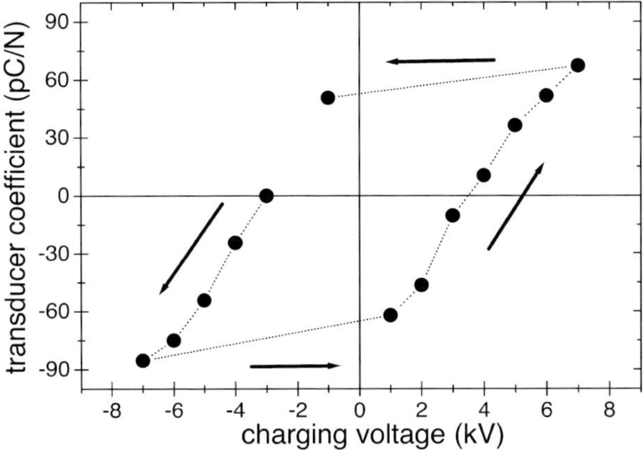

Fig. 7 Electromechanical activity of cellular PP after charging as a function of the voltage applied during charging.

applied voltage of 1 kV, will not destroy the internal polarization responsible for the electromechanical activity. The internal polarization will switch if the applied electric field is in the range of the threshold. The switching of the internal polarization was also measured in-situ during the charging process by determining the thickness change as a function of the applied field. A combination of quasi-piezoelectric and electrostrictive responses was found upon application of charging fields higher than the threshold [31].

2.3.5 Electroluminescence During Charging

Microdischarges in thin voids are usually combined with electroluminescence. Recently, electroluminescence was observed during charging of cellular PP films in direct contact (Fig. 8, left) [32] as well as with a corona discharge (Fig. 8, right) [29]. The electroluminescence observed during charging of cellular PP and of other porous polymers such as porous PTFE is a clear proof that the void charging is based on internal electrical breakdown.

2.3.6 Influence of the Ambient Gas During Charging

Beside the investigation of suitable charging parameters at normal pressure and humidity, the ambient gas in the sample chamber and in the sample itself was exchanged in order to influence the breakdown and charge-trapping processes. By means of charging in gases with higher dielectric breakdown strength, the applicable corona voltage could be increased [19, 33]. The resulting microdischarges at the higher field lead to higher charging of the voids. After a slight discharge during the first week after charging, the remaining deep charge-trapping levels were stable at normal conditions. The higher charging within the voids also leads to larger transducer effects. In particular, charging of cellular PP in nitrogen gas

 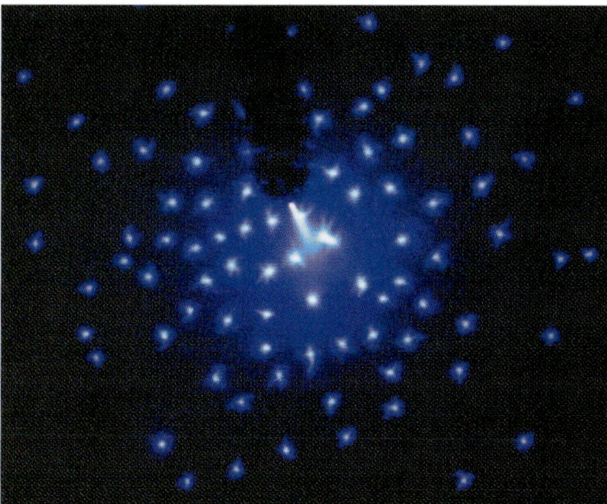

Fig. 8 Electroluminescence during charging in direct contact (left) [32]; by means of a corona discharge (right) [29].

leads to transducer coefficients of up to 790 pC/N. Furthermore, charging experiments performed in gases containing larger molecules such as SF_6/Ar result in lower film thicknesses after the gas-pressure treatment and in smaller transducer coefficients. Both results demonstrate that the gas-pressure treatment influences the charging inside the voids.

2.3.7
Summary of Charging Evaluation

In summary, we found:
- a threshold for the initiation of microdischarges,
- electromechanical activity already after short charging times,
- no differences between the transducer coefficients after charging with various methods,
- switching of the internal polarization,
- electroluminescence during charging, and
- a strong dependence of the transducer coefficients on the gases present in the voids of the cellular PP during charging.

These results show that the electromechanical activity of cellular PP is based on the mechanical properties of polymer and voids (not discussed here) and on the electrical charging within the voids. Therefore, in cellular PP, the symmetry breaking occurs on a µm scale and the resulting electromechanical effect is based on the change of the large dipoles. Therefore, suitable voided space-charge electrets with electromechanical properties should be classified as *quasi-ferroelectric* materials or as *ferroelectrets* [34].

3
Applications

3.1
Proposed Electromechanical and Electroacoustical Transducer Concepts

For electromechanically active polymers, numerous applications were suggested, demonstrated and commercially realized [2, 3]. The "classical" non-voided polar piezoelectric polymers, such as PVDF and P(VDF-TrFE), are for instance used in ultrasonic sensors for materials testing and in several medical applications [35, 36]. Owing to their low acoustic impedance and good matching to air, the voided electromechanically active polymers will usually be suitable for other application areas than non-voided piezoelectric polymers. A further advantage for use in medical applications is the rather small pyroelectric activity of voided space-charge electrets. Temperature oscillations caused by close contact of sensors to human or animal bodies will not much influence the sensor signal, in contrast to the "classical" piezoelectric polymers or ceramics. Various applications are proposed such as the monitoring of pressure distributions in shoe soles as well as quantitative dynamic-force measurements on dog limbs where the sensors were attached with elastic bandages [37]. For orthopedic diagnostics, also pressure monitoring on seats and backrests is very interesting. The results could lead to further optimization of office chairs or of seats in cars, trains, and aircraft. Because of the high sensitivity of the electromechanical films, even small motions can be detected. If the transducers are located in rat cages, the respiration of resting rats can be monitored [38]. The sensitivity is also high enough for recording the respiration of human patients even if the transducers are not directly attached to the body of a patient [39].

Beside the recording of pressure distributions outside human or animal bodies, the new thin cellular polymer films allow for measuring pressure distributions within the body itself. A current field of development is the sensing of pressure distributions between vocal cords [40].

Using larger sensor areas below floors, a surveillance of rooms and of the surroundings of machines is possible [41]. Small-area sensors are interesting for control panels and keyboards as described below.

For electroacoustical applications, microphones and loudspeakers were suggested [24, 42–44]. Microphones are already on the market, produced by the Finnish company Emfit Ltd. For loudspeakers, two different concepts were developed for employing cellular electret materials. In order to generate small sound-pressure levels, the intrinsic electromechanical effect of the cellular polymer films can be used [45]. A larger elongation of the transducer films and therefore higher sound-pressure levels can be obtained if the charged, non-metallized films are inserted in a specially prepared electrode structure as discussed earlier [16, 45]. The last-mentioned application relies on the electret effect of charges trapped on the outer surfaces. Owing to the high charge levels obtained with corona charging [29], cellular PP films are very attractive for such applications. Up to now, thin large-area loudspeakers are still under development. Simultaneously, the concept will be developed in order to control or cancel sound [45]. In order to increase the displacement amplitudes, the folding and bending of cellular PP films was also investigated [46, 47].

In the following, we briefly demonstrate investigations and prototypes of additional applications, such as control panels containing thin push buttons as well as new concepts for vibration analysis and active noise control.

3.2
Control Panels with Pushbuttons Made of Cellular PP Electrets

Very thin keyboard pushbuttons and control panels can be prepared with electromechanically active polymer films such as cellular PP. Furthermore, the electromechanically active polymers can be placed behind thin metal plates in order to achieve vandal-proof control panels. Based on soft and flexible polymer films, the construction of flexible keyboards is also possible if the electronics are separated from the sensor film and connected via (flat) cables.

We prepared a three-pushbutton test keyboard in order to stimulate further development and to demonstrate the requirements on the read-out electronics and the further processing of the electrical signal from the cellular PP sensor film. On

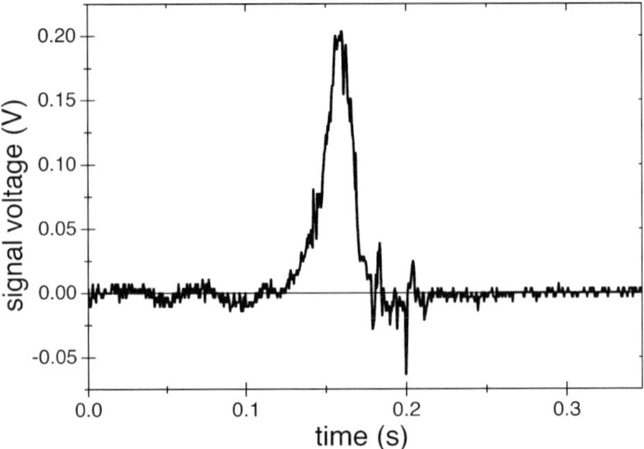

Fig. 9 Voltage–time signal during activation of a sensor push-button in a control panel.

the upper surface of the 100 μm thick cellular PP film, a pushbutton pattern was metallized. The lower surface was metallized completely. Changing the mechanical pressure on one of the pushbuttons with a load of 200 g leads to a typical sensor signal, as shown in Fig. 9. The signal with a maximum voltage of around 200 mV and a rise time of 30 ms can be easily detected with standard electronics.

3.3
Concept for Vibration Control

Electromechanically active polymers are suitable for the vibration control of different kinds of surfaces, for instance cabin walls of aircraft or roofs of cars and trains. Especially non-cellular piezoelectric polymers are often used, because of their significant piezoelectric 31-effect. However, electromechanically active cellular polymers with low electromechanical activity within the film plane, but very high activity across the film thickness are also attractive for vibration control. For employing the strong 33-effect of cellular polymers, the films must be mechanically clamped between the vibrating surface and a stiffer electrode. As an example, we clamped a thin metal stripe (300×4 mm) on one side and inserted the cellular sensor film on a stiffer back electrode behind the clamping. Then the free end of the metal stripe was vibrated. With this arrangement, a clear detection of the vibration was possible as shown in Fig. 10, where the deflection amplitude represents the vibration applied to the metal stripe and the sensor-signal voltage shows the output of the sensor. In order to enhance the sensor signal during vibration-control measurements, we developed multi-layer systems, such as the four-layer system shown in Fig. 11. Because of the small thickness of the individudal sensor films, multi-layer stacks of electromechanical films are still relatively thin. As can be seen in Fig. 10, the four-layer system shows a sensitivity nearly four times as high as that of a single sensor film.

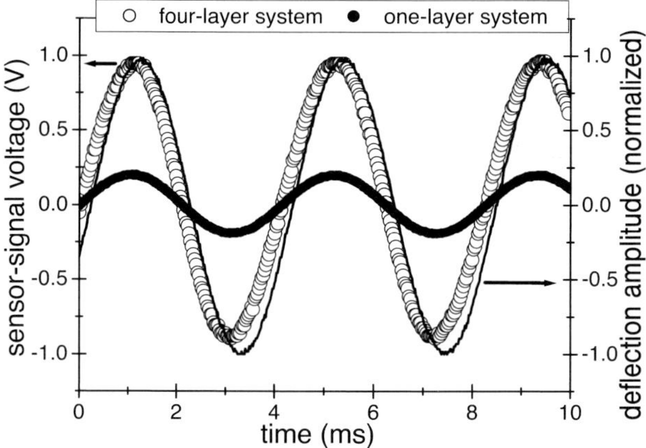

Fig. 10 Vibration control of a metal stripe by means of cellular PP films with an individual sensor film (●) as well as with a four-layer sensor system (○).

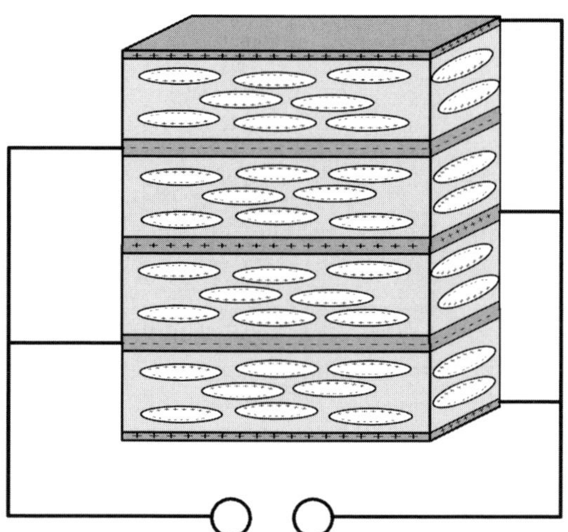

Fig. 11 Multi-layer transducer with four electrically charged, voided space-charge electrets.

3.4
Concept for Active Noise Control

Significant research and development is being devoted to active noise control in the cabins of cars, trains, or aircraft. With electromechanical polymers such as cellular PP, different concepts are feasible: The noise-generating vibration may be detected by means of electromechanically active polymer layers, in order to determine the vibration frequency and amplitude. Knowing these parameters, a sound

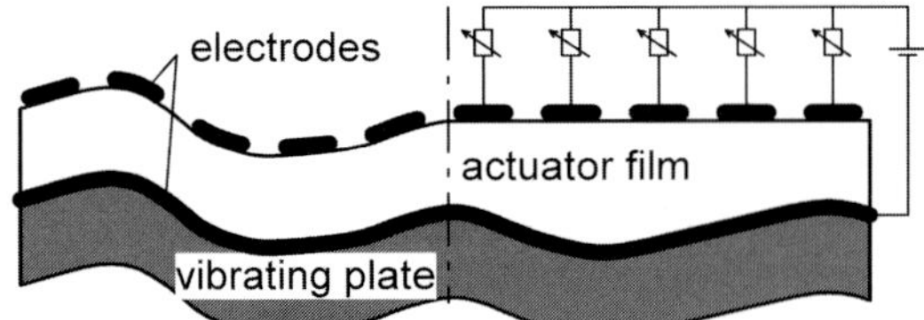

Fig. 12 Concept for vibration canceling or active noise control by means of electromechanically active films.

wave with optimal frequency and amplitude (anti-sound) may be generated and focused so that the noise is cancelled locally, that is at the human ear [45]. With the available charging possibilities a suitable pattern may be prepared on the polymer film. The use of a Fresnel zone plate for directional sound radiation was already demonstrated [48].

Another concept consists of two parts: First, vibration detection, for example by means of electromechanically active polymer layers covering a larger surface or by means of other systems for vibration detection. Second, canceling of the surface vibration at the inner surface of the room or cabin. In order to realize this concept, an electromechanical-active film with a suitable electrode pattern must be attached to the vibrating surface as shown in Fig. 12 (left). From the detected noise within the room or cabin, the surface vibration can be determined. Based on this, the necessary thickness change of the electromechanical film must be calculated and an appropriate voltage must be applied to the cellular film in order to generate the desired thickness variations within the film area (Fig. 12, right). After optimization, there will be no vibration of the outer wall surface facing the room or cabin and therefore no noise generation resulting from the vibration.

To demonstrate this concept, we used two electromechanically active cellular PP films with a thickness of around 70 μm. After metallization of round electrodes with a diameter of 6 cm, the samples were fixed together by means of double-sided adhesive tape on top of a circuit board that connects the films electrically. In order to measure the noise, acoustical near-field measurements were performed. A sinusoidal voltage of 100 V amplitude was applied and the sound-pressure level was recorded. The measured sound-pressure levels as a function of the frequency of the applied voltage are shown in Fig. 13. Curves (a) and (b) represent the sound-pressure levels during excitation of the individual upper (a) and lower film (b), respectively. The calculated transducer coefficients are 108 pC/N and 95 pC/N for the upper and lower film, respectively. The difference may be resulting from non-optimized film attachment. In order to demonstrate the concept of active noise control, the lower film was used as the vibrating surface. Then the upper film was driven with the same voltage amplitude, but applied with a phase shift of π (inverted). Curve (c) shows the resulting sound pressure level, which demonstrates a significant noise reduction over the investigated frequency range.

However, if a single electromechanically active polymer film is used this concept will only work for small displacements of the vibrating surface: up to thick-

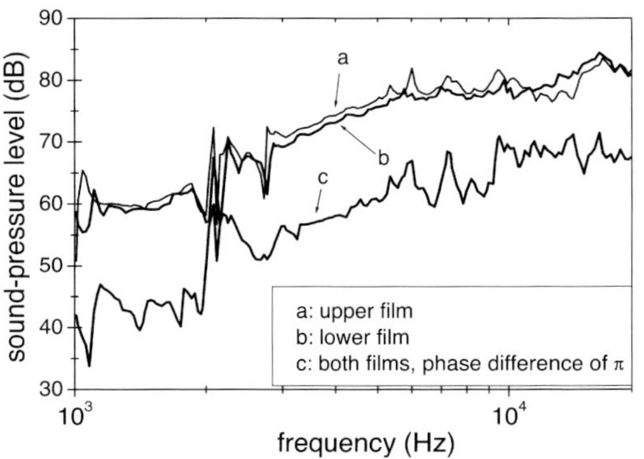

Fig. 13 Measured sound pressure level as a function of the frequency of the applied voltage (sinusoidal, amplitude 100 V) of a system containing two electromechanically active polymer layers fixed on a circuit board.

ness vibrations of around 30 nm. In order to cancel larger displacements, multilayer actuator systems as described above must be used.

4
Conclusions

Suitably voided space-charge electrets represent a new class of electromechanically active materials. Several procedures for preparing thin voided films and for their electrical charging were available. Prerequisites for electromechanical activity in cellular PP are suitable mechanical properties of the heterogeneous structure and macroscopic electric dipoles generated by means of charge separation within the voids and charge trapping at opposite surfaces.

The change of dipole sizes during mechanical or electrical excitation causes the electromechanical activity of voided space-charge electrets. The completely different microscopic process in voided polymers, when compared to that of non-voided piezoelectric polymers, leads to similar electromechanical properties. As a consequence, the macroscopic electromechanical effect may be explained in a similar manner for both (different) classes of polymers. Because of the close relationship to non-voided, polar ferroelectric polymers, voided space-charge electrets with electromechanical properties may be called *ferroelectrets*.

Due to their high electromechanical effect and their excellent acoustical coupling to air, cellular electromechanically active polymers have a good potential for use in applications such as control panels, vibration-control devices, and active noise control concepts.

Acknowledgments

The authors are indebted in particular to Prof. Reimund Gerhard-Multhaupt, Dr. Axel Mellinger, Dr. Wolfgang Künstler and Andreas Pucher (University of Potsdam, Germany), Manfred Kornelson (Potsdam, Germany), Prof. Siegfried Bauer and Dr. Simona Bauer-Gogonea (Johannes Kepler University, Linz, Austria), Prof. Gian Carlo Montanari (University of Bologna, Italy) and Dr. Mika Paajanen (VTT Processes, Tampere, Finland).

References

1 H. S. NALWA (Ed.), *Ferroelectric Polymers*, Marcel Dekker, New York, USA **1995**.

2 R. GERHARD-MULTHAUPT (Ed.), *Electrets*, 3rd edn. Vol. 2, Laplacian Press, Morgan Hill, CA, USA **1999**.

3 R. GERHARD-MULTHAUPT, *IEEE Trans. Dielect. Electr. Insul.* **2002**, *9*, 850–859.

4 A. MELLINGER, M. WEGENER, W. WIRGES, and R. GERHARD-MULTHAUPT, *Appl. Phys. Lett.* **2001**, *79*, 1852–1854.

5 R. KACPRZYK, *Proc. 11th Int. Symp. Electrets*, Melbourne, Australia **2002**, 207–210.

6 R. GERHARD-MULTHAUPT, W. KÜNSTLER, T. GÖRNE, A. PUCHER, T. WEINHOLD, M. SEISS, Z. XIA, A. WEDEL, and R. DANZ, *IEEE Trans. Dielect. Electr. Insul.* **2000**, *7*, 480–488.

7 M. WEGENER, W. WIRGES, K. RICHTER, W. KÜNSTLER, and R. GERHARD-MULTHAUPT, *Proc. 4th Int. Conf. Electric Charges in Non-Conductive Materials (CSC'4)*, Paris, France **2001**, 257–260.

8 S. BAUER-GOGONEA and S. BAUER, *Physik J.* **2003**, *2*, 41–46, in German.

9 J. HILLENBRAND, Z. XIA, X. ZHANG, and G. M. SESSLER, *Proc. 11th Int. Symp. Electrets*, Melbourne, Australia **2002**, 46–49.

10 R. KACPRZYK, E. MOTYL, J. B. GAJEWSKI, and A. PASTERNAK, *J. Electrostatics* **1995**, *35*, 161–166.

11 R. KACPRZYK, A. DOBRUCKIAND, and J. B. GAJEWSKI, *J. Electrostatics* **1997**, *39*, 33–40.

12 G. M. SESSLER, J. E. WEST, and R. L. WALLACE jr., *IEEE Trans. Commun.* **1973**, com-21, 61–65.

13 K. KIRJAVAINEN, US Patent No. 4654546, **1987**.

14 A. SAVOLAINEN and K. KIRJAVAINEN, *J. Macromol. Sci. Chem.* **1989**, *A26*, 583–591.

15 J. I. RAUKOLA, *VTT Publications* **1998**, Vol. 361.

16 M. PAAJANEN, J. LEKKALA, and K. KIRJAVAINEN, *Sens. Actuators* **2000**, *84*, 95–102.

17 J. RAUKOLA, N. KUUSINEN, and M. PAAJANEN, *Proc. 11th Int. Symp. Electrets*, Melbourne, Australia **2002**, 195–198.

18 M. PAAJANEN, H. MINKKINEN, and J. RAUKOLA, *Proc. 11th Int. Symp. Electrets*, Melbourne, Australia **2002**, 191–194.

19 M. PAAJANEN, M. WEGENER, and R. GERHARD-MULTHAUPT, *Conf. Electrical Insulation and Dielectric Phenomena CEIDP, Annual Report* **2001**, 24–27.

20 W. KÜNSTLER, Z. XIA, T. WEINHOLD, A. PUCHER, and R. GERHARD-MULTHAUPT, *Appl. Phys. A* **2000**, *70*, 5–8.

21 R. SCHWÖDIAUER, G. S. NEUGSCHWANDTNER, K. SCHRATTBAUER, M. LINDNER, M. VIEYTES, S. BAUER-GOGONEA, and S. BAUER, *IEEE Trans. Dielect. Electr. Insul.* **2000**, *7*, 578–586.

22 M. PAAJANEN, H. VÄLIMÄKI, and J. LEKKALA, *J. Electrostatics* **2000**, *48*, 193–204.

23 R. KRESSMANN, *J. Appl. Phys.* **2001**, *90*, 3489–3496.

24 J. BACKMAN and M. KARJALAINEN, *Proc. IEEE Int. Conf. Acoustics, Speech, and Signal, Processing*, IEEE Service Center, Piscataway, NJ, USA, **1990**, *2*, 1173–1176.

25 R. KRESSMANN, *J. Acoust. Soc. Am.* **2001**, *109*, 1412–1416.

26 H. OHIGASHI, *J. Appl. Phys.* **1976**, *47*, 949–955.

27 A. MELLINGER, *Dielectric Newsletter* **2002**, *4*, 1–3.

28 G. S. NEUGSCHWANDTNER, R. SCHWÖDIAUER, M. VIEYTES, S. BAUER-GOGONEA, S. BAUER, J. HILLENBRAND, R. KRESSMANN, G. M. SESSLER, M. PAAJANEN, and J. LEKKALA, *Appl. Phys. Lett.* **2000**, *77*, 3827–3829.

29 M. Wegener, M. Paajanen, W. Wirges, and R. Gerhard-Multhaupt, *Proc. 11th Int. Symp. Electrets*, Melbourne, Australia **2002**, 54–57.

30 R. Gerhard-Multhaupt, M. Wegener, W. Wirges, J. A. Giacometti, R. A. C. Altafim, L. F. Santos, R. M. Faria, and M. Paajanen, *Conf. Electrical Insulation and Dielectric Phenomena CEIDP, Annual Report* **2002**, 299–302.

31 M. Lindner, R. Schwödiauer, M. Dansachmüller, S. Bauer-Gogonea, and S. Bauer, *Proc. 11th Int. Symp. Electrets*, Melbourne, Australia **2002**, 106–109.

32 M. Lindner, S. Bauer-Gogonea and S. Bauer, *J. Appl. Phys.* **2002**, *91*, 5283–5287.

33 M. Paajanen, M. Wegener, and R. Gerhard-Multhaupt, *J. Phys. D: Appl. Phys.* **2001**, *34*, 2482–2488.

34 S. Bauer, *Abstract Book, 12th Rolduc Polymer Meeting* **2003**.

35 P. M. Galletti, D. E. De Rossi and A. S. DeReggi (Eds.), *Medical Applications of Piezoelectric Polymers*, Gordon and Breach, New York, USA **1988**.

36 F. S. Foster, E. A. Harasiewicz, and M. D. Sherar, *IEEE Trans. Ultrasonic. Freq. Contr.* **2000**, *47*, 1363–1371.

37 L. M. Heikkinen, H. E. Panula, T. Lyyra, H. Olkkonen, I. Kiviranta, T. Nevalainen, and H. J. Helminen, *Scand. J. Lab. Animal Sci.* **1997**, *24*, 85–92.

38 L. Räisänen, R. Pohjanvirta, M. Unkila, and J. Tuomisto, *Pharmacol. Toxicol.* **1992**, *70*, 230–231.

39 J. Siivola, K. Leinonen, and L. Räisänen, *Med. Biol. Eng. Comput.* **1993**, *31*, 634–635.

40 P. Mirow, Mirow Systemtechnik GmbH, private communication.

41 M. Paajanen, J. Lekkala and H. Välimäki, *IEEE Trans. Dielect. Electr. Insul.* **2001**, *8*, 629–636.

42 J. Backman, *J. Audio Eng. Soc.* **1990**, *38*, 364–371.

43 J. Lekkala, *Industrial Horizons*, December **1997**, 12–13.

44 M. Antila, T. Muurinen, L. Linjama, and H. Nykänen, *Proc. Active 97*, Budapest **1997**, 607–618.

45 H. Nykänen, M. Antila, J. Kataja, J. Lekkala and S. Uosukainen, *Proc. Active 99*, Fort Lauderdale **1999**, 1159–1170.

46 M. K. Hämäläinen, J. K. Parviainen, and T. Jaaskelainen, *Rev. Sci. Instrum.* **1996**, *67*, 1598–1601.

47 L. Räisänen, K. Kirjavainen, K. Korhonen, and J. Sarlin, *High Technology in Finland* **1995**, 164–165.

48 H. Hoislbauer, R. Schwödiauer, S. Bauer-Gogonea, and S. Bauer, *Proc. 11th Int. Symp. Electrets*, Melbourne, Australia **2002**, 58–61.

Subject Index

a

absorption band 108, 116
– carbon dioxide 233
absorption coefficient 267
ACCVD process 62
acoustic coupling coefficient 182 f, 192
action teams 28
active noise control 313
agglomerates 234
air, optical properties 268
air quality monitoring (AQM) 227–238
AlCoStruct 273 f
Alq$_3$ thin films on Au 113
aluminum oxide 263 ff
aluminum nitride (AlN) 182
amorphous electrically conducting
 materials 239–248
amorphous phase, titanium dioxide 218
amorphous silicon
– pin diodes 124 f
– thin film transistors 130
anatase 207
annealing
– Ta–Si–N films 240
– temperatures 170 ff
antifog functionality 56
antireflective surfaces imprinting 263–280
application-specific lab-on-a-chip (ALM)
 119–138
applications 225–317
– automobile 55
– customer related 14
– glass coatings 55
arrayed waveguide gratings (AWG)
 207
Arrhenius plot 294
Asia–Pacific nanotechnology 35–48
ASIC metal 120 ff
aspect ratio 266

b

atomic force acoustic microscopy (AFAM)
 89 ff
atomic force microscopy (AFM) 89 ff
– Ta–Si–N films 243
– TiAl$_6$V$_4$ 159
– titanium dioxide 208
Australia, R&D 37, 42
automatic coating line 54
automobile applications, glass coatings 55
automobile fuel 62

band-gap effect 141
band structure 207
bandwidth 183
barrier layers 197, 264 ff
batteries 293 ff
benchmarking 17
Bio Nanotech Research Institute (BNRI) 61
biocompatibility 152 f
bioengineering 72
biohybrids 154
bio-inspired antireflective surfaces 263–280
biometric recognition system 74
biomimetric titanium structures 151–162
blades, diamond 254 f
blood clotting 153
blue luminescence 174
Boehmite aluminum oxide 264 f
bottom-up approach 79
bow measurements, silicon wafer 94
Bragg law 217
BRITE-EURAM III 227
brookite 207
bulk starting materials 80 f
bulk structures 140
bulk–liquid eutectic temperature 198
bump measurements 97

The Nano–Micro Interface: Bridging the Micro and Nano Worlds
Edited by Hans-Jörg Fecht and Matthias Werner
Copyright © 2004 WILEY-VCH Verlag GmbH & Co. KGaA, Weinheim
ISBN: 3-527-30978-0